THE FIRST ELECTRONIC COMPUTER

John Vincent Atanasoff

THE FIRST ELECTRONIC COMPUTER

The Atanasoff Story

Alice R. Burks and Arthur W. Burks

Ann Arbor   The University of Michigan Press

Copyright © by The University of Michigan 1988
All rights reserved
Published in the United States of America by
The University of Michigan Press
Manufactured in the United States of America

1991   1990   1989   1988   4   3   2   1

Library of Congress Cataloging-in-Publication Data

Burks, Alice R., 1920–
   The first electronic computer.

   Bibliography: p.
   Includes index.
   1. Electronic digital computers—History.
2. Atanasoff, John V. (John Vincent)   I. Burks,
Arthur W. (Arthur Walter), 1915–   . II. Title.
QA76.5.B825   1987   621.39′09   87-25547
ISBN 0-472-10090-4 (alk. paper)

Preface

This book has grown, literally, over a period of five years. It began as an article describing the computer John V. Atanasoff invented at Iowa State College in the late 1930s. Our idea was to provide a full description of what we regarded as the first electronic computer, and to do so in a way that would bridge the gap from the vacuum-tube technology of the early electronic computers to that of solid-state transistors, microchips, and semiconductors. We then decided to include an evaluation of Atanasoff's contribution to the overall history and also a more technical description of his electronic switching circuits. Finally, we decided to include an examination of the patent litigation that ended in the finding by Judge Earl R. Larson that Atanasoff's computer was an "automatic electronic digital computer" prior to the ENIAC, and that, in fact, the ENIAC—itself unveiled as the first electronic computer in 1946—had been derived from it.

We would like to trace this evolution of our interest in Atanasoff's work and to explain some of the circumstances that motivated us in what we realize is a highly controversial undertaking.

Arthur first heard of Atanasoff from John W. Mauchly when he was working with him and J. Presper Eckert on the design of the ENIAC, at the Moore School of Electrical Engineering, University of Pennsylvania. The impression was not strong, and Arthur had no reason to suspect any Atanasoff influence on the ENIAC. He did not hear of Atanasoff again until after the ENIAC patent issued in 1964; then, as a consultant to Honeywell in its efforts to avoid the very high royalties Sperry Rand was asking on that patent, he began to learn more of Atanasoff's work and of its bearing on the ENIAC patent.

The 1973 decision in the *Honeywell v. Sperry Rand* case, which invalidated the ENIAC patent and brought Atanasoff to the attention of the world, surprised us, however, as it did most of the computing community. It also further stimulated Arthur's interest in Atanasoff. He was starting to teach a course in the history of computers at the University of Michigan, and he wrote a paper for a 1976 Los Alamos conference in which he made his first statement of an Atanasoff-ENIAC link (Burks 1980). He contacted Atanasoff in connection with this paper, after overcoming some initial resistance owing to Atanasoff's identification of him with Eckert and Mauchly.

He planned next to write a longer article on the ENIAC, from his first-hand experience with it, and to follow that with a similar article on the first generation of stored-program computers, basically the EDVAC and the Institute for Advanced Study machine (which he had also helped design). He soon had Alice involved in the ENIAC paper as both sounding board and editor. She had been a "computer" of firing tables for the Army Ordnance at Aberdeen and the Moore School in 1942–43, working on the very tables the ENIAC was conceived and built to compute, but then entered the College for Women at the university to complete her undergraduate degree in mathematics. She had later assisted Arthur in his writings in philosophy and computer science.

Her interest in the ENIAC paper, especially with regard to the role of Atanasoff in the conception of the ENIAC, led finally to her becoming a coauthor (Burks and Burks 1981). That paper developed this role considerably, bringing to bear the 1941 Mauchly-Atanasoff correspondence, on the one hand, and, on the other, the relevance to the ENIAC of both Atanasoff's computer and an idea he had given Mauchly for a more sophisticated computer.

It was at this point that the Atanasoff book actually got started, as Alice decided to write a description of the Atanasoff machine similar to the one we had provided of the ENIAC, and Arthur began to integrate Atanasoff's achievements into his course in the history of computers. It was also at this point that, upon the appearance of our ENIAC article, we became aware of the reluctance of much of the computing community to accept Judge Larson's finding on Atanasoff.

Although we were somewhat surprised at this reluctance, we could see on reflection how deeply entrenched the popular belief in the ENIAC—and in Eckert and Mauchly—had become with the passage of some thirty-five years, as against almost no public recognition of Atanasoff and complete silence on his part concerning any contribution he had made. But we found no examination of the court proceedings in the literature, despite a considerable volume of dispute over the correctness of the decision. Historians and others cited the existence of the huge store of evidence and testimony, then went on to criticize the outcome, often giving what we considered either erroneous or irrelevant arguments against it.

We had felt that with Judge Larson's decision the burden of proof shifted to the other side, especially since there had been no appeal. We ourselves had been satisfied with the decision, bolstered by the correspondence from the period. Ultimately, of course, our curiosity led us over the threshold into the vast expanse of trial material, where we found not just the historical protocols that only the trial process can produce but a story rich in human interest. We believe the portions of this material that we present show how well, contrary to popular opinion, a court of law can handle complicated technical issues.

We should mention one particular event that contributed to our curiosity about the trial proceedings. When our ENIAC article was published in the *Annals of the History of Computing,* interested parties were invited to respond to it in the same issue; Presper Eckert and Kathleen Mauchly, John Mauchly's widow, joined in a short criticism of our treatment of the Atanasoff-ENIAC connection and promised to write their own article of refutation. Mrs. Mauchly alone published an article in the *Annals* that we take to be her answer, although she did not address our arguments directly (Mauchly 1984). She presented Mauchly letters and artifacts not introduced at trial, or, in her opinion, not adequately exploited at trial; and she quoted testimony of both Mauchly and Atanasoff to support her argument that Mauchly learned nothing from Atanasoff, indeed, that he was well on his way to inventing the ENIAC before he met Atanasoff in late 1940.

Mrs. Mauchly's article, like the ENIAC trial records, has not been subjected to close scrutiny in the literature. But, unlike Judge Larson's decision, neither has it been challenged. For this reason, and because we consider her presentation the best case to be made for the Mauchly-Eckert side, we have included a response to that presentation as an appendix.

Finally, we should note in our brief account of the motivation behind this book that Atanasoff has enjoyed increasing recognition of his achievements in electronic computing in the last few years, and that there has been increasing acceptance in the literature of Judge Larson's decision. Moreover, Atanasoff has published his own long article in the *Annals,* providing a much needed balance of personal accounting (Atanasoff 1984).

As to our actual writing of the book, it was very much a joint effort. Alice was able to work full-time, with very little interruption for other projects, whereas Arthur had teaching and other research responsibilities. But he provided the added resource of his own involvement in the technological history, together with his study of that history, over a period of forty years. Thus, while Alice did most of the writing and a large part of the research into the trial records, she consulted with Arthur on a continuing basis. He also wrote the first several drafts of the more technical appendix A, and provided long tracts of material for chapter 5 on Atanasoff's place in history. In short, Arthur is responsible for the engineering content, but both are responsible for the other technical content and arguments, as well as for the analyses of events.

We want to thank John Atanasoff for his many memory searches, at Arthur's instigation, to bring out some of the more elusive details of his computer. We hasten to add, however, that we never relied entirely on his recollection for our presentation of these details, but always checked to see not only that they squared with his 1940 article describing the computer and with his courtroom testimony about them, but also that the engineering was sound. We should add, too, that we never had the impression that Atanasoff wanted to distort these details. He was comfortable with the essentials of his computer, all of which were clearly documented anyway.

We also want to thank Alice Atanasoff for her hospitality during our several visits to interview her husband, and for her persistence in locating and supplying particular documents and photographs from among his countless papers and memorabilia.

We are indebted to others, too, in the writing of this book. Philosopher Carl Cohen did us a great service with his thorough criticism of an early draft of our machine description. Sociologist Jacqueline Scott gave us her impressions of the entire manuscript in its penultimate form, impressions that were both reassuring and helpful in subsequent revisions. Charles G. Call, chief patent attorney for the Honeywell suit against Sperry Rand, also read a late draft of the entire manuscript and affirmed its authenticity with regard to the issues and proceedings of the trial. We are grateful to these three friends; at the same time, we remain responsible for all particulars of our book.

We want to thank Arthur's doctoral student Christopher Langton and our son Douglas for their comments on appendix A. We also want to acknowledge the more indirect influence of Richard V. Burks, Arthur's brother, whose views on what constitutes history—imparted over the years—inspired Alice's decision to describe Atanasoff's computer in the context of his own chronology of its conception. We want to remember the late Henry Halladay, attorney in charge of the Honeywell suit, for supplying several documents that we requested during and after the trial, and to thank archivist Bruce H. Bruemmer and the Babbage Institute for supplying requested documents in recent years. Finally, we want to thank Judge Larson for guiding us to the repository of trial documents in Chicago, where we found some otherwise unavailable material.

Leslie R. Thurston is to be credited with most of the drawings, and Derwin L. Bell with earlier versions of figures 7, 16, and 18. Figures 2–6, 9, 16, 17, and 20–23 are based on photographs and diagrams introduced at the ENIAC patent trial. Figure 15 is reproduced by courtesy of the *Des Moines Register;* the frontispiece and figures 1, 8, and 19, by courtesy of John Atanasoff. Appendix A figure A.20 was drawn from figures in the Jevons references; figures A.22, A.24, and A.26 are by courtesy of the Princeton University Library, as is also the quotation from a Peirce letter in appendix A. The quotations in appendix B from the article Kathleen R. Mauchly published in the *Annals of the History of Computing* are reprinted by permission of the American Federation of Information Processing Societies, which holds the copyright.

We are grateful to the National Science Foundation for research support under grants SES 82-18834 and DCR 83-05830.

Contents

Illustrations

Introduction

John Vincent Atanasoff initiated the computer revolution with his invention of the world's first electronic computer. This special-purpose digital machine, completed in the spring of 1942 in Atanasoff's Iowa State College laboratory at Ames, not only proved the feasibility of electronic computation, but also contributed basic features to nearly all subsequent computers. It led directly to the ENIAC, the world's first general-purpose electronic computer, through a series of contacts between Atanasoff and John W. Mauchly, one of the two principal designers of the ENIAC.

These contacts took place from December, 1940, to October, 1941, and included both frequent correspondence and a visit by Mauchly to Iowa to see Atanasoff's computer. They conveyed critical information as to the innovative technology of that machine, and also as to possibilities for future machines.

The ENIAC was built at the Moore School of Electrical Engineering, University of Pennsylvania, by a small team headed by Mauchly and the other principal designer, J. Presper Eckert. It was begun in 1943 and completed in 1946. The programmable ENIAC led in turn to the first generation of stored-program computers: the EDVAC of the Moore School, the IAS computer of the Institute for Advanced Study, and others that adopted the basic design concepts of those two. Like Atanasoff's machine and the ENIAC, these all computed with vacuum tubes. There followed in due course the replacement of vacuum tubes with transistors, and then the replacement of entire circuits with computer chips.

Thus the causal linkage from Atanasoff's computer to the computers of today is a straightforward one. We wish to make clear at the outset, however, that each new machine in the causal chain far outstripped its predecessors even as it arose from them. In particular, the ENIAC was a very different computer from Atanasoff's. It constituted a great advance over it, and it embodied much inventive substance beyond concepts derived from him.

Our purpose here is to tell the story of Atanasoff's invention from a historical perspective. We want to describe the machine itself not just in terms of its final form, but in terms of its development. Moreover, we want to present it within the broader history of computing, that is, to place it in the context of the technologies from which it grew and the new technology it launched. This latter goal, as it happens, entails perhaps our most important task of all, namely, establishing that Atanasoff's computer *was* the first electronic computer and *was* causally related to the ENIAC and the machines that followed. For, while we find this part of the

Atanasoff story straightforward—indeed, indisputable—it has actually been engulfed in controversy of the most vehement sort since it was first uncovered some twenty years ago.

At the heart of this controversy is the issue of whether Atanasoff actually influenced the conception of the ENIAC by Mauchly and Eckert, and whether, in fact, Atanasoff's machine was even an electronic digital computer. Strangely, this issue was not laid to rest by a federal court's unappealed decision, in 1973, that his machine was such a computer and that key claims of the ENIAC patent were derived from it. Rather, that decision has been largely dismissed by the computing community, on the word of persons who are in responsible positions but have made no serious effort to address the trial proceedings on which it was based.

Yet the records of this *Honeywell v. Sperry Rand* suit are a researcher's dream: an immense public treasure of evidence and testimony, scrutinized under the most stringent courtroom conditions by highly skilled attorneys engaged in a contest over many millions of dollars. They unfold an intriguing drama, not only of corporate maneuverings, but of individual strife and human foibles, in a landmark trial that ran more than nine months and consumed more than five years of intensive study by both sides. This case provides its measure of irony, too, most notably in that, except for it, Atanasoff's role in electronic computing might never have come to light, and certainly could never have gained credence if it had come to light much later than it did.

We ourselves have felt a sense of urgency in writing this account, because all too often an invention unrecognized in its own day has not received due credit when affirmed very many years after the fact. The commonly accepted version becomes entrenched, for one thing, and, for another, the original device becomes harder and harder to understand as it is further and further removed from the current practice. In the case of Atanasoff's computer, for example, vacuum-tube technology is already foreign to today's younger engineers.

Our purpose in this book, then, is really twofold: first, to present Atanasoff's electronic digital computer as it evolved in his mind from the existing technologies; second, to defend its priority and its causal linkage to later computers.

We begin with the computer itself (chap. 1), explaining how and why Atanasoff conceived it, how he went about building it, what its main components were, and how it was to work. Our basis for the developmental aspect of this descriptive portion is the inventor's own manuscript, written in 1940 as he was starting to build. Details of the machine are drawn from that manuscript and also from the extended explanations he gave in court; Arthur Burks's personal experience with vacuum-tube computing has, of course, facilitated our comprehension of these two sources. This chapter serves not only as a full accounting of the first electronic computer but as a foundation for the balance of the book. The main features of Atanasoff's computer figure both in the arguments over his influence on Mauchly and in the evaluation of his contribution.

We follow this presentation of Atanasoff's work with a parallel one of John Mauchly's achievements at Ursinus College from 1936 until late 1940, at which time he first met Atanasoff and began his exchanges with him about electronic digital computing. This portion of the book (chap. 2) is based on Mauchly's own testimony in the ENIAC trial, together with the physical evidence, notes, and documents he produced under subpoena. It demonstrates how little he was concerned with digital computing and, contrariwise, how much he was involved in analog computing in connection with a meteorological project he was conducting at Ursinus College during that period.

We begin chapter 3 with a summary of the ENIAC project at the Moore School of Electrical Engineering, University of Pennsylvania, and then tell the story of Atanasoff's influence on the conception of that computer. We present and analyze much evidence from the ENIAC patent trial here: Mauchly's notes from his first encounter with Atanasoff, on December 28, 1940; his notes and drawings on computing from January 1, 1941, through the next ten months; letters he and Atanasoff exchanged between January 19 and October 30, 1941; and an exchange of letters with a third party on the subject of computing in the summer of 1941. It also includes the testimony of both Mauchly and Atanasoff at that trial, plus Mauchly's testimony in deposition for another case that scarcely saw trial, and the testimony of two others taken in deposition for that same earlier case.

Next we look at the role of Atanasoff in the *Honeywell v. Sperry Rand* case (chap. 4). We start with an overview of the trial issues and their resolution, proceed to a presentation of Atanasoff as a witness, and conclude with an examination of Judge Earl R. Larson's findings on the influence of Atanasoff and his computer on the invention of the ENIAC (and also of the EDVAC, as it happened). This chapter balances our greater concentration on Mauchly's courtroom performance in the preceding two chapters and, in fact, draws a sharp contrast between the two men. It does so, however, in the context of several issues not treated earlier but of significance to the case in their own right.

The last chapter of this book is devoted to an evaluation of Atanasoff's contribution to computing (chap. 5). We outline the mechanical, electromechanical, and electronic digital computing technologies; defend the judge's decision that Atanasoff's machine was a prior "automatic electronic digital computer"; and place that computer in a causal chain of invention. This last assessment includes: a listing and characterization of the novel aspects of Atanasoff's computer; a consideration of its patentability, had a patent been applied for; an analysis of the utilization of Atanasoff's ideas in the ENIAC and the EDVAC, with particular attention to patent claims; John von Neumann's concepts of a variable-address program language and a random-access store; and, finally, the transition from the EDVAC and the IAS computers to the computers of today.

There are two appendixes. Appendix A gives a general account of computer switching and spells out Atanasoff's pioneering role in its history. As will be apparent from the main text, Atanasoff's invention of electronic switching was one of two basic contributions, the other having been the more easily grasped invention of the regenerative memory.

Appendix B is our response to what is clearly the best case yet made for

Mauchly's independence of Atanasoff, as presented by his widow, Kathleen R. Mauchly, in "John Mauchly's Early Years" (Mauchly 1984). It explores Mauchly's pre-Atanasoff digital devices in the light of her new interpretation of them, and it examines letters found after his death that she maintains show significant progress toward an electronic digital computer, indeed, toward the ENIAC, prior to any association with Atanasoff.

CHAPTER 1

A Computer in the Making

Biographical Sketch

Let us begin the story of John Vincent Atanasoff with a very brief biography. He has given a fuller account in a recent article, "Advent of Electronic Digital Computing" (Atanasoff 1984).

Atanasoff was born in Hamilton, New York, on October 4, 1903. When he was eight years old, his family moved to central Florida, where he lived through his public school years. His father was an electrical engineer for a phosphate-mining company, having learned his engineering by correspondence after taking a bachelor of philosophy degree at Colgate University. John Atanasoff's abilities in science and mathematics were recognized early, and he was encouraged and stimulated in various ways by both his father and his mother.

By the age of fourteen or fifteen, Atanasoff had decided that he wanted to become a theoretical physicist. He did his undergraduate work at the University of Florida in electrical engineering, however, as a program he considered closer to theoretical physics than any other available there at that time. After receiving his bachelor's degree in 1925, he went on to Iowa State College to earn a master's in 1926, with a major in mathematics and a minor in physics. He had also worked as a teaching fellow that year, and now he continued at Iowa State as an instructor in mathematics.

In March of 1929, he enrolled at the University of Wisconsin as a doctoral student in theoretical physics. He had done some studying on his own while at Iowa, and though he entered courses well into Wisconsin's second semester, he succeeded in gaining credit for a full year's work in all of them by the end of that term. He took mathematical physics from Warren Weaver and elasticity from H. W. March, both in the mathematics department, and theoretical physics from John Van Vleck, who was to be the 1977 Nobel laureate. In Van Vleck's course, only five of the twenty-five or so students even attempted the final examination!

During the academic year 1929–30, Atanasoff did some teaching and wrote his thesis, "The Dielectric Constant of Helium," receiving his Ph.D. in the summer of 1930. He returned to Iowa that fall as an assistant professor, but still in the mathematics department. He continued to study on his own, this time delving into the relatively new field of electronics in order to fill what he considered the one serious deficiency remaining in his training. When, a few years later, Iowa promoted him to associate professor of both mathematics and physics, Atanasoff shifted his base permanently to the physics department (fig. 1).

This arrangement obtained until the fall of 1942, at which time Atanasoff left Iowa to direct war research unrelated to computing at the United States Naval Ordnance Laboratory in Washington, D.C. Although he was made a full professor in absentia, and was offered the chairmanship in physics, he never returned to Iowa State—or to teaching anywhere, for that matter. When World War II ended, he stayed on at Naval Ordnance, leaving in 1949 for a year as scientific advisor to field forces at United States Army Fort Monroe, then returning to Naval Ordnance for two more years as director of the Navy Fuze Program.

In 1952, Atanasoff formed his own research and development company, Ordnance Engineering Corporation. He sold this company to Aerojet Engineering Corporation in 1956, becoming a vice-president of Aerojet and serving in that capacity until 1961, after which he retired to pursue private interests. Atanasoff holds about twenty patents. He lives today with his wife, Alice, on a small farm near Frederick, Maryland.

Defining His Task

John Vincent Atanasoff did his pioneering work on the use of vacuum-tube circuits for computing while he was at Iowa State College. With a need to solve a wide range of physical problems numerically, he had by 1935 begun to think seriously about methods of mechanizing digital calculation; by early 1938, he had conceived the general electronic and logical design of an automatic digital computer for solving large sets of simultaneous linear equations. He explained the many possible applications of his machine after it was well along:

> In the treatment of many mathematical problems one requires the solution of systems of linear simultaneous algebraic equations. The occurrence of such systems is especially frequent in the applied fields of statistics, physics and technology. The following list indicates the range of problems in which [their solution] constitutes an essential part of the mathematical difficulty:
>
> 1. Multiple correlation.
> 2. Curve fitting.
> 3. Method of least squares.
> 4. Vibration problems including the vibrational Raman effect.
> 5. Electrical circuit analysis.
> 6. Analysis of elastic structures.
> 7. Approximate solution of many problems of elasticity.
> 8. Approximate solution of problems of quantum mechanics.
> 9. Perturbation theories of mechanics, astronomy and the quantum theory.

Fig. 1. John Atanasoff as a young man

> This list could be expanded very considerably, for linear algebraic systems are found in all applications of mathematics which possess a linear aspect. . . . The writer is of the opinion that . . . approximate methods using large systems of linear algebraic equations constitute the only practical method of solving many problems involving linear operational equations. (This general type of equation includes [some] differential and integral equations as special cases.) . . .

This passage is from Atanasoff's proposal, "Computing Machine for the Solution of Large Systems of Linear Algebraic Equations" (pp. 305–16), that he completed in August, 1940 (Atanasoff 1940/1973).

The need for such a computer had actually been recognized in England as early as 1879, when, eight years after the death of inventor Charles Babbage, a government-commissioned investigating committee recommended against funding construction of his analytical engine. Whereas the committee found that undertaking too expensive and too risky, it suggested as a worthwhile alternative a machine to evaluate determinants for the solution of simultaneous equations (Merrifield 1879).

Atanasoff spent what time he could from late 1935 to early 1939 studying the existing analog and digital devices, conducting experiments, and determining both the design principles and the physical components of his proposed computer. Satisfied at last with his plan, he hired a graduate student, Clifford E. Berry, to assist him. Berry joined him in September, 1939, and before the end of that year the two had built and demonstrated a model of the machine's central computing apparatus, an electronic (vacuum-tube) arithmetic unit interacting with a rotating electrostatic (capacitor) memory (see fig. 5).

Stated briefly for our present purpose, Atanasoff's idea was the following: he would solve a large set of equations by eliminating a designated variable from successive (overlapping) pairs, thereby generating a new set in one fewer variables, then repeating the process for the new set, and so on, until finally a single equation in a single variable emerged; he could then find single equations in all the other variables, as well, and so calculate the value of every variable. He would have a memory separate from the arithmetic unit, in the form of two drums turning on a common axle, each drum large enough to store the coefficients of one equation in capacitor elements. The coefficients of any given pair to be processed would be fed simultaneously into the electronic arithmetic unit and operated on to eliminate a designated coefficient from one of them. The new equation thus formed would be recorded on a card as one of the next smaller set, to be reentered in the next round of eliminations. All of this is explained in detail when we describe the machine.

Construction of the computer itself began in early 1940 and ended in April or May of 1942 (fig. 2). Test runs showed that the full arithmetic unit and the two-drum store worked perfectly, as did most of the supporting components of the input-output system. Only an electronic card-writing and -reading method, which Atanasoff had invented for intermediate storage of the newly computed sets of equations, failed sporadically. Its failure rate of less that one in ten thousand instances was still high enough to spoil results for large sets of equations. As it happened, United States entry into World War II in December, 1941, caused both Atanasoff and Berry to leave Iowa State before the difficulty with this particular procedure could be resolved. This matter, too, is fully explored later.

In describing Atanasoff's computer, we present his design challenges from his perspective but also relate them to modern computing concepts in modern terminology. All quotations are from his August, 1940, manuscript cited above (Atanasoff 1940/1973).

Preliminary Decisions

During his initial investigations, Atanasoff made four decisions of a critical nature. The first was to compute in the digital mode of such devices as keyboard desk calculators (Fridens, Marchants, Monroes) and punched-card machines, rather than in the analog mode of, say, the differential analyzer. For he realized he could not accurately solve the large sets of equations he needed to solve with analog technology.

The second settled the exact size of these sets of simultaneous linear equations. In his manuscript, he discussed the current state of the art as to number of unknowns, difficulty of solution, computing time, and accuracy of results. He then noted that solving systems with more than ten unknowns was rarely attempted at

Fig. 2. Top view of finished computer as of May, 1942. One of the two electrostatic storage drums at the rear of the computer is exposed here, and the other, to its right, is covered. To its left is a decimal-binary base-conversion drum, with bundles of wires to carry its signals to other parts of the machine. The large tray in the front left corner received blank binary cards to be passed through a recording mechanism at the extreme left, and the tray just behind it received the recorded cards to be read back into the computer by a similar mechanism. A keyboard of manual controls is partly visible at the top right. Between it and the binary-card recording tray is a decimal-card reading tray, somewhat elevated. On the front face below that binary tray are thyratrons (gas tubes) used in processing the binary cards. Vacuum tubes of the electronic arithmetic unit occupied the balance of the front of the machine; some of them are visible in this photograph. Atanasoff's computer was about as large as an office desk, though taller.

that time, and yet was precisely what was "needed to make approximate methods more effective in the solution of practical problems" (p. 306). He himself intended to solve sets of up to twenty-nine equations in twenty-nine unknowns, with each of the thirty coefficients (including constant term) of each equation having about fifteen decimal places. Such a goal was very ambitious for the late 1930s. It was especially ambitious for Atanasoff, who had both teaching and other research responsibilities, could not expect much research assistance, and had severely limited financial resources.

His third decision was his choice of a method of solving large sets of equations, the elimination process outlined in the previous section. He considered this "the one practical method," as against one using determinants, whose evaluation "is as difficult a problem as the solution of the original system of equations" (p.

306). What he had in mind was evidently the standard Gaussian elimination algorithm used then with desk calculators; he referred to "the well known process" (p. 306) and he himself had been working with calculators. Or, rather, it was some form of that algorithm, for at some point, perhaps even before he began to design the computer's components, he developed an original adaptation of Gauss's algorithm that avoided certain complex equipment. We present this modified algorithm in the next section.

Before giving Atanasoff's fourth preliminary decision, let us note the implications of the first three for machine design. He would have to provide read-write storage capacity for two equations of twenty-nine variables (thirty coefficients) and arithmetic capability for combining a given pair of equations to eliminate a designated coefficient; notice, however, that neither the type of storage facility nor its final relation to the arithmetic facility was as yet determined. He also would have to provide an input-output system capable of entering data initially, extracting data intermediately (as new sets of equations were computed) and reinserting them later, and of extracting final results; it will be clear from Atanasoff's comments on his fourth decision that he was already anticipating the use of punched cards throughout his input-output system.

That fourth and last precondition was use of the base two for his arithmetic operations, a decision that proved prophetic for the computer revolution. He had tried, once he formulated his general procedure, to adapt the current punched-card tabulating devices, or to design a decimal computer of his own along conventional lines, without satisfaction. When ultimately he saw that he would have to construct a machine that went well beyond the existing ones, he considered computing in some base other than ten. Of the several he investigated, the base two seemed best, for both speed and simplicity.

The advantage for simplicity was particularly striking to Atanasoff. The binary system not only allowed much simpler ongoing storage and computation mechanisms than any other but also facilitated the use of cards for storage of intermediate results: "Now at each spot on the card there are two possibilities; either there is a hole or there is not a hole. This corresponds exactly to the use of base-two numbers and greatly simplifies the mechanism for punching the card and reading it" (p. 307). He was, of course, adding two further processes to the input-output system, namely, converting 29×30, or 870, decimal coefficients of about fifteen places, at the beginning of a problem, into binary coefficients of about fifty places and, at the end of a problem, performing the necessary binary-decimal conversions.

Atanasoff's Elimination Algorithm

The standard Gaussian algorithm can be thought of as having a "forward" part and a "backward" part. In the forward part, the elimination process is carried out between successive (overlapping) pairs of equations until a single equation in a single variable emerges; that equation is then solved for that variable. In the backward part, successive equations in new single variables are generated by the process of back-substitution and each newly generated equation solved for its variable.

The process of eliminating a designated variable from a given pair of equations calls for dividing the variable's coefficient in the second equation by its coefficient in the first equation, multiplying the first equation through by that quotient, and subtracting the new equivalent of the first equation from the second equation. The equation in one variable that ultimately results from repeated application of this process is then solved by dividing its constant term by the coefficient of its variable and affixing the proper sign. All successive equations, each in a new variable, generated by the back-substitution process are solved in the same way. As each new variable is solved, it is substituted, along with the values of all the others so far obtained, in one equation of the next larger set, to obtain an equation in another new variable.

Thus the standard Gaussian algorithm required not only many additions and subtractions, but many multiplications and a lesser number of divisions. Naturally, the latter two operations would be much harder to execute automatically than the former. And so Atanasoff, to avoid the complications that multiplication and division would entail in a new machine, conceived a variant for which only addition, subtraction, and shifting were required.

His variant of the standard algorithm called for regarding equations as vectors, in effect, and repeatedly subtracting one vector from the other in a given pair—or adding one to the other, depending upon their signs—with appropriate shifting of the first, until the coefficient of a designated variable was eliminated. Completion of the elimination at each digit place of that coefficient would be signaled by a change in its sign, actually indicating an overdraft; compensation would then be made through the reverse operation.

As in the forward part of the Gaussian algorithm, Atanasoff's procedure would be applied to successive pairs of equations in the original set until a new set of one fewer equations in one fewer variables was generated, then to pairs in that set, and so on. The following simple set of three decimal equations in three variables, contrived to avoid shifting, illustrates vector addition and subtraction in the forward part of Atanasoff's algorithm.

$$(1) \quad x + 3y - 2z - 1 = 0$$
$$(2) \quad 2x + 5y - 2z - 6 = 0$$
$$(3) \quad 4x + 13y - 9z - 3 = 0$$

$$(4) \quad -y + 2z - 4 = 0$$
$$(5) \quad 3y - 5z + 9 = 0$$

$$(6) \quad z - 3 = 0$$

Equation (1) was subtracted twice from equation (2) to eliminate x and yield equation (4); similarly, equation (2) was subtracted twice from equation (3) to yield equation (5). Equation (4) was added thrice to equation (5) to eliminate y and yield equation (6).

This procedure circumvented the one division and the many multiplications required by the Gaussian algorithm for each pair of equations in its forward portion. But Atanasoff went further. He continued to apply this same elimination procedure to selected pairs of equations in the backward part of his algorithm, thus also circumventing the many individual multiplications required by back-substitution. Here the procedure called for not immediately solving the equation in one variable that ultimately emerged in the forward part, but pairing it with an equation from the prior set of two equations in two variables and, by repeated subtraction or addition, eliminating the shorter equation's variable from the longer equation. The result would be a new equation in a second single variable. Next, this new equation and the original one in a single variable would be paired successively with one from the set of three equations in three variables, and a third equation in a new (third) variable obtained. At each subsequent step, all of the single-variable equations thus far accumulated were to be paired successively with one (and the same) equation from the next larger set, to form still another equation in still another single variable. This part of the algorithm would finally produce an equation in one variable for each of the variables of the original set of equations. Its application to our preceding example is as follows.

$$(6) \qquad z - 3 = 0$$
$$(4) \quad -y + 2z - 4 = 0 \qquad \qquad (7) \quad -y + 2 = 0;$$

$$(6) \qquad\qquad z - 3 = 0 \qquad (7) \qquad -y + 2 = 0$$
$$(1) \quad x + 3y - 2z - 1 = 0 \qquad (8) \quad x + 3y - 7 = 0 \qquad (9) \quad x - 1 = 0.$$
$$\therefore \ x = 1, y = 2, z = 3.$$

And so Atanasoff, through this algorithm consisting in both forward and backward application of his vector procedure, succeeded in limiting his machine's arithmetic operations to addition and subtraction. Only one division for the solution of each equation in a single variable was required, and he planned to do that on a desk calculator after converting its two coefficients to decimal. He did have to provide for sensing the changes in sign of a particular coefficient of a particular pair of equations being processed, and also for shifting its vector. We explain these aspects of the procedure when we describe the arithmetic unit of his computer. We should note in passing, however, that while avoiding the operations of multiplication and division with his variant of Gauss's elimination algorithm, Atanasoff still reserved the option of incorporating them in his machine at some future time. Indeed, with the capability of adding and subtracting, of shifting numbers, and of sensing sign changes, he had already resolved the major design problems of automatic multiplication and division.

Original Storage Design

Atanasoff contemplated the design of his computer from the aspect of the two interacting functions to which all others would be auxiliary, storage and arithmetic (fig. 3). He considered many possibilities for each, conducting numerous tests and

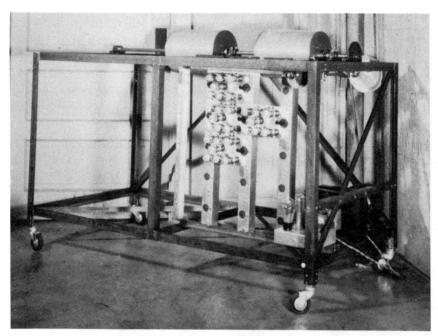

Fig. 3. Front view of computer as of August, 1940. This photograph, taken for inclusion in Atanasoff's 1940 proposal, provides a good perspective on the orientation of all other systems toward the store and the arithmetic unit. The thirty add-subtract mechanisms of the latter had already been built, though only four are plugged in here; the vertical racks mounted on the angle-iron frame could receive fifteen from the front and fifteen from the back. The two storage drums are in place on their common axle, with space to the far left for the base-conversion drum yet to be installed. At the lower right is a voltage-regulated power supply, with its filament transformer just behind it. Behind that is a temporary direct-current motor to turn the drum axle; this motor would be replaced later by a synchronous (alternating current) motor.

experiments; in the end he chose separate units, as we have indicated, a rotating electrostatic store based on capacitors and an electronic arithmetic unit based on vacuum tubes. We follow the order of presentation of Atanasoff's proposal, discussing storage first and then arithmetic, but we speculate at times on details he omitted and also on reasons for certain choices. It should be borne in mind, though, that as he designed any one component of his computer, he was always cognizant of the needs of every other component. For example, his storage choices were guided both by his stringent budget and by his intention from the start to compute electronically, and his calculating speed was paced to match that of his input-output procedures.

Jogging the Memory

Atanasoff's computer had to provide storage for two large blocks of numbers, each representing a linear equation with thirty binary coefficients. With each such coefficient conveniently represented as 49 digits and a sign, he would need a memory capacity of $2 \times 30 \times 50$, or 3,000, binary digits. He therefore set out to find a simple, inexpensive two-state element. One state would represent a "0," the other a "1"; similarly, one state would represent a plus, the other a minus, in the sign position.

Atanasoff thought briefly of using vacuum tubes, but because of their cost these were out of the question for him. He gave more serious consideration to magnetic recording on small pieces of ferromagnetic material, but they would require amplification and so, again, more tubes than he could afford. Finally, after weighing various forms of these and other elements, testing many, he arrived at his decision to use small paper condensers, or capacitors. The great and apparent advantage of condensers is that they can transmit signals directly to vacuum tubes without amplification. As he explained (p. 308),

> . . . The magnitude of the charge on small condensers fits in naturally with the use of thermionic tubes since this charge is large enough to control the grids of these tubes but small enough to be taken from the plate circuits of the tubes. . . .

Their serious disadvantage for holding numbers more than a few minutes is that their charge leaks off. Atanasoff, having excluded condensers from consideration while experimenting with other elements, finally decided to attack their leakage problem. And in time he found a way to stabilize their states for however long was necessary. He would have them recharged at regular, very short intervals through interactions with vacuum tubes in his arithmetic unit.

Some of these interactions would be in the normal course of computation; for the output tubes of the computing mechanisms would, in the process of sending signals to the storage condensers, automatically recharge them. The greater need would be to hold condenser states during certain intervals when they were *not* involved in computation. The initial input process of converting decimal coefficients to binary form and entering them into the memory was one case in which the charges on some condensers would have to be retained for a relatively long time while subsequent condensers were receiving their charges. Another was the final output process of converting a binary solution to decimal form, where, again, charges on the condensers not involved in the current conversion would need to be held for subsequent use. Still another was simply any time the operator had to leave the computer running idly, as when attending to a malfunction or just taking a break.

In studying this problem, Atanasoff realized that with some further interconnections and switches, and possibly a few more tubes, he could arrange for all the condensers in his 3,000-element read-write store to be recharged at the same regular intervals, regardless of whether they were engaged in computing.

Atanasoff described this recharging process as follows (p. 308):

> . . . Then the idea occurred to the writer that it might be possible to so arrange the [computing] mechanism that the machine would jog its memory

at short intervals, and this idea has been incorporated along with the use of condensers as [memory] elements in the present design. The condensers will retain their memory (i.e., their charge) to a sufficient degree for five minutes but they have it jogged (i.e., they are recharged in their original direction) at intervals of perhaps a second.

Today we refer to his "jogging" process as restoring or refreshing the memory, and to the "jogged memory" as a regenerative or a dynamic memory.

Atanasoff's discovery of this process of restoring the charges on storage capacitors represented one of two highly significant and influential inventions in electronic computation made by him in late 1937 or early 1938. (The other was his adding circuit.) The concept of a regenerative memory found application in the first stored-program computers. And regenerative memories based on capacitors for storage, with refreshing, are in wide use today, in the form of very tiny and sophisticated memory chips (Dynamic Random Access Memory chips, or DRAMs).

Basic Design of the Memory

While considering capacitors, Atanasoff must also have conceived the rotating drum memory. He would have realized that transmitting signals between a stationary arrangement of 3,000 capacitors and their corresponding arithmetic mechanisms would require so many switching circuits, electronic and electromechanical, as to defeat their low cost advantage. But a separate cyclic memory in the form of a rotating drum, or cylinder, would afford relatively simple communication with the arithmetic mechanisms, whatever form these might take.

Atanasoff's was the first computer to have a drum store. This kind of memory was reinvented about ten years later, in the form of a magnetic drum, and was common for another twenty years or so. Although drum memories have now been completely replaced by disk memories, these latter operate on a similar principle. Atanasoff's was also the first electronic computer to have a separate internal store, either stationary or cyclic.

Very probably at this point Atanasoff also decided to store a pair of equations on two separate drums rotating on a common axle. He would mount 1,500 condensers inside each drum and wire them to contacts set in its surface; they could then be read or written on from the exterior by brushes as the drums rotated, their charges being restored with each cycle. He then faced two further considerations. The first was how fast to rotate the drums. The second was how to orient the array of contacts representing an equation on a drum, that is, whether there should be fifty bands of thirty contacts around the drum or thirty bands of fifty contacts.

Atanasoff arrived at a drum rotation rate of once a second, not for computational reasons but because of the requirements of his intermediate input-output system. He meant to use electric sparks to punch the binary cards as newly formed sets of equations were read from the drums, and to time these sparks with sixty-cycle (per second) alternating current. His thinking here, as elsewhere, actually worked both ways, since the decision to punch with sparks arose from his desire to

Fig. 4. Storage drum. The arrangement of thirty bands of metal contact studs, 50 per band, around the circumference of the drum is shown in this view. Each band stored one of the thirty binary coefficients of one of a pair of equations to be combined arithmetically; two extra bands were provided to allow for condenser failure. Inside the drum, arranged radially to correspond with the 1,600 studs, were thirty-two bands of condensers, their outer wires connected to the studs, their inner wires joined and connected to the power supply system through an end mounting plate. Notice that the studs do not completely encircle the drum; a 60-degree blank space was left so that in a given rotation there would be time for decisions and control operations for the next rotation. The two drums were made of a compound similar to Bakelite, with removable aluminum end plates, and measured eleven inches in length and eight inches in diameter.

punch at a faster rate than was possible with the usual mechanical means. Nevertheless, the fact remains that once he had decided on his electrical method of punching the binary cards, as he did at an early stage in the design, he was locked into the rate of sixty punches, or one drum rotation, per second.

As to the orientation of the coefficients of an equation on a drum, his choice again arose from a preliminary decision, this time his decision to treat equations as vectors in some form of the Gaussian algorithm. Because vector operations lend themselves much more readily to what is today called word-parallel bit-serial processing, he chose to run the coefficients around the drum (fig. 4). He would have thirty parallel bands of fifty contacts each, so that as a drum rotated, successive binary digits (bits) of all thirty coefficients (words) would be sent in parallel to the arithmetic unit for processing. Notice that this was a marked departure from the processing mode of the digital devices he had studied, desk calculators and punched-card machines, which added and subtracted successive decimal numbers in the digit-parallel mode.

As it happened, this arrangement of coefficients in bands of fifty bits around a drum was nicely accommodated by the rotation rate of once a second. The rotation through sixty cycles, with certain activities timed off the cycles, required sixty equally spaced stations around the drum for each band. With fifty of these occupied by a coefficient, ten remained vacant, allowing a sixth of a second in which commands could be issued for the activities of the next rotation. This fifty-ten ratio of computation to control made for an efficient use of each second's rotation, and was much better than the thirty-thirty ratio that would have ensued from the alternative arrangement of the coefficients on the drum.

We are uncertain as to just when Atanasoff made the switch to his own modified version of Gauss's algorithm. It is probable, however, that he had had it in mind for some time, because he had been exploring ways of adapting other machines to do this same task (see chap. 4). On the other hand, he probably did not make a final decision as to the exact form of the algorithm until he had determined the features of his memory.

Interaction with the Arithmetic Unit

The immediate implication of the memory design was that, in the execution of the algorithm, the two storage drums would be read from and written on in unison; but that, while identical in construction, they would have different functions, with the numbers held on one drum repeatedly subtracted from or added to those held on the other (see fig. 6). The implication for the arithmetic unit was that it should have thirty identical "computing mechanisms" (p. 308) working in parallel. For a given pair of equations being processed, each such mechanism would receive a pair of corresponding coefficients from the two drums, digit pair by digit pair, combine these by subtraction or addition as ordered by the controls, and send the result back to the drum keeping the successive tallies. This step would be repeated over and over until that drum recorded "0's" for all the digits in the coefficient that had been designated for elimination. The remaining coefficients then represented a new equation in the next smaller set being generated.

Atanasoff called the bands of coefficients around his storage drums *abaci*. The original abacus was essentially a storage device, a decimal register, actually, with each wire representing a decimal position and the beads so placed on the wire as to register a decimal value for that position. Likewise, one of Atanasoff's storage drum bands was a binary register, with each capacitor representing a binary position and the charge on the capacitor registering a binary value for that position. Today we would call Atanasoff's "abacus" a fifty-bit serial register.

Notice that we distinguish a register from an accumulator. Although usage varies for these terms, as for many others, we reserve the latter for devices that accumulate totals by adding and subtracting, with carrying and borrowing. We will explain this accumulator principle in the next section.

Atanasoff named the two drums themselves after parts of the decimal desk calculator he associated with them, the "counter" and the "keyboard." In performing a division on a desk calculator, the operator entered the dividend on the set of dials (each one a counter in the decimal system), punched the divisor on the keyboard, and proceeded to subtract (add) the latter from (to) the former. Correspondingly, Atanasoff called the drum holding the vector from (to) which the other vector was to be subtracted (added) repeatedly the *counter abaci cylinder*. And he called the drum holding the vector to be subtracted from (added to) the other repeatedly the *keyboard abaci cylinder*. Finally, he used the terms *counter abaci* and *keyboard abaci* to distinguish the abaci of the two drums.

Now while "abacus" was an appropriate designation, "keyboard" and "counter" were less so. The keyboard drum was not an input device, and the counter drum did not count. Rather, they constituted a read-write memory unit, separate from the arithmetic unit, that performed functions analogous to the memory functions incorporated in a desk calculator. In short, these two drums received, held, and returned numbers to be processed, but did not participate in the processing. Atanasoff's usage here reflects the degree to which he continued to look to the desk calculator as a digital model, despite the several departures he had made from it in his memory design. In the next section, we will see that this same very natural, but ultimately frustrating, association carried over to his efforts to design an electronic computing mechanism.

Vacuum-Tube Logic

Atanasoff's chief task in designing his arithmetic unit was to design the thirty computing mechanisms, or rather, to design one of these. Clearly, these mechanisms would be electronic; he had firmed that commitment when he chose capacitors as memory elements, with "jogging" from the arithmetic unit. Yet he arrived at the general principle behind these computing circuits only after lengthy experimentation—through a truly dramatic insight, in fact, which resulted in original logical design that is in standard use to this day.

Problems with Electronic Counters

Atanasoff had first planned to base his computing mechanism on the electronic scale-of-two counters used by scientists at that time to count physical phenomena. That is, he meant to convert such a binary counter into a circuit that would function, in conjunction with his counter drum, in a fashion analogous to that of a decimal mechanical accumulator of a desk calculator. He would take a pair of numbers (coefficients) from corresponding bands of his two storage drums into a given mechanism and add or subtract them on this circuit, with his answer accumulating on the band of the counter drum.

This effort met with failure, as Atanasoff noted in his 1940 manuscript (pp. 308–9):

. . . The first plans made were to use the circuit of the scale-of-two counters but after months of experimental work this idea was abandoned because of the inherent instability of the circuits. At times these circuits could be made to work but obscure factors strongly influenced their operation. . . .

It is hard to know the precise nature of his problem with these counters. As we explain in chapter 2, on John Mauchly's efforts during this period, it was not a simple matter to move from vacuum-tube circuits that could only count to ones that could add. The scale-of-two device was similar to an odometer: it received a single stream of pulses in the units place and passed along a carry pulse at each successive place as needed. Adding two many-place numbers was a much more difficult undertaking.

We believe, however, that the crucial problem for Atanasoff, in his attempt to construct his computing mechanism from these scale-of-two counters, was the unsuitability of the accumulator principle to the digit-serial processing mode he needed for his algorithm.

The standard decimal desk calculators to which he looked for a model operated their accumulators in the *digit-parallel* mode. Such an accumulator can be visualized as a set of counters arranged from right to left in ascending place order, with a carry mechanism between each counter and the one to its left. In the case of a series of additions, the successive numbers are "counted" onto the counters, all digits in parallel. For each new addend, the carry mechanisms "remember" the instances of counters passing to "0" and add "1's" to the next higher counters, again in parallel; they also remember and add on any subsequent (delayed) carries. At the conclusion of each complete addition, then, the counters hold the sum *accumulated* thus far.

Now to apply the accumulator principle to the addition of two numbers in the *digit-serial* mode would entail a single counter, with an associated carry mechanism, that could receive and add two numbers in the order of their ascending places. Granted that Atanasoff meant to operate his counter in conjunction with his counter drum, on which the sum would be continuously accumulating, and that he was working in the base two, the procedure would still be extremely complex. In the first place, since a counter cannot receive two digits simultaneously, the two "streams" of digits would have to be coordinated through spacing delays. In the second place, a third stream of digits, that generated by the carry mechanism, would also have to be coordinated with the other two. Such an accumulator would be clumsy indeed.

We are of the opinion, then, that the essence of Atanasoff's difficulty with electronic counters lay in this inherent unsuitability of the accumulator principle for the bit-serial processing mode of his vector algorithm. We should stress, of course, that he was in entirely uncharted waters, working with electronic circuits a decade before logical design problems had been separated from electronic design problems. That is, he probably thought he had a standard electronic design problem when in actuality he had such a problem compounded by a novel logical design problem.

An Electronic Breakthrough

In any event, as he continued to pursue this accumulator approach to his computing mechanism, he grew increasingly frustrated, almost to the point of despair. Then, suddenly, he saw that he could dispense with counters altogether. The traditional accumulator, he realized, performed two distinct and separable functions: addition of successive addends and storage of the accumulating totals. But he had already provided the necessary storage, in the form of the drum on which the totals were to accumulate, his counter drum. It remained only to provide for the addition. And that could be done with logic!

Of this momentous discovery, made in late 1937 or early 1938, Atanasoff wrote in his 1940 proposal (p. 309):

> . . . At last the writer hit upon another type of circuit that in the end proved very stable and entirely satisfactory in other ways. This circuit operates upon new principles in the computing art, principles that are rather analogous to the function of the human brain in mental calculation. The circuit takes cognizance of what is in a given [memory] element, what is to be added into or subtracted from the element, and from a [third] memory device it receives a signal indicating carry over from the previous place. Having been taught by a man with a soldering iron it selects the right answer and replaces what is in the [accumulating drum] by this result. . . . At the same time, the over-all complexity and cost of the computing machine is greatly reduced.

He meant to replace the accumulator principle with the principle of logical switching, that is, to replace the electronic counting circuits he originally planned for his mechanisms with electronic switching circuits. The "man with the soldering iron" was the builder of these circuits, who would "teach" the mechanisms to produce the correct sum or difference for any pair of arriving bits, taking into account the carry or borrow bit from the addition or subtraction of the preceding pair. The process would be "analogous to the function of the human brain" in that our memories hold the addition and subtraction tables that enable us to avoid counting out every number we wish to add to or subtract from another number.

The binary system now proved highly advantageous, because its addition and subtraction tables were much simpler than those for the decimal system. More fundamentally, though, the bit-serial processing of numbers required of his individual computing mechanisms was ideally suited to switching circuits, because switches could handle the three bits at once, and could also generate the carry and borrow bits, with relative ease. Lastly, as he noted, the required circuitry would be much simpler, and so cheaper, for switching than for counting mechanisms.

Atanasoff had now crowned his invention of the regenerative (drum) memory with an invention of still greater significance and influence: the first electronic switching adder—indeed, the first electronic switching circuit of any complexity.

And he had shown how to use vacuum tubes in the digital or discrete mode for computing, as contrasted with their almost exclusive prior use in the analog or continuous mode for radio and other forms of communication.

Variants of his switching circuit and switching adder are common in modern hardware design, just as the regenerative memory, using capacitors, is the essence of the modern dynamic memory chip. And just as that memory played a critical role in starting the computer revolution—in the EDVAC—so did that circuit and that adder play critical roles—the one in the ENIAC, the other in the EDVAC. We present these derivations from Atanasoff in chapters 3 and 4, on the Atanasoff-ENIAC connection and the patent litigation involving his inventions.

The logical switching insight was the culmination of about a year of study of all the different ideas Atanasoff had had for processing his pairs of equations. Actually, every prior decision had remained tentative until, at last, with this new idea for a switching adder, the general design of the entire computer took shape. Another year was then required to refine the logic and electronics of the arithmetic unit, to determine the specifics of the memory, and to devise a system of communication between the two.

In the spring of 1939, Atanasoff received a small grant from Iowa State College to cover the cost of hiring an assistant and building a model to test the operation of these two units (fig. 5). Clifford Berry, a first-year graduate student, began work in September, and the model was successfully demonstrated before the end of the year. Construction of the machine itself, together with various test setups, now commenced and continued for over two years, although Atanasoff took time in the first months to write the proposal we have been citing. In that manuscript, completed on August 14, 1940, he not only described the current state of progress, but also explained his plans for the entire computer. His immediate purpose in writing it was to secure a grant to complete the project; indeed, he received $5,000 from the nonprofit Research Corporation, of New York City. His longer-term purpose was to lay the groundwork for a patent application; unfortunately, no application was ever filed.

The Memory

Atanasoff and Berry, as would be expected, built the memory and the arithmetic unit first. The keyboard and counter cylinders, or drums, mounted on a common axle, were each eleven inches long and eight inches in diameter. Their surfaces were an insulating material, paper-reinforced plastic, similar to Bakelite in appearance, and their end mounting plates were aluminum. Penetrating the surface of each drum from the outside were 1,600 brass contact studs, arranged in thirty-two bands of 50. (Notice the increase of 100 studs, or two bands, over the 1,500 needed to accommodate an equation of thirty coefficients of fifty binary places each—a safety allowance for condenser failure.)

The studs of these bands were spaced 6 degrees apart, occupying 300 degrees

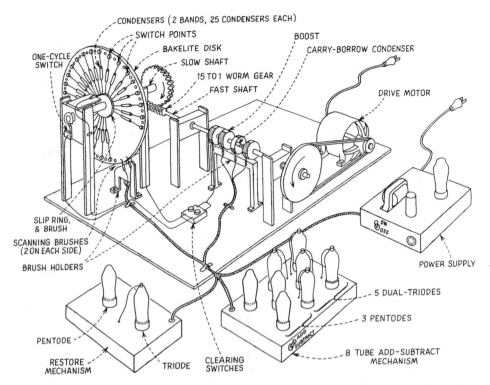

Fig. 5. Drawing of 1939 model. This drawing (now slightly modified) was made for court proceedings about thirty years after the fact. The model had two storage bands, each with twenty-five condensers, on the outer faces of a large disk; one band represented an abacus or coefficient of the counter drum, the other that of the keyboard drum. It had an add-subtract mechanism, served by a single carry-borrow condenser, to add or subtract the two abaci, and also a mechanism to perform the restore, but not the shift, function. The model proved that Atanasoff's principles of electronic computation were practicable. On the basis of this success and the manuscript he finished in August, 1940, he received a grant to complete his computer.

of circumference and leaving 60 degrees blank for control purposes. Immediately inside each drum was a thick layer of moisture-proof wax, within which the 1,600 condensers were embedded, spokelike in correspondence with the studs; their outer sides were wired individually to the studs, their inner sides wired together and brought outside through a slip ring fixed to that plate but insulated from it.

Which binary digit a given condenser stored at a given time depended on whether its voltage was relatively high or relatively low: a high voltage on the side wired to a stud corresponded to "0," a low voltage to "1." In turn, the actual high-low voltage levels that represented these two memory bits at a given time depended on the role the storage drums were playing in the computation—whether they were being read from or written on. During the reading (output) phase, the levels were a high of about +30 volts for the digit "0" and a low of about −40 volts for the digit "1." For the writing (input) phase, these levels were momentarily elevated by 90 volts to a high of +120 volts and a low of about +50 volts, so as to achieve compatibility with the voltage levels of the writing (output) vacuum tubes of the arithmetic unit. This elevation was accomplished by arranging to connect the slip ring mentioned above to a power supply of +90 volts during the writing phase only, and to 0 volts (ground) during the reading phase and also an interim storage phase. We will return to this arrangement, which Atanasoff called "boosting," when we present the electronics of the arithmetic unit.

Each coefficient stored on a drum abacus ran from its least to its most significant digit, the order in which serial additions and subtractions must be carried out. Negative numbers were stored in two's complement form, though the subtractions were executed with borrowing. Forty-nine of the fifty places represented digits (bits) and the last, most significant place, the sign; there "0" meant a positive number, "1" a complement.

The two drums, the counter drum holding one vector and the keyboard drum another, were rotated at the rate of once a second, their common axle driven by a synchronous motor operating on sixty-cycle alternating current. On a given rotation, fixed rows of brushes ranged along the two drums read the charges on the studs, thirty bits at a time from each drum, and sent the corresponding pairs over wires to the thirty corresponding adding and restoring mechanisms of the arithmetic unit.

In executing Atanasoff's variant of the Gaussian elimination algorithm, these mechanisms combined the paired streams of incoming bits, one pair per mechanism, by addition or subtraction, always adding the keyboard vector to or subtracting it from the counter vector. They then sent to the counter drum the streams of sum or difference bits, but returned to the keyboard drum the same streams of bits it had just contributed, refreshed ("jogged") for use on the next rotation. This transmission, too, was via rows of brushes ranged along the drums.

We describe the steps and timing of this transmission procedure in detail in the next section, accounting there for the complicating factor of shifting, and also explaining how carry or borrow bits were generated and integrated into the computation.

The drum-rotation time of one second and the 6-degree spacing of studs in the abaci meant a *pulse time* of one-sixtieth of a second (the time required for the parallel processing of thirty pairs of bits), and an *addition time* of one second (the time required to combine two vectors by thirty complete additions or subtractions). During passage over the fiftieth (sign) position and over the 60-degree blank span of the drum's surface, general determinations were made by the controls and orders issued for the next rotation. Thus in each addition time, five-sixths of a second was spent on calculation and one-sixth of a second on control (fig. 6).

We estimate that the time needed by Atanasoff's computer to reduce each pair of equations to an equation in one fewer variables would be approximately two minutes—a very slow process by today's standards, of course, but still an astounding rate of computation for the late 1930s.

The Arithmetic Unit

Atanasoff's arithmetic unit consisted of his thirty computing mechanisms together with several control mechanisms. We present these here in terms of their roles in

Fig. 6. Rear view of computer as of August, 1940. This close-up shows the two storage drums in place on their axle, the keyboard abaci cylinder on the right and the counter abaci cylinder on the left, and also the "undersides" of the few add-subtract mechanisms plugged in at this time. Above these are three terminal blocks to which the plug-in sockets of the add-subtract mechanisms are connected. As the axle rotated at the rate of once every second, the vector on the keyboard cylinder was repeatedly subtracted from or added to the vector on the counter cylinder, with appropriate shifting of the keyboard abaci whenever there was a change in sign of the designated coefficient on the counter cylinder, until the coefficient was eliminated. Each vector addition or subtraction was executed by the arithmetic unit in a single rotation as the digits of the two drums were read off by brushes and sent to that unit's add-subtract mechanisms.

executing the elimination algorithm, leaving their roles in the various input-output procedures to the relevant sections. Let us begin with an overview of the entire unit, and then examine the different components and their functions more closely (fig. 7).

Each of the computing mechanisms, which Atanasoff had expected to be electronic counters with some kind of carrying arrangement, now had three interconnected components. That is, the thirty computing mechanisms now consisted of thirty electronic add-subtract mechanisms (to use his term); thirty other primarily electronic mechanisms called (by us) restore-shift mechanisms; and the thirty electrostatic bands of a carry-borrow drum. From the standpoint of interaction with the two storage drums, then, for each pair of abaci on the counter and keyboard drums, respectively, the arithmetic unit had an add-subtract mechanism, an associated restore-shift mechanism, and an associated band on the carry-borrow drum.

We saw in the preceding section that, during the execution of the elimination algorithm, each computing mechanism performed serial additions or subtractions on streams of bits arriving from its associated counter and keyboard abaci. We can now say that these basic operations were performed by the add-subtract mechanism, which also incorporated in them a third stream of bits arriving from its band on the carry-borrow drum.

The carry-borrow drum was a long, slender cylinder driven by the same synchronous motor as the two storage drums, but mounted on a second axle perpendicular to (and geared to) theirs. Atanasoff found it convenient to rotate this drum fifteen times a second and process four bits per rotation to achieve the overall rate of sixty bits per second that it also required. Each of its thirty bands was made up of four successive strips of metal that served as contacts to be written on and read from, all wired to one end of a single condenser inside the drum whose other end was connected to a slip ring.

An add-subtract mechanism not only processed the incoming streams of bits from its band on the carry-borrow drum, but sent back a newly generated stream of carry or borrow bits to that band. Of course, it also sent the stream of sum or difference bits to its counter abacus. It did not, however, return the stream of bits that had arrived from its keyboard abacus to that abacus. Rather, its associated restore-shift mechanism performed this function.

The restore-shift mechanism actually intercepted the stream of bits entering the arithmetic unit from the keyboard abacus and electronically routed it on two different paths: one going forward to the add-subtract mechanism, to be added to or subtracted from the stream arriving directly from the counter abacus, the other going back to the keyboard abacus. The purpose of this intercession was twofold: to send the bits back to their abacus over and over, for repeated participation in the arithmetic process; and to refresh them, that is, to restore the charges on the condensers of the keyboard abacus, so that they would retain their original "0" or "1" values for use on each succeeding drum rotation. This task, accounting for the *restore* of *restore-shift mechanism,* was accomplished automatically as the stream of signals was passed through an output vacuum tube of the mechanism. As the name also implies, the restore-shift mechanism had the further function of shifting the keyboard abacus when ordered to do so.

The controls for the arithmetic unit sensed the situation as the fifty addition or subtraction steps of a given rotation were finished, ordered the shift, and dictated the next operation, whether addition or subtraction. They also sensed when the designated coefficient was completely eliminated, and ordered a cessation of the computation for that pair of equations. These and other controls were largely electromechanical.

Atanasoff himself did not think in terms of an arithmetic unit, and it is true that the add-subtract and restore-shift mechanisms also performed the memory restoration function. He did, however, think of his add-subtract mechanisms as agencies of coaction between the keyboard and counter abaci (p. 310), and these mechanisms did operate along with the restore-shift mechanisms and the carry-borrow drum to perform the computer's arithmetic functions. It is therefore reasonable to view all of these devices collectively as an arithmetic unit.

Figure 8 shows Atanasoff teaching physics during this period when he was designing his electronic computing circuits.

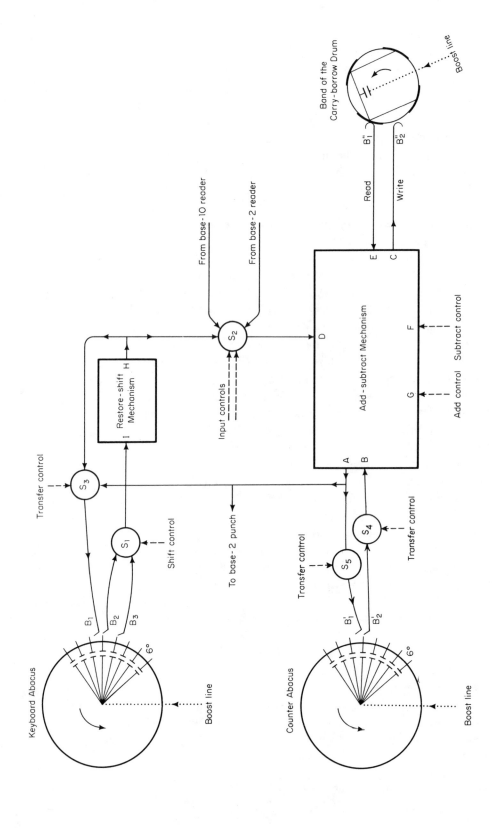

Fig. 7. Cross section of store and arithmetic unit. One abacus each of the keyboard and counter drums, and one band of the carry-borrow drum are shown in cross section, together with their brushes and brush wires. The add-subtract and restore-shift mechanisms to and from which these wires transmitted signals are represented by rectangles. It can be seen that, in the addition or subtraction of the two abaci, as the counter abacus rotated, brush B_2' contacted a given stud at the beginning of a pulse time and sent its bit to input point B of the add-subtract mechanism, and, 3 degrees later in the rotation (also one-half pulse time later), brush B_1' contacted the same stud and deposited the sum or difference bit sent from output point A.

As the keyboard abacus rotated during that same one-half pulse time, if no shift had been ordered the shift control operated by switch S_1 caused a bit to be read from a stud by brush B_2 and sent to input point I of the restore-shift mechanism; there it was restored and routed to output point H, from which it was sent, respectively, to input point D of the add-subtract mechanism via switch S_2 and to brush B_1 via switch S_3. Alternatively, if a shift had been ordered the shift control caused a bit to be read from a stud by brush B_3 and sent to input point I of the restore-shift mechanism, there to be processed in that same manner. Notice that although brush B_3 was positioned earlier in the rotation, it was read at the same time brush B_2 would have been read, on the no-shift alternative, because it was a full 6 degrees earlier.

The carry-borrow drum rotated fifteen times as fast as the storage drums, but it was alternately read from and written on at half-pulse-time intervals in phase with theirs. Brush B_1'' read a bit from a strip at the start of a pulse time, and brush B_2'' wrote the new carry or borrow bit on the same strip one-half pulse time later. At the start of the next pulse time, brush B_1'' read that new carry or borrow bit from the next strip, and brush B_2'' wrote the next carry or borrow bit on that strip. A bit could be written on one strip and read from the next because of the connection of all strips to a single condenser inside the drum; the connections were made as indicated.

Shown here also are the various lines of the input-output system, whose interactions with the arithmetic unit and the store were effected by the switches S_1 to S_5, and the "boost" lines to all drum condensers. The latter raised, from ground to +90 volts, the voltage level on the inner side of each condenser for the writing phase (when its outer side was written on via its stud), as a necessary adjustment to the incoming voltage levels; it then returned the inner side to 0 volts, allowing the stud voltage to drop 90 volts to its proper value for the reading phase of the next rotation. (See also fig. 18; fig. 7 is an abacus cross section of part of fig. 18.)

Fig. 8. Atanasoff in the late 1930s

Adding and Subtracting

As we have indicated, an add-subtract mechanism received, at the start of a pulse time in the elimination process, one bit from its counter abacus directly, one bit from its keyboard abacus via its associated restore-shift mechanism, and either a carry or a borrow bit from its carry-borrow band. It combined these three bits to produce a sum or difference bit and a new carry or borrow bit, consuming roughly half a pulse time in doing so. The sum or difference bit was returned to the *same* contact stud from which a bit had just come, ready to be read on the next *addition time*. The next carry or borrow bit was also returned to the *same* contact strip from which the earlier one had come, but to be read from the *next* strip on the next *pulse time;* thus the carry-borrow drum was a half-pulse-time store for the serial borrow and carry bits. It was possible to read from one strip a bit written on another because of the connection of all strips in a given band to a single condenser: to write on one was to write on all, and a reading could be taken from any one; it was just necessary to adjust the lengths of the strips so that writing and reading could not coincide (see fig. 7).

The reading and writing exchanges between add-subtract mechanisms and the counter and carry-borrow drums were accomplished through two parallel rows of thirty brushes stationed along the length of each drum, with the individual brushes wired to their particular add-subtract mechanisms. In both cases, the two rows were spaced one-half pulse time apart (3 degrees of rotation for the counter drum, 45 degrees for the carry-borrow drum); and the reading row was earlier in the rotation than the writing row, the reading occurring at the start of a pulse time, the writing at the halfway point. These and other reading and writing brushes were made of wire bristles.

Shifting

As a restore-shift mechanism transmitted its keyboard bit to its associated add-subtract mechanism, it also sent it back to the keyboard abacus, to be held until the next rotation. The keyboard condenser to which this bit was sent corresponded to the counter condenser that had just contributed a bit and was now receiving another; but only if a shift was not in progress was this the condenser from which the keyboard bit had come. If a shift was in progress, the restore-shift mechanism had arranged for the reading of that condenser to be superseded by the reading of the one following it on the drum. These reading and writing exchanges were accomplished, in the case of no shift, by two rows of brushes stationed 3 degrees apart as they were for the counter drum; in the case of a shift, by the same writing row and by an alternate reading row stationed 6 degrees earlier in the rotation than the other. Notice that whenever a shift was executed, the least significant digit on each keyboard abacus dropped away.

The control mechanisms determined when a shift was required and issued orders accordingly. A control circuit sensed whether the sign bit of the coefficient designated for elimination had changed (from "0" for plus to "1" for minus, or vice versa) as compared to that of the previous addition time. If no change had occurred, it ordered the restore-shift mechanisms to read keyboard bits from the row of brushes preceding the writing row by 3 degrees (not to shift), and the add-subtract mechanisms to repeat the operation they had just performed. If a change had occurred, it ordered the restore-shift mechanisms to read from the row preceding the writing row by 9 degrees (to shift), and the add-subtract mechanisms to reverse operations. These orders were given in the 60-degree blank portion of the rotation, and were carried out through 300 degrees of the next rotation.

Atanasoff's adaptation of the Gaussian elimination algorithm thus had much in common with the *non*restoring variant of the method of division by repeated subtraction. That is, when a change in sign (an overdraft) occurred, he did not restore the subtrahend or addend (thereby restoring the sign as well) before shifting; rather, he simply shifted and commenced "restoring" in the next binary position. He did not, however, record the number of operations at each position, and so formed no quotient.

The controls also determined whether the initial operation should be addition or subtraction, and when the operations for a particular pair should be terminated.

Restoring

As we have indicated, the stream of bits that a restore-shift mechanism returned to its associated keyboard abacus, for further use in the elimination process, was refreshed as it passed through an output vacuum tube of that mechanism.

The restore-shift mechanisms also kept their keyboard capacitors recharged at other times, as when the computer was idling or when the keyboard drum, at least, was idling. Instances of this latter phenomenon, when the machine was engaged in activities not involving this drum, will be evident in our presentation of the various input-output procedures.

Notice that we reserve the term "restore" (or "refresh," "regenerate") for cases in which streams of bits read from the drum capacitors were being returned without change. We do not consider, for example, that the add-subtract mechanisms were "restoring" the charges on the counter drum capacitors when they sent them streams of newly computed sums or differences, or on the carry-borrow drum capacitors when they sent them carry or borrow bits.

There were times when the bits on the counter drum abaci did pass repeatedly through the add-subtract mechanisms and back without entering into computation—in other words, when they were refreshed. As with the keyboard drum, one such instance was when the entire machine was idling. At least some of the add-subtract mechanisms were involved, however, in every computing activity. In the terminal binary-decimal conversions, for example, one counter drum abacus at a time was converted; in each such instance, the signals of the other twenty-nine were repeatedly passed through their add-subtract mechanisms and were refreshed on each passage.

In this regard it may be noted that regeneration occurred at times when there was no need to retain the values on the storage capacitors. On the other hand, the values were always available if they were needed, so that it was unnecessary to provide for distinguishing the two situations.

Electronic Design of the Add-Subtract Mechanism

Having explained in the previous section the *functions* of the basic components of the arithmetic unit, we now turn to the *design* of the add-subtract mechanism and, incidentally, the restore-shift mechanism.

Arithmetic, Logic, and Electronics

Atanasoff's design of the circuitry for his thirty electronic add-subtract mechanisms implied a two-stage transition, from arithmetic to logic and from logic to electronics. The first stage required translating the steps of binary addition and subtraction into logical operations suited to the use of vacuum tubes in their new discrete, on-off, mode. The second required devising electronic switching circuits to perform these operations (fig. 9).

Today the translation from arithmetic to logical operations would be accomplished by logical switching theory, a post–World War II development of Boolean algebra and truth-function theory. The electronic circuits to perform these logical

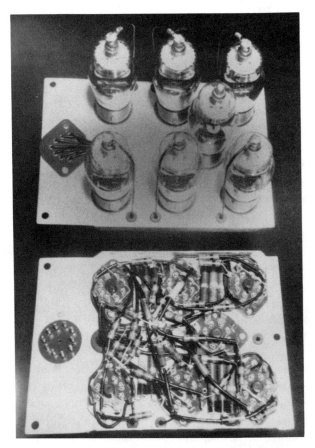

Fig. 9. Add-subtract mechanism. These photographs are top and bottom views of a serial add-subtract mechanism, the basic component of the parallel arithmetic unit. The seven vacuum tubes were double triodes, or envelopes, containing fourteen triodes, of which thirteen were used. The entire mechanism measured only five by seven inches. Note that it was a module, easily plugged in and removable for inspection, repair, or replacement.

operations would be devised by translating a logical switching diagram into an electronic circuit diagram (see app. A). Atanasoff knew of Boolean algebra and knew that he was doing logic, but he did not know of truth-function theory or even of truth tables as such. Nor did he know of the development by Claude Shannon, published in 1938, of a general method for applying this theory to the synthesis of electromechanical relay switching circuits. No one had ever applied Boolean algebra in any form to electronic circuitry, of course, and Atanasoff's synthesis was more a matter of trial and error than of formal methodology.

Thus he did not, when presenting the add-subtract mechanism in his 1940 proposal, either spell out the logical operations of the first design stage or supply the details of the electronic operations of the second stage. Instead, he did three things. First, he provided his electronic circuit diagram for the mechanism (p. 313).

Second, he provided the equivalent of a truth table for the sixteen cases of addition and subtraction this circuitry was to perform (p. 314); his table, though, translated the initial "0's" and "1's" into "high" and "low" voltage levels, respectively, to represent those of the drum capacitors and of the input and output tubes of the add-subtract mechanism. Third, he provided a brief explanation of how the circuits in the diagram would perform the operations in the table (p. 314).

For our purpose, it is better to present Atanasoff's electronic switching circuits more nearly as they would be developed today. Accordingly, we first give the truth table for binary addition and subtraction—his table, except that we will keep to "0's" and "1's" throughout.

We then derive from that table four general rules that express the contents of the table. These rules give the correct sum and carry bits for every addition, and the correct difference and borrow bits for every subtraction. Neither Atanasoff nor Berry, who contributed much to the details of this design, mentioned such rules in their writings, and we doubt that they formulated any of their own. Still, it seems likely that they recognized overall patterns in the table and were guided by them at the electronic stage.

We next turn to the electronics, presenting Atanasoff's adaptation of the vacuum tube for use in the binary mode and his arrangements for communicating among capacitors and tubes in the arithmetic process. In the matter of the circuitry itself, here again Atanasoff's procedure was not formal. He did, however, in his explanation of his circuit diagram, state a general principle that defined the basic circuits shown there. We give this principle and then use it to explain the circuits, along with their corresponding logical primitives. Finally, we give an example of synthesis, by showing how one rather complex primitive *could be* applied to execute one of *our* four general rules, namely, the rule for producing the carry bit in addition.

Our presentation here is a simplified version of the circuits of the add-subtract mechanism, showing the basic principles of their functioning. A more complete and precise analysis is provided in appendix A.

Four General Rules for Binary Addition and Subtraction

Table 1 is our adaptation of the table given in Atanasoff's proposal. From the addition table we can derive the following two rules. An odd number of inputs of "1" yielded a sum of "1." At least two inputs of "1" yielded a carry of "1." (All other cases yielded "0's.")

Our two rules for subtraction parallel the addition rules. In fact, the rule for the difference is the same as the rule for the sum, and the rule for the borrow *would be* the same as the rule for the carry if every counter bit were reversed (if every "0" from the counter drum were "1" and every "1" were "0"). Thus the same circuitry *could be* used for both addition and subtraction, so long as the counter bit was reversed in each determination of a borrow bit.

TABLE 1.   Truth Table for Binary Addition and Subtraction

Addition					Subtraction				
Bits in			Bits Out		Bits In			Bits Out	
Counter	Keyboard	Carry	Sum	Carry	Counter	Keyboard	Borrow	Difference	Borrow
0	0	0	0	0	0	0	0	0	0
0	0	1	1	0	0	0	1	1	1
0	1	0	1	0	0	1	0	1	1
0	1	1	0	1	0	1	1	0	1
1	0	0	1	0	1	0	0	1	0
1	0	1	0	1	1	0	1	0	0
1	1	0	0	1	1	1	0	0	0
1	1	1	1	1	1	1	1	1	1

Adaptation of the Vacuum Tube for Binary Arithmetic

In this and the next two subsections, we discuss the electronic details of Atanasoff's arithmetic circuits in terms of voltage levels, resistor values, and tube characteristics. Although we do not have access to precise figures, such as would be contained on circuit diagrams, we employ some plausible values that explicate the general principles in a realistic manner. As it happened, resistors and tubes varied significantly from one to another, and Berry determined the actual resistances for the add-subtract mechanisms by laboratory measurement.

We saw earlier that for a storage capacitor Atanasoff chose to have a relatively high voltage represent a "0," and a relatively low voltage represent a "1." We also saw that this association of a high voltage with "0" and a low voltage with "1" carried over to the vacuum tubes of the add-subtract mechanism.

The vacuum tubes Atanasoff selected for his binary arithmetic operations were triodes (plate, grid, cathode). He connected their plates to a direct-current supply of +120 volts through load resistors, most *but not all* of their grids to a supply of −120 volts through bias resistors, and their cathodes to ground (0 volts) as a reference voltage. Thus *input was to the grids,* and *output was from the plates.* These tubes were actually twin triodes, of type 6C8G, so that he needed only half as many envelopes per add-subtract mechanism as individual triodes; as it was, the complete network required thirteen triodes, or seven envelopes.

To give the high-low input and output voltage levels for these tubes, we must distinguish three types of application in the mechanism. There were three *input triodes,* driven by drum capacitors and driving other triodes. There were seven *internal triodes* driven by and driving other triodes. And there were three *output triodes,* driven by other triodes and driving drum capacitors. Notice that there were three triodes in this last category, even though only two output bits were produced for each arithmetic operation of one pulse time's duration. One output triode sent the sum or difference bit to the counter drum; it was driven by other triodes and drove a drum capacitor. The other two output triodes shared the sending of the carry or borrow bit to the carry-borrow drum; that is, during addition one triode was operative and sent the carry bit to a drum capacitor, while during subtraction the other was operative and sent the borrow bit to that same capacitor. This objective of a joint output was accomplished by having a common load resistor on the plates of these two output triodes.

In the following discussion, we give the voltage levels for just one of these applications, that of the internal triode, in order to elicit the basic features of the tube. That purpose is best served, in fact, if we limit ourselves to the simplest case of a triode driven by just one prior triode and driving just one subsequent triode (fig. 10). We will give the levels for the two applications involving communication between tubes and capacitors in the next subsection, after which we will move on to the design of basic circuits to execute the desired logical primitives.

At *input*, this triode's high voltage level representing "0" was 0 volts *or higher,* and its low voltage level representing "1" was −10 volts *or lower.* The high voltage level was determined by the fact that for a tube whose cathode was grounded, 0 volts at the grid would ensure that the tube was *on,* and the low voltage level by the fact that −10 volts would ensure that it was *off.* Notice that these input "levels" were really very wide ranges, which allowed considerable variation in the output levels of the driving tubes. And notice that a 10-volt *difference* in voltage at input sufficed to distinguish the on and off states of the tube.

At *output,* the high voltage level of this internal triode was about +100 volts, and the low voltage level was about +35 volts. Both levels were determined by the amount of voltage drop across its load resistor as current flowed from the power supply of +120 volts; each level was equal to +120 volts less that voltage drop, which, of course, depended primarily on whether the tube was on or off at the time. The voltage at the tube's plate was low when the tube was *on,* because a large current flowed from the +120 volt supply through the tube to the 0 volt supply. The voltage at the tube's plate was high when the tube was *off,* because with no current flowing through the tube only a small current flowed from the +120 volt supply to the plate through its load resistor, and on to the grid of the succeeding tube through an input resistor and to the −120 volt supply through the bias resistor on that grid. In each case, the amount of drop across the tube's load resistor depended on the value of the load resistor relative to the values of the two resistors on the grid of the driven tube, and also on the inherent characteristics of this tube Atanasoff had selected.

Notice now that the output voltage levels were appreciably higher than the input voltage levels. This discrepancy meant that one triode could not be wired directly to another that it was to drive, since there would have been no switching action; both a relatively high and a relatively low voltage at the plate of the first would have produced a relatively high voltage at the grid of the second, and so assured the on state only. It was therefore necessary to shift the voltage levels at the driving tube's plate downward to match those of the driven tube's grid. The input resistor mentioned above, running from that plate to that grid, had this purpose, in conjunction with the bias resistor on that same grid; the shift was accomplished by

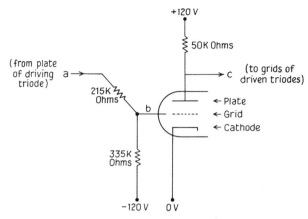

Fig. 10. Use of the vacuum tube in binary mode, NOT circuit. This figure shows a triode adapted for binary computation, specifically, a single internal triode that was driven by one prior triode and drove one or more further triodes. Three fixed voltages came from the direct-current supply: +120 volts to the plate load resistor, 0 volts to the cathode, and −120 volts to the grid bias resistor. Point *a* represents the input voltage to the circuit, coming from the plate of the driving triode. Point *b* represents the voltage on the grid of the (driven) triode. Point *c* represents the output voltage from the circuit, coming from the plate of the (driven) triode.

A voltage on the grid of −10 volts or lower ensured that the triode was off, so that no current flowed through it and the plate output became about +100 volts. A voltage on the grid of 0 volts or more ensured that the triode was on, so that current flowed through the plate resistor (50K ohms), and the plate output became about +35 volts.

Atanasoff adopted the convention that a high voltage at the plate or the grid signified a "0," a low voltage a "1." The binary electronics and its logical/arithmetic translation are shown in the following tables.

Input Voltage *a*	Grid Voltage *b*	Output Voltage *c*
+100	+15	+35
+35	−25	+100

Input Bit *a*	Output Bit *c*
0	1
1	0

Atanasoff's arrangement of his triode clearly caused it to execute the logical switching function NOT, the circuit he designated (1,1), meaning that it was a single-triode circuit driven by one triode, and that the plate voltage of the driving triode had to be low to turn the driven triode off (and high to turn it on).

Notice that Atanasoff's triode also performed the electronic function of signal amplification. A voltage difference of 10 volts at the grid was amplified to a voltage difference of about 65 volts at the plate.

assuring that the value of the input resistor bore the appropriate ratio to the value of the bias resistor. Thus it was that the choice of load and bias resistors for Atanasoff's triode and of input resistor to connect a driving triode to a driven triode was a critical factor in his circuitry. We return to the choice of resistor values later. Appendix A explores this topic in depth.

Notice also that whereas at the grid a high voltage representing "0" caused the tube to be on, the tube's being on caused a low voltage representing "1" at the plate; alternatively, a low voltage at input produced a high voltage at output. This triode, then, arranged as it was, actually performed the logical operation of negation. (It would not have negated if output had been at the cathode instead of at the plate.) Obviously, this feature of Atanasoff's triode was another critical factor in his circuitry.

Notice finally that the high-low voltage *difference* at output, about 65 volts, was up to nearly seven times that at input. This increase occurred because the vacuum tube was an amplifier, and when arranged so that output was from the plate, it was a *voltage* amplifier—a feature that made possible the use of capacitors for storage. The amplification process was reversed when the output levels of a tube were brought down to match those of a succeeding tube, but the remaining difference in high-low voltage levels was always more than sufficient to switch the driven tube off and on.

Communication between Vacuum Tubes and Drum Capacitors

Atanasoff had chosen capacitors as storage elements because they could communicate directly with vacuum tubes; that is, they did not require amplification on entry to the arithmetic unit. And he had seen that he could resolve their leakage problem by having output vacuum tubes restore their charges once every drum rotation.

In this subsection, we explain the arrangements for transmitting signals from the drum capacitors to the input triodes of the add-subtract mechanisms, where the arithmetic operations would be performed, and from the output triodes of those mechanisms back to the drum capacitors. We commence this cycle from the output triodes, however, as a more convenient approach.

But first let us take a brief look at the electronic circuitry of the restore-shift mechanism associated with each add-subtract mechanism. The add-subtract mechanism had three input triodes, to receive the three bits shown entering the arithmetic process in the truth table (see table 1). It also had three output triodes, one to send the resulting sum or difference bit to the counter drum, the other two to send (jointly) the carry or borrow bit to the carry-borrow drum. As was explained in the previous section, the restore-shift mechanism mediated the bit arriving from the keyboard drum, sending it both on to an input triode of the add-subtract mechanism and back to the keyboard drum.

The restore-shift mechanism had three triodes of the same type as those of the add-subtract mechanism. Two were input triodes, one to receive bits when a shift was in effect and one to receive bits when no shift was in effect. The third was an output triode. A separate output triode was required because of the negating action of the input triodes; it reversed again the bit it had received from one of these, returning it to its correct value (polarity). We do not give the communication details of this mechanism, since they are readily surmised from those of the add-subtract mechanism.

Communication from Output Triodes to Drum Capacitors

For convenience, we treat the three ouput triodes of the add-subtract mechanism as two, taking the pair that sent out bits jointly as one, although the resistor networks between the prior driving triodes and the grids of that pair were obviously more complicated than the network on the grid of the single output triode. The output triodes differed from the internal triode described earlier in that they were driven by more than one prior triode, as the previous sentence indicates, but that difference need not concern us here. They also differed in a highly relevant way, however, namely, in that while they drove only one subsequent element, that element was a drum capacitor rather than another triode.

Atanasoff wanted to make this communication as simple and direct as possible, while using capacitors of the size he needed. Because of the amplification mentioned above, he had to make an adjustment between the output voltage levels of the triode and the lower levels of the drum capacitor. But he did not want to insert an input resistor, as for connecting two triodes, because this would cause a loss of signal, and so of reliability. He therefore chose to make the difference in levels at the plate of his triode equal to the difference in levels in his drum capacitor, and to make an absolute voltage adjustment between the two during the actual signal transfer. Let us see how he achieved each of these conditions.

The decision not to insert an input resistor meant that the only resistor involved in the transfer of signals was the load resistor of the output triode, since of course there was no bias resistor. Moreover, with no subsequent triode, the high voltage level at output was the full +120 volts across that load resistor, because when the tube was *off*, there was no drop. The low voltage level (its *on* state) now depended on the value of the load resistor, and so did the difference in voltage levels. The difference in voltage levels for the drum capacitor was about 70 volts (high, about +30 volts; low, about −40 volts). Accordingly, Atanasoff set the value of the load resistor to achieve a low output voltage level for his output triode of about +50 volts, thus achieving the desired difference.

He now had triode output levels differing by the same 70 volts as the drum capacitor levels, but 90 volts higher (+120 to +30, and +50 to −40). He could match the two pairs of levels during each signal transfer either by momentarily adjusting the triode plate levels downward 90 volts or by momentarily adjusting the capacitor levels upward 90 volts. He chose to do the latter, by a process he termed a "boost" (p. 312). He achieved this once-per-pulse-time voltage elevation by installing a commutator system on the "fast" axle, that of the carry-borrow drum—in fact, contiguous to that drum.

The output from this commutator system went to the slip rings of all three drums, each slip ring being connected to the inner sides of its capacitors. It held the slip rings at 0 volts during the reading phase and an interim storage phase, but connected them to a supply of +90 volts for the writing phase. That is, the inner sides of the capacitors began receiving the 90-volt boost just before the moment of writing, they were at the full 90 volts during the writing, and they were returned to ground immediately after the writing (fig. 11).

This boost of 90 volts on the *inner* sides of the capacitors meant that the voltage on the *outer* sides, and so on the external studs, was raised by 90 volts at the moment of writing, either from about +30 to +120 volts (in the case of a "0") or from about −40 to about +50 volts (in the case of a "1"). The arrival now of a signal, via wire and brush, from an output triode of an add-subtract mechanism caused a given capacitor to acquire the voltage transmitted by that triode. In each of the four possible conditions (high or low at the plate of the output triode, to high or low at the capacitor), the capacitor was either charged or discharged until it came to have the plate voltage.

The immediate drop to 0 volts on the slip rings brought the voltage on the outer side of a given capacitor down again to either about +30 volts or about −40 volts, depending on whether a "0" or a "1" had been sent by the corresponding output triode. Having now received a bit from an output triode, the capacitor was ready for the reading phase of the next second's addition or subtraction operation. See appendix A for more on the boost.

Notice in passing that the "overwrite" feature in the writing phase precluded the need to clear the storage drum capacitors prior to each new addition or subtraction. It was necessary to clear the capacitors of the carry-borrow drum to "0's," however, so that "1's" left over from one operation would not enter the units place addition or subtraction of the next operation. This clearing was effected by pulses from a commutator system on the "slow" axle between the two storage drums.

Communication from Drum Capacitors to Input Triodes

The reading phase was much more easily accomplished than the writing phase, because a direct voltage level match already existed between capacitors and input triodes. The capacitor levels were a high of about +30 volts and a low of about −40 volts; the input (grid) levels of the input triodes were a high of 0 volts *or higher* and a low of −10 volts *or lower,* just as for the internal triode presented in a previous subsection. Since this triode was to send its output signals on to other triodes, it could be arranged as given there, except that this was the one case where the grid was not connected to the supply of −120 volts through a bias resistor. Instead, the grid of an input tube was connected to a very small capacitor within the add-subtract mechanism, and on to ground (see fig. 11).

Each grid capacitor received the signal from its particular drum capacitor,

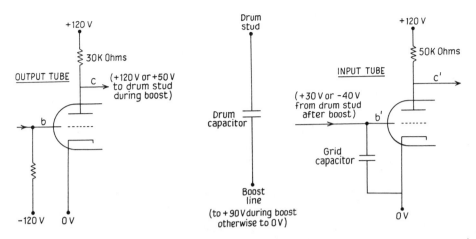

Fig. 11. Boost circuit. Shown in cross section are an output tube of an add-subtract mechanism, a capacitor and stud of an associated drum abacus, and an input tube of that same mechanism. Also shown is the boost line connecting the drum capacitor to the commutator system on the fast shaft of the computer.

The boost occurred during the second half of a pulse time, as the bit sent from the output tube was being written on the stud of the drum capacitor; it raised the voltage levels on the stud (about +30 volts, about −40 volts) by 90 volts, to those of the plate of the output tube (+120 volts, about +50 volts), so that the bit represented by the voltage at that plate could be recorded on the stud. In order to achieve this voltage boost, the boost line, whose reference connection was to ground, was connected to +90 volts via the commutator system just before the writing brush engaged the stud. It was returned to 0 volts, again via the commutator system, immediately after the writing phase; the stud voltage then dropped 90 volts to one or the other of its reading values (about +30 volts, about −40 volts), whichever represented the bit value just received from the output tube.

via the wire from the brush that read the capacitor, and it held this signal until the outputs of the add-subtract mechanisms were recorded on the drums (about half a pulse time). There was some reduction in voltage during the transmission from drum capacitor to grid capacitor, but the high and low levels remained such that the signal for "0" was always positive and the signal for "1" was well below −10 volts. Also, because the grid capacitors were considerably smaller than the drum capacitors, it was not necessary to clear them between successive bit transmissions. Appendix A provides a fuller account of the situation between pulse times.

Communication among Vacuum Tubes

Our example of a simple internal triode of the add-subtract mechanism, one driven by one prior triode and driving one subsequent triode, was given to elicit the basic features of the triode Atanasoff used throughout the mechanism (and also in the restore-shift mechanism). We saw that he arranged to bring the output voltage levels of the driving triode down to match the input levels of the driven triode by inserting an input resistor from the plate of the first to the grid of the second, and choosing appropriate values for this resistor, the bias resistor of that same grid, and the load resistor on the plate of the driving triode. This arrangement on the grid of a driven tube constituted a simple *resistor network*. The driven tube, together with the resistor network on its grid, constituted a simple *switching circuit*. It was a *one-triode* circuit, in that it was a circuit in which a single triode was driven by *one or more* prior triodes.

We also saw in that discussion that a single internal triode executed the logical operation of negation. A one-triode circuit, then, with one driving triode constituted a NOT circuit (see fig. 10). Atanasoff's add-subtract mechanism had two such NOT circuits among his internal triodes. We should keep in mind, though, that the three input triodes were single triodes driven by one *source,* a drum capacitor, and they, too, negated the bits they received at their grids. Thus, while the "networks" on their grids differed from those of the internal triodes, the input triodes, together with those arrangements, also constituted simple switching circuits.

But the binary additions and subtractions Atanasoff had worked out in his truth table called for more complex logical operations than negation, and so for more complex switching circuits to perform them. Most of these were one-triode circuits with two or three driving triodes. In these cases, an input resistor was inserted between the plate of each individual driving tube and the grid of the driven tube. The two or three input resistors on a given grid were of equal value, a value chosen along with that of the grid's bias resistor to achieve the necessary voltage match from driving tubes to driven tube.

Of course, an arrangement also had to be made for performing the desired logical operation in each case. Atanasoff saw that he could choose his resistor values in such a way as to attain the two goals simultaneously; that is, he could achieve the desired operation at the same time that he achieved the voltage match. The details of this phenomenon become quite technical. We give just a brief explanation here, before turning to the logical primitives executed by the add-subtract mechanism, and to the principle by which basic circuits that executed them were defined.

Because of the great leeway afforded by the input requirements of the input triodes (0 volts or higher, -10 volts or lower), there existed several alternative ways to set the values for the resistors of a given circuit: several different pairs of operative high-low voltage levels could be calculated that would be acceptable at the grid of the driven triode. Of these pairs, one could be found that would also serve to accomplish the logical operation desired of a particular circuit. Such a pair, for example, might cause a low voltage at the grid of the driven tube under certain conditions at the plates of the driving tubes, and a high voltage at that grid in the absence of those conditions. See appendix A for a detailed description of these resistor networks.

Notice in passing that the concept of a circuit used here—and the one Atanasoff used in this defining formula—focuses on the *driven* tube and the number of tubes (or other sources) that drove it. It does not address the number of tubes (or other elements) each such driven tube might in turn have driven. The seven internal triodes drove up to four triodes each. As for the input triodes, each of them drove two others. The two outputs of the (three) output triodes, of course, each drove one drum capacitor.

Logical Primitives and Basic Circuits

Atanasoff, in his 1940 manuscript, formulated six basic one-triode circuits, by stating the principle defining them (p. 314) and marking the instances of each in his circuit diagram of the add-subtract mechanism (p. 313) (fig. 12). Rather, his principle defined six such circuits, but he found need for—and marked in his diagram—only five. Each of these basic circuits consisted of the driven triode and its resistor network connecting one, two, or three prior (driving) triodes to it. And each executed a particular logical primitive—although, as we have said, Atanasoff did not go through the process of translating his electronics into logic.

We now state Atanasoff's defining principle and show how his three simplest basic circuits executed the primitives, NOT, NOR, and NAN (NOT-AND), respectively. We also comment on the three more complex circuits covered by the principle, give the primitives they *could* execute, and note what use if any Atanasoff assigned each of them in his diagram. We then conclude our presentation of the electronic design of the add-subtract mechanism with an example of synthesis, showing how one of these more complex circuits *could be* combined with a NOT circuit to execute the carry rule we formulated in connection with table 1.

Atanasoff expressed the principle of his basic one-triode circuits in terms of the notation (p,q). Here p was the number of driving triodes in the designated circuit; and q was the minimum number of those driving triodes that had to have *low* plate voltages in order to turn off the driven triode (p. 314), that is, in order to ensure a *low* input voltage at the grid of the driven triode and a *high* output voltage at its plate. Notice that this principle logically defined just one circuit to be driven by one triode, two circuits by two triodes, and three circuits by three triodes, for a total of six possible circuits; it was one of the three in the last category that Atanasoff dispensed with.

The simplest circuit to which this principle applied was, of course, that for which (p,q) was $(1,1)$. It executed the logical primitive NOT (see fig. 10). As we have seen, it was driven by one triode, and the plate voltage of that triode had to be low to turn off the driven triode (ensure a low voltage at its grid). There were two $(1,1)$ circuits in the add-subtract mechanism, both of them internal.

The next simplest circuits were the two driven by two prior triodes. These executed the primitives NOR and NAN, respectively. This two-input NOR circuit—as distinguished from the three-input NOR circuit, $(3,1)$, also covered by the defining principle—had a (p,q) of $(2,1)$ (fig. 13). Its resistor network was designed so that the plate voltage of at least one of the driving triodes had to be low to turn off the driven triode. The add-subtract mechanism had just one $(2,1)$ NOR circuit, an internal one.

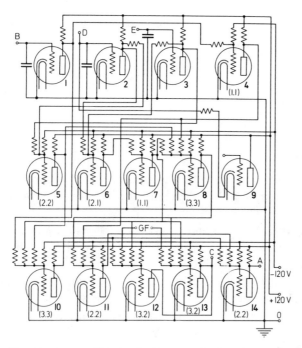

Fig. 12. Atanasoff's circuit diagram of add-subtract mechanism. In this diagram from Atanasoff's 1940 manuscript (p. 313), *B* (upper left) represents the input point that receives a bit from an abacus of the counter drum and passes it to input triode 1, and the short double line is that tube's small grid capacitor. Similarly, *D* represents the input point that receives a bit from a keyboard abacus and passes it to input triode 2, and *E* the input point that receives a bit from a band of the carry-borrow drum and passes it to input triode 3. (These and other letters correspond to those used in fig. 7.)

The (*p,q*) designations of one-triode circuits as defined by Atanasoff are given for triodes 4 through 14, with the exception of 9, the inactive extra triode of the seven envelopes of each add-subtract mechanism. All plates are connected to the +120 volt power supply through load resistors, all grids to the −120 volt supply through bias resistors, and all cathodes to ground. The connections from triode to triode, including resistor networks, can be traced; they are obviously very complex, with one triode driving as many as four others. (The logical structure of this circuit is shown in app. fig. A.17.)

A (lower right) represents the output point that passes the sum or difference bit from output triode 14 to the counter drum abacus, and *C* represents the output point that passes the carry or borrow bit from paired output triodes 12 and 13 to the band of the carry-borrow drum. Recall that whether the pulse-time operation has been addition or subtraction, the resulting sum or difference bit is the same for any given combination of three input bits. Accordingly, no distinction need be made between the two operations to produce the bit to be sent to the counter drum by output triode 14. The distinction must be made, however, for the carry or borrow bit to be sent by paired triodes 12 and 13.

G and *F* represent relays that together signal output triodes 12 and 13 as to whether their joint output to the carry-borrow drum (joint because their plates have been connected to a common load resistor) is a carry bit or a borrow bit. If *F* is high ("0") and *G* low ("1"), triode 13 sends a carry bit, to be treated as such in the next pulse time, while triode 12 is inactive; if *F* is low ("1") and *G* high ("0"), triode 12 sends a borrow bit, while triode 13 is inactive. Thus, whereas Atanasoff has designated both triodes 12 and 13 as one-triode circuits with (*p,q*)s of (3,2), they really constitute a two-triode circuit with six driving sources. (See also app. figs. A.16 and A.17.)

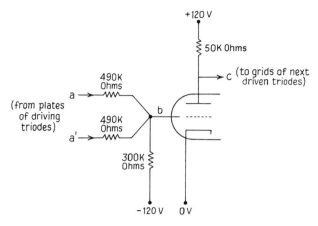

Fig. 13. Two-input NOR circuit. Atanasoff's (2,1) circuit executed the logical primitive NOR. Its voltage table and truth table are as follows.

Input Voltages		Grid Voltage	Output Voltage		Input Bits		Output Bit
a	a'	b	c		a	a'	c
+110	+110	+5	+35		0	0	1
+110	+35	−15	+110		0	1	0
+35	+110	−15	+110		1	0	0
+35	+35	−35	+110		1	1	0

The two-input NAN circuit had a (p,q) of (2,2) (fig. 14). Its resistor network was designed so that the plate voltages of both driving triodes had to be low to turn off the driven triode. There were three (2,2) NAN circuits in the add-subtract mechanism. Two of these were internal, and the third was that of the output triode which sent the sum or difference bit to the counter drum capacitors.

The three remaining circuits covered by Atanasoff's defining principle were those driven by three prior triodes. Their (p,q)s were (3,1), (3,2), and (3,3), respectively, and they were quite complex. One of them, (3,1), was a three-input NOR circuit; it was also the one circuit of the possible six that Atanasoff did *not* include in his design. Another of them, (3,3), was a three-input NAN; he had two of these, both internal.

Atanasoff marked two triodes (3,2), namely, the two output triodes whose plates shared a common load resistor to produce a single output bit for the carry-borrow drum (see fig. 12). But this designation has two difficulties. First and more important is the fact that their operation was mutually exclusive: one of them produced the carry bit, the other the borrow bit. Their joint output depended on their having a common load resistor, which performed an OR function. Second, each of these two output triodes was actually driven by two triodes and a relay,

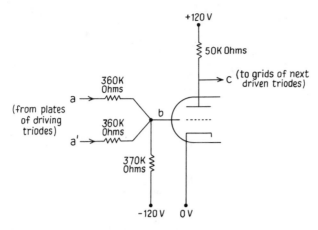

Fig. 14. Two-input NAN Circuit. Atanasoff's (2,2) circuit executed the logical primitive NAN (NOT-AND). Its voltage table and truth table are as follows.

Input Voltages		Grid Voltage	Output Voltage		Input Bits		Output Bit
a	*a'*	*b*	*c*		*a*	*a'*	*c*
+110	+110	+35	+35		0	0	1
+110	+35	+10	+35		0	1	1
+35	+110	+10	+35		1	0	1
+35	+35	−15	+110		1	1	0

instead of by three triodes. The role of the relays was to signal whether the output bit in a given instance was to be added (a carry bit) or subtracted (a borrow bit): *either* the relay that drove one output triode signaled addition *or* the relay that drove the other output triode signaled subtraction, as appropriate for that instance.

Thus these paired output triodes really constituted a two-triode circuit, with six driving sources and a single output. We will not discuss them further. However, we have still to present the basic one-triode (3,2) circuit of the usual arrangement—with a standard plate output and with three driving triodes—in order to give the promised illustration of synthesis.

First, though, let us note that we have accounted for all thirteen triodes of Atanasoff's add-subtract mechanism. There were seven internal one-triode circuits; two were (1,1) NOT circuits, one was a (2,1) NOR circuit, two were (2,2) NAN circuits, and two were (3,3) NAN circuits. There were three one-triode input circuits, each of which was a NOT circuit (driven by a capacitor on one of the three drums). There was one one-triode output circuit, a (2,2) NAN circuit (that drove a counter drum capacitor). And there was a two-triode output circuit (that drove a capacitor of the carry-borrow drum); it did not execute any of the commonly recognized logical primitives. (The output triode of the restore-shift mechanism drove a capacitor of the keyboard drum.) Appendix A gives a much fuller account of the logical and electronic design of the add-subtract mechanism.

Carry Rule: A Synthesis of Primitives

The rule we abstracted from the truth table for addition (table 1) stated that at least two inputs of "1" yielded a carry of "1," and all other cases yielded ·0's." Now this happens to be an instance of a positive logical primitive known as a "threshold-two element with three inputs."

In Atanasoff's add-subtract mechanism, a one-triode circuit with a (p,q) of (3,2) would have had its resistor network designed so that the plate voltage of at least two of the three driving triodes had to be low to turn off the driven triode. If we substitute arithmetic values for electronic, this translates into: At least two of the bits from the three driving triodes had to be "1" to produce a "1" at the grid of the driven triode. Because of the negating action of this tube, this translation is equivalent to: At least two of the bits from the three driving triodes had to be "1" to produce a "0" at the plate of the driven triode. And so the logical primitive executed by Atanasoff's (3,2) circuit is the negation of the positive primitive of which our carry rule is one instance. By the law of double negation, then, our carry rule could be executed by negating a (3,2) circuit as defined by Atanasoff.

Atanasoff *did not* execute our carry rule, although, as we have seen, he used his (3,2) circuit (driven by two triodes and a relay) in producing both his carry and his borrow bits. We have given the circuitry for our rule here as he *could have* executed it, to provide an example of synthesis of primitives. Appendix A has more on the circuitry for carrying and borrowing.

The three primitives NOT, NOR, and NAN that we presented earlier were negative; and we have just seen that the primitive with a (p,q) of (3,2) was also negative. In fact, because of the negating effect of Atanasoff's vacuum tube as he had arranged it, all his primitives had to be negations; positive primitives like AND and OR were attainable only by further negation.

Concluding Comments

It was through a very efficient electronic design that Atanasoff—with Berry's assistance—limited the number of triodes in each add-subtract mechanism to 13. The restore-shift mechanisms associated with each added another 3 triodes, for a total of 16 for each pair of mechanisms. All of these triodes were housed in envelopes of 2. Because the thirty add-subtract mechanisms were mounted individually on small boards (five by seven inches), one mechanism required seven envelopes, with 1 triode standing idle. The restore-shift mechanisms, on the other hand, were not mounted separately, and so each one required one and one-half envelopes, sharing its odd envelope with a neighboring mechanism (fig. 15). There were thus a total of 390 triodes *used* in the add-subtract mechanisms, and a total of 90 in the restore-shift

Fig. 15. Clifford Berry holding restore-shift mechanisms

mechanisms, for a grand total of 480 triodes used by the arithmetic mechanisms proper; in terms of envelopes, there were 210 + 45, or 255, envelopes.

As for tubes in other parts of the computer, Atanasoff used a few triodes in connection with his electromechanical controls, and another 31 in his binary-card input-output system. We should perhaps mention also the 31 thyratrons (gas tubes) in that system. These bring the total for the entire machine to a figure very close to 525 triodes, or 280 envelopes, plus 31 thyratrons.

We hope our detailed presentation of Atanasoff's add-subtract mechanism has conveyed not only an understanding of the device itself, but also an appreciation for the elegance of his conception and design. It was a highly significant contribution to the invention of today's electronic computer, and a most extraordinary achievement for the year 1938.

Timing and Control

Before moving on to the input and output procedures of Atanasoff's computer, we now pause to explain how the activities of the two systems described thus far, the memory and the arithmetic unit, were synchronized. In fact, as we shall see, the timing and control arrangements these systems entailed applied as well to those input-output procedures. For all incoming and outgoing information either passed through or was processed by the add-subtract mechanisms, and the associated base conversions were also processed by them.

All of the transactions between the add-subtract and restore-shift mechanisms of the arithmetic unit, on the one hand, and the storage and carry-borrow drums, on the other, were ultimately timed from the synchronous motor. This was true for such transactions whether the machine was actually executing a procedure involving at least some of the storage drum capacitors or was just running idly, so that at any given time all storage drum capacitors either were receiving new values or were being continually refreshed.

As we have seen, the synchronous motor, through its gear drive, rotated the computer's "slow" axle at the rate of 60 revolutions per minute, its "fast" axle at the rate of 900 revolutions per minute. Some things were effected simply by the rotation of the drums on these axles, as readings from the drums' contacts were taken by brushes. Others were effected by the rotation on each axle of a commutator system, made up of commutators and their associated slip rings.

The fast commutator system, contiguous with the carry-borrow drum, operated on a pulse-time basis. Its sole function was the one presented previously, to emit a "boost" signal during the second half of each pulse time for the recording of bits from the add-subtract and restore-shift mechanisms onto all three drums.

The slow commutator system, which we call the timing drum, was positioned between the two storage drums. It operated on an addition-time basis, during the 60-degree blank portion of each storage drum rotation; or, more specifically, during the period from the last bit position (the sign position) of a given vector addition or subtraction through that blank portion. It had two functions. One of these, also presented previously, was to emit the pulses that cleared the capacitors of the carry-borrow drum for each new addition or subtraction. The other was to time the various control operations. Let us see how this latter function was executed during the elimination process.

As the sign bits left the add-subtract mechanisms, the timing drum sent a signal to a vacuum-tube switch which passed the sign bit of the designated coefficient on the counter drum to a control mechanism. This electromechanical control then interpreted the sign bit and issued the appropriate orders to the add-subtract mechanisms (to add or to subtract) and the restore-shift mechanisms (to shift or not) on the next rotation. For all except the first rotation in the processing of a pair of equations, this interpretation depended on whether the sign of the designated coefficient had changed. If it had changed, there would be a shift and a change in arithmetic operation; otherwise, no shift and no change in operation. For the first rotation in a given series, there would be no shift, and the arithmetic operation ordered would depend on whether the sign of the designated coefficient on the counter drum was the same as that of the corresponding coefficient on the keyboard drum. That is, the keyboard equation would be subtracted from the counter equation if their signs were the same, and added to it if they were different.

Besides this activity during the blank portion of each drum rotation, another control mechanism counted the shifts as they occurred, by means of a relay counter. Because a shift corresponded to the elimination of a digit from the designated coefficient on the counter drum, a count of forty-nine shifts indicated that the entire coefficient had been eliminated. At this point, the control ordered a halt to the series of add-shift-subtract operations.

There was also a one-cycle switch to advance an operation just one addition time, for checking and demonstration purposes.

Atanasoff had intended, as late as August, 1940, when he finished the manuscript describing his computer—already under construction—to have his controls be electronic. "In designing these controls, free use has been made of vacuum tubes" (p. 319). We take this statement to mean, not that he had completed their design, but that he believed he knew how he could do it. Electronic controls would have entailed switches, flip-flops, and counters. He had already developed switches for his add-subtract mechanisms, and flip-flops were so simple as to present no problem. The counters to count the shifts would have been harder, but still considerably easier than the ones on which he had tried to base his computing mechanisms, since the required operation was both much simpler and much slower.

In the end he used electromagnetic relays, probably at least partly as an economy measure. Whereas he had to pay about four dollars for vacuum tubes, he could buy relays for about one dollar apiece, and although the overall savings would be just a few hundred dollars, this was a significant percentage of his budget. He did not need the electronic speed in these control operations, of course, because of the time factor mentioned before, namely that the control mechanisms had 60 degrees of rotation (a sixth of a second) in which to function.

We should note that Atanasoff also mentioned manual controls in his manuscript, in reference to the entire computer: "power switches, a keyboard, switches for controlling the start of various operations (such as card punching or reading), routing switches to select the particular abacus on which any number is to be placed, and a flexible arrangement of plugs and jacks to provide for special setups" (p. 319).

Decimal Input-Output and Base Conversion

As we have seen, the solution of a set of simultaneous linear equations on Atanasoff's computer required four different input-output procedures. Two of these, the initial input of decimal data and the final output of decimal results, had an external connection. The two intermediate procedures, recording binary equations from the read-write memory as they were formed in the elimination process, and reentering them in that memory as they were needed for further processing, were completely internal.

Atanasoff designed and installed a different mechanism for the execution of each procedure. For the initial input he devised a decimal-card reader, and for the final output he devised what may best be described as a decimal accumulator. For the two intermediate procedures he devised, respectively, a binary-card "punch" and a binary-card reader.

The first and last procedures were greatly complicated by the necessity for base conversion, for which he devised a decimal-binary conversion drum. The conversions were computed either onto or off of the counter drum, through the add-subtract mechanisms. In the first instance, this was conversion from decimal to binary of the original data, namely, up to thirty coefficients for each of up to twenty-nine equations in the original set to be solved. In the last, it was conversion from binary to decimal of the two coefficients of each single-variable solution equation. (Recall that Atanasoff meant to use a desk calculator at the end to divide one such converted coefficient by the other for the ultimate solution of each variable.)

The present section is devoted to explaining this complementary pair of "external" procedures, taken now to include the base conversions with which they were integrated. The other complementary pair of "internal" procedures, which in the end gave Atanasoff trouble, is explained in the next section. They too involved the add-subtract mechanisms, and although they were not complicated by base conversion, they were still highly complex. There was no precedent for recording and reentering binary data, and Atanasoff invented an entirely original means of "punching" and reading cards.

For this presentation, we draw on an untitled memorandum written by Clifford Berry in late 1941 or early 1942 for patent application purposes, as well as on Atanasoff's 1940 proposal (ENIAC Trial Records).

Decimal to Binary

The counter drum played its usual accumulator role as the decimal-binary base conversions were computed onto it through the add-subtract mechanisms. The usual readings from the keyboard drum, however, were replaced by binary readings from the special base-conversion drum. This drum held a read-only decimal-binary conversion table, and it rotated in unison with the counter and keyboard drums. (It was not located on their axle, however, as Atanasoff had originally intended; to make room for this drum's reading brushes and several large bundles of wires leading from those brushes, he placed it on an axle parallel to but set back a few inches from the longer one and, of course, geared to it.)

The conversion drum and its reading brushes provided, in parallel, 135 streams of binary signals, one for each decimal digit 1, 2, . . . , 9 in each of the fifteen positions of a possible decimal coefficient. All these streams went to the decimal-card reader, which selected certain streams from them according to the decimal digits punched in the card and sent the selected streams on to the arithmetic unit. There the binary conversion of a given coefficient was computed onto a counter abacus by its associated add-subtract mechanism, which combined each successive partial conversion, from greatest to least, into the total already registered on the counter abacus. Thus with each drum rotation, an add-subtract mechanism combined a newly arriving partial conversion with the current total arriving from the counter abacus, and returned a new total to that abacus.

Notice that this decimal-binary conversion was made in the order: most significant decimal digit to least. It could have been made in the opposite direction, but the chosen order was that required by the binary-decimal conversion at the end of the computation. Notice also that no binary representations of the decimal "0" were provided from the drum; instead, the absence of a stream of signals indicated a "0."

Table 2 shows the successive binary transmissions for the decimal number 00. . .0875. After a series of 00. . .0000000000 entries (binary equivalents of $0 \times 10^{14}, 0 \times 10^{13}, \ldots, 0 \times 10^{3}$), the binary equivalent of 8×10^{2} was entered onto the counter drum; then the binary equivalent of 7×10^{1} was added to that, and the sum entered onto the counter drum; lastly, the binary equivalent of 5×10^{0} was added to that sum, and the new sum, the desired binary equivalent of 875, entered onto the counter drum.

The binary equivalents of particular decimal numbers, each a multiple of a power of ten out to 9×10^{14}, were represented by parallel bands of studs around the circumference of the conversion drum, with the presence of a stud signifying "1," the absence of a stud signifying "0," just as the absence of an entire band signified repeating "0's." Like the storage drums with which it was rotating, this drum also had a blank portion of 60 degrees that could be used to prepare for the operations of the next rotation. Brushes stationed along the length of the drum *read* all of the binary numbers as it turned, but for a given coefficient only the partial conversions that matched its particular digits were passed to an add-subtract mechanism, and so to a counter abacus.

As we said earlier, this selection of partial conversions was accomplished by a decimal-card reader that filtered signals from the conversion drum to the add-subtract mechanisms through the holes in a decimal card. All the coefficients of all the equations of the original set to be solved were punched onto standard IBM cards of eighty ten-digit columns. With each coefficient occupying fifteen columns, one card could hold five coefficients as five successive matrices, and six cards were required for one equation with thirty coefficients. The sign of each coefficient was recorded in an extra row of the first-read column of each matrix.

The tray of the card reader onto which cards were to be placed, one by one, consisted in an array of contacts that matched the hole positions of a card. The brushes at the conversion drum were wired to these contacts, and so, with a card on the tray, a sequence of signals representing the appropriate partial conversion could pass through the hole in each column, provided a means existed for transmitting the sequences to the add-subtract mechanisms (fig. 16).

Atanasoff arranged to read the coefficients from a given card by means of columns of brushes mounted on a rack that was to be lowered over the tray of the card reader. There were five such columns, with ten brushes each (one for reading the sign), and they were spaced so that if one was over, say, the hundreds column of a card, the other four were over the other hundreds columns (see fig. 2). With this

TABLE 2.   Example of Decimal-Binary Conversion

Drum Decade	Decimal Digits	Binary Equivalents	Counter Drum Abacus
10^{14}	00. . .0000	00. . .0000000000	00. . .0000000000
. . .	. . .	. . .	. . .
10^2	00. . .0800	00. . .1100100000	00. . .1100100000
10^1	00. . .0070	00. . .0001000110	00. . .1101100110
10^0	00. . .0005	00. . .0000000101	00. . .1101101011
Decimal Total:	00. . .0875	Binary Total:	00. . .1101101011

arrangement, he could read the successive digits of the five coefficients of a given card in unison; that is, he could read all five columns in the fifteenth place in parallel, then all five in the fourteenth place, and on down to the hundreds, tens, and units columns. Connection of each column of brushes to its assigned add-subtract mechanism was made by hand-plugging a cable at the time a new card was inserted in the reader. Entry to the mechanism was directly to the input triode that normally received bits from the keyboard drum via a restore-shift mechanism; that is, the restore-shift mechanism was now bypassed.

The reading and conversion procedure was as follows. With a card on the tray and the five columns of reading brushes plugged into their assigned add-subtract mechanisms, the rack was lowered over the tray with its brushes engaging the most significant (fifteenth-place) columns. Five streams of bits from the rotating conversion drum passed through the punched holes of those columns and on to the add-subtract mechanisms. As the drum rotated during its blank portion, the rack of brushes was advanced automatically to the fourteenth-place columns, which were then read into the add-subtract mechanisms, and so on down to the units columns. In this way, the five coefficients of one card were converted in fifteen rotations, or fifteen seconds, and one complete equation of thirty coefficients was entered on the counter abaci in ninety seconds of machine time.

Positive and negative coefficients had to be handled differently, since the latter had to be entered on the counter abaci in two's complement form. The sign of each coefficient was punched in the zero row of the most significant column, and it was sensed during the 60-degree control portion of the drum rotation. According to the result, the controls of the corresponding add-subtract mechanism were set to "add" or to "subtract." For a positive sign, each binary number received from the conversion drum was added to the sum being accumulated on the counter drum. For a negative sign, each binary number received from the conversion drum was subtracted from the number being accumulated on the counter drum.

We have now completed our explanation of Atanasoff's initial input procedure, except for a description of his base-conversion drum. Because this drum was used in the final output procedure, as well, we treat it in a separate subsection and then go on to that last procedure.

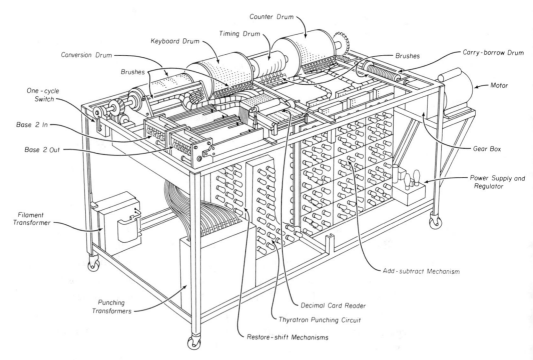

Fig. 16. Drawing of computer. Like figure 5, this drawing (with minor modifications) was made for use in court.

Design of the Base-Conversion Drum

The read-only decimal-binary conversion drum was of about the same length as the read-write storage drums, but of about half their diameter. As we have said, the binary representations of the nine multiples of the powers of ten out to 9×10^{14} $(1, 2, \ldots, 9; 10, 20, \ldots, 90; 100, 200, \ldots, 900; \ldots; 1 \times 10^{14}, 2 \times 10^{14}, \ldots, 9 \times 10^{14})$ were available from this drum, in the form of bands of studs around it, with a stud signifying "1," no stud signifying "0."

Wires from the inner ends of the studs were connected to a negative direct-current power source on the outside through a slip ring fixed to, but insulated from, an end mounting plate. This negative voltage would have sufficed to transmit a "1" to an input triode of an add-subtract mechanism whether the triode was already off or on. However, the absence of a signal from a position on the conversion drum would transmit a "0" only if the triode was already on, that is, only if the input was normally at ground. This could have been arranged by connecting a bias resistor between each line from the decimal-card reader and ground, so that the input to each add-subtract mechanism being used for conversion would be 0 volts in the absence of a signal from the reader. Berry, in his memorandum on the computer's specifications mentioned above, suggested that better tolerances could have been achieved by connecting the lines into these triodes to ground each pulse time, through a commutator system on the fast axle.

Table 3 shows how these bands of studs would be formed for the first three decades, that is, for the units, tens, and hundreds decades.

Sometime before the drum was built, Atanasoff and Berry realized that the number of bands could be reduced much further, by taking advantage of the fact that certain binary sequences were products of others by powers of two, and so could be obtained by shifting. Berry explained this principle in his memorandum.

The band in the third place of each decade would yield a reading for the sixth place if an extra reading brush was stationed one stud position (6 degrees) later in the drum rotation: the band at 3 could be made to give the conversions for both 3 and 6; at 30, for both 30 and 60; . . . . This is easily confirmed from table 3, where we see the following binary equivalents in the first two decades:

3:	. . . 000000000011	If read 6 degrees later, the band for 3
6:	. . . 000000000110	would be the same as the band for 6, and
30:	. . . 000000011110	the band for 30 the same as the band for
60:	. . . 000000111100	60.

In the same way, the band in the first place of each decade would yield a reading for the second, the fourth, and the eighth places, if an extra reading brush was stationed one, two, and three positions (6, 12, and 18 degrees) later in the drum rotation: the band at 1 could be made to give the conversions for 1, 2, 4, and 8; at 10, for 10, 20, 40, and 80; . . . . But except for the first decade, the band in the first place would also yield a reading for the fifth place in each *previous* decade, if an extra brush was stationed one position (6 degrees) *earlier* in the drum rotation. These multiple readings can also be confirmed from table 3, where we see that the band for 10 yielded the binary equivalents of five different decimal numbers:

5:	. . . 000000101	If read 6 degrees *earlier,* the band for 10
10:	. . . 000001010	would be the same as the band for 5. If read 6,
20:	. . . 000010100	12, and 18 degrees *later,* it would be the same
40:	. . . 000101000	as the bands for 20, 40, and 80, respectively.
80:	. . . 001010000	

In the fifteenth decade, of course, a band to yield 5×10^{14} did have to be provided, since there was no succeeding (sixteenth) decade from which the reading could be taken.

All of this brings to sixty-one the total number of bands needed on the conversion drum. The number of reading brushes, however, remained at 135, and each brush had to be wired to five contacts in the tray of the card reader. The wire from the brush reading the binary conversion of the decimal 3, for example, was fanned to transmit this sequence to the contact below the "3" position in the units column of each of the successive matrices; the wire from the

TABLE 3.   Pattern of Decimal-Binary Conversion Drum

The array continues to the left for twelve more decimal digits.

Hundreds (×10²)									Tens (×10¹)									Units (×10⁰)									
9	8	7	6	5	4	3	2	1	9	8	7	6	5	4	3	2	1	9	8	7	6	5	4	3	2	1	
																											2^{48}
																											•
																											•
																											•
																											•
																											2^{10}
●	●	●	●																								2^9
●	●			●	●	●																					2^8
●		●		●	●		●																				2^7
			●	●			●	●	●	●	●																2^6
	●	●		●		●		●				●	●	●													2^5
		●	●	●	●				●	●		●	●		●	●											2^4
		●	●			●	●		●			●		●	●		●	●	●								2^3
●		●		●		●		●			●	●			●	●				●	●	●	●				2^2
									●		●		●		●		●			●	●			●	●		2^1
																		●		●		●		●		●	2^0

A large dot indicates the presence of a stud, for a "1."
A blank cell indicates the absence of a stud, for a "0."
Compare table 2 and its representation of 875.

brush reading the binary conversion of the decimal 800 was fanned to the contact below the "8" position in the hundreds column of each of the five matrices (see figs. 2 and 16).

Binary to Decimal

The decimal-binary *input* procedure already described had to be reversed to a binary-decimal *output* procedure at the conclusion of the internal solution process. The elimination algorithm for a full set of equations ended with twenty-nine equations in one variable (each in a different variable). The solution for each variable could now be found by dividing one coefficient (the constant term) by the other (the coefficient of the variable). Atanasoff had not arranged for this machine to divide two binary numbers, and even if he had, he would still have had to convert each resulting quotient to the decimal answer he needed. He therefore decided to convert each of the two coefficients of each binary equation in one variable to decimal form, and then divide the one decimal coefficient by the other on a desk calculator, thereby achieving the final decimal solution to each such equation. This procedure meant a total of fifty-eight binary-decimal conversions, as against the twenty-nine he would have required had he been able to divide on his computer; but it also meant a great saving in not having to provide for full division.

We noted earlier that he used the same drum for these final binary-decimal conversions as for the initial decimal-binary conversions, but with a different procedure. There he had accumulated the partial binary conversions of a decimal coefficient on a counter abacus, decade by decade, progressing from most to least significant decimal digit, or decade. In the binary-decimal conversions, on the other hand, he started with a binary coefficient on a counter abacus and successively subtracted (added) partial binary conversions, decade by decade, until the coefficient was eliminated, again progressing from most to least significant decade. There he had used a decimal-card reader to select the decimal number in each decade whose binary equivalent was to be contributed to the accumulating total. For the reverse process, he used the decimal-card reader to obtain the binary equivalents of 10^{14}, 10^{13}, . . . , 10^1, 10^0, in succession, and a set of decimal dials to store the answer.

As described in Atanasoff's manuscript (pp. 318–19), his "base-ten dials" constituted an electromagnetic accumulator, with carry connections between adjacent dials. Each time a binary equivalent of 10^n was subtracted from (added to) the remainder of the binary number being converted, the dial for n was advanced (retracted) a step. In this way the decimal solution accumulated on the decimal dials. (There was, however, no problem of delayed carry, because mechanical pulses were fed into one dial at a time.)

The conversion process was as follows. A decimal card with 1111 . . . to fifteen places punched in one of the five fields was placed in the decimal-card reader, and the output of this field was routed to the add-subtract mechanism being used for the conversion. The binary equivalents of 10^{14}, 10^{13}, . . . , 10^1, 10^0 were obtained by scanning this IBM card, as in the case of decimal-binary conversion.

If the number (on a given counter abacus) to be converted was positive, the procedure was to *subtract* the binary equivalent of 10^{14} repeatedly until, on a particular drum rotation, the sign indication on the abacus changed to negative, whereupon the subtraction was stopped; meanwhile, the decimal dial in the most significant position had counted up the number of subtractions required to create the overdraft. Next the binary equivalent of 10^{13} was *added* repeatedly until the sign returned to positive; meanwhile, the dial in the second most significant position had counted down (subtracted) the number of additions required to restore the sign, in the process borrowing from the number recorded on the first dial and so correcting its reading. Addition and subtraction continued to alternate in this fashion until the binary number on the abacus was eliminated. The decimal output accumulator now held on its dials the converted decimal form, with a "+" in the sign position.

If the number on the counter abacus was in two's complement form, the binary equivalent of 10^{14} was *added* repeatedly, that of 10^{13} was *subtracted* repeatedly, and so on, in each instance until the sign changed; meanwhile, the most significant dial counted up the number of additions, the next dial counted down the number of subtractions (borrowing from the first dial), and so on until the entire number on the abacus had been eliminated. The decimal output dials now held the negative decimal equivalent of the binary number whose complement had been on the abacus.

Berry, in his memorandum, described a better arrangement for the output dials. The carry mechanisms between dials were omitted, and the method of advancing (retracting) a dial was modified so that the dial was moved one fewer steps than the number of subtractions or additions. Thus if a binary conversion number was subtracted m times before the sign of the counter number changed, the corresponding decimal dial was advanced from position 0 to position $m - 1$, while if a binary conversion number was added m times before the sign of the counter number changed, the corresponding decimal dial was retracted from position 9 to position $9 - (m - 1)$.

Similarities will have been noticed between these binary-decimal procedures and the elimination algorithm procedure; in particular, Atanasoff's use here again of a method akin to the nonrestoring method of division by repeated subtraction. This method was much more efficient, for a machine, than the restoring method, and called for simpler controls of the successive operations. And now there is the added feature of an actual means of *counting* the number of subtractions being made—which we remarked earlier was missing from Atanasoff's otherwise complete provision for the option of incorporating division (and multiplication) in his computer at some future time.

Intermediate Binary Input-Output

The central and most basic operation of Atanasoff's computer was the elimination of a variable from a pair of binary equations on the storage drums. Integral to this operation were the procedures of reading two equations in and of writing a reduced equation out; that is, of reading one equation onto the counter and another onto the keyboard drum, and then of writing a new equation out from the counter drum on which it had been formed.

In the next and final section of chapter 1, we trace through all the steps in the solution of a full set of 29 equations, in order to estimate the time required to solve such a set. Taken together, those figures reveal that a total of 463 different equations were generated and had to be recorded (29 converted from the original set, 406 formed in the forward portion of the elimination algorithm, and 28 more formed in the backward portion); and that there were 1,247 instances in which an already recorded equation had to be read, or reread, into the machine (812 in the forward portion and 434 in the backward portion).

Thus in terms of sheer volume of activity, these two procedures constituted a much greater problem than the initial input and final output procedures. Moreover, the only fast way of reading data into and writing data out of a machine at that time was with punched cards. Yet no means of punching cards in the binary system (or adaptable components thereof) existed; all punched-card machines were decimal.

Atanasoff recognized the importance of this binary input-output system for his computer at the same time that he recognized its clearly auxiliary role. "Strictly speaking, such a mechanism is only an auxiliary part of the machine, and not a part of the actual computing devices; however, it is an essential auxiliary if an over-all speed of operation consistent with the speed of computing is to be obtained" (pp. 314–15). This overall speed had, of course, been tentatively set very early, in terms of one drum rotation (one addition time) per second with sixty stations (sixty pulse times) circling the drum. That is, Atanasoff had in mind a new and faster "punching" method that would record on and read from cards at this same rate of sixty impressions per second.

A New "Punching" Method

He envisioned an electronic spark-recording method of marking cards, together with its card-reading counterpart. He would represent a "1" at a given position on the grid of a card by charring that spot, instead of by punching a hole. And he would represent a "0" with the absence of a char.

Atanasoff saw that he could record an entire binary equation on a single card, for two reasons. First, only one grid position per digit was needed for binary representation, as against the ten required for a decimal digit in the standard notation. Second, even though there would be a much greater frequency of weakening "punches" than in the decimal mode, their spacing would not have to be increased proportionately, because the spark method would impact on the card less than the usual mechanical punching did. Thus he found that he could safely store his largest binary equation of 1,500 bit positions on an 8½-by-11-inch card, with the thirty coefficients running the length of the card on a grid of thirty rows by fifty columns.

His method was to record or *write* "1's" on a card with electric sparks, the arcs passing through the card between pairs of electrodes and carbonizing it at the appropriate positions. Because a carbonized spot offered much less resistance to electric current than a noncarbonized spot, the card could be *read* later by applying a lower voltage—to test for "1's," as it were. Accordingly, the writing voltage was applied only to those grid positions where a "1" was to be recorded. But the reading voltage was applied to all positions: if current flowed, a "1" was signaled, and if no current flowed, a "0" was signaled. Considerable experimentation showed that 5,000 volts would produce the desired degree of charring on the kind of card material Atanasoff planned to use. The reading voltage then had to be low enough that the card would not be charred by the reading process, yet high enough that a charred spot was reliably detected. A reading voltage of one-quarter to one-third the writing voltage proved satisfactory.

The thirty coefficients of an equation would be read from, or written on, a card in parallel, from least to most significant digit as always, because they were to be passed to and from the storage drums via the add-subtract mechanisms. As to the card handling, a card to be "punched" or read would be placed on the appropriate tray by hand, and at the push of a button, two "fingers" would pass the card into the mechanism lengthwise, to be drawn through by rollers (fig. 17).

Fig. 17. Front view of computer as of fall, 1941. The add-subtract mechanisms are all in place in this photograph, with fifteen visible on the front right face. The tubes to their left are the vacuum tubes of the restore-shift mechanisms, and just above and behind these may be seen thyratrons of the binary-card processing circuits. The synchronous motor is now in place (driving the fast axle, whereas the direct-current motor had driven the slow axle). The keyboard of manual controls is on the top front right, to its left is the decimal-card processing mechanism, and farther to the left are trays of the binary-card writing and reading devices (the writing tray in front). Finally, two covers across the rear top conceal, on the left, the base-conversion drum and the keyboard drum, and, on the right, the timing drum and the counter drum.

Binary-Card Writing

The writing mechanism consisted in two sets of thirty tungsten electrodes, positioned one directly above the other in straight lines, the electrodes of each set matching the thirty bit positions in a column of a card. The card was passed between the two sets, so that as column after column of bit positions moved through, each position in each column was momentarily aligned between two electrodes, one above it and one below. At each column's "moment," thirty signals (one from each of thirty streams) arrived from the counter drum abaci via the add-subtract mechanisms; for each bit position where a "1" was indicated, a sparking circuit now applied 5,000 volts across the associated pair of electrodes, to produce an arc and leave a small round charred spot on the card at that position.

With all thirty abaci recorded in parallel, one in each row of a card, bit by bit, an entire equation was transferred from the counter drum to a card in a single operation. This writing operation was, of course, executed in one drum rotation, or one second (one addition time); each column of bits on a card was recorded in one-sixtieth of a second (one pulse time). The procedure was essentially a vector addition, with the circuitry so arranged that the add-subtract mechanisms received only

"0's" on the keyboard input. (We should note also that, although equations usually came from the counter drum, where they had been formed in the elimination process, Atanasoff did arrange for writing them from the keyboard drum as well, via the restore-shift mechanisms. This transfer was useful for checking the computer's operation [see fig. 7].)

For the sparking circuit that would cause the arcing between a given pair of electrodes, Atanasoff experimented at length with different possibilities, including "various modifications of vacuum-tube-controlled spark coils, radio-frequency-initiated arcs, and thyratron-controlled discharges" (p. 315). He concluded that current discharges produced by thyratrons (gas tubes) were most suitable for writing at the required speed, and just as suitable for reading later. He arranged for the firing of each writing tube to be under the control of a vacuum tube: a series of "1's" and "0's" arriving from an add-subtract mechanism passed through a vacuum tube which allowed its associated thyratron to fire whenever a "1" was presented.

Thus his sparking method called for the following sequence of events, occurring in parallel for each of the thirty coefficients being recorded from a storage drum: the reading of a stream of pulses (bits) from a drum abacus; their transmission through the associated add-subtract mechanism and on to a vacuum tube in control of the sparking circuit; the passage by that vacuum tube to its associated thyratron of every "1" pulse presented; the discharge by that thyratron of 5,000 volts of current to its associated pair of electrodes; the arrival of the corresponding grid position of the binary card being "punched" at the moment of this discharge; the arcing between the two electrodes of the 5,000 volts; the charring of the card at that point.

Obviously, a great deal of coordination, or synchronization, was required for the success of this system. But much of it had already been achieved, as it had with other systems of the computer, by the simple measure of gearing axles one to another. For, ultimately, the rotations of the storage drum axle timed all the systems we have described so far; they even timed the two commutator systems that timed certain essential activities, since these rotated on their axle and that of the carry-borrow drum geared to it. Atanasoff now extended this basic timing method to coordinate his thyratron firing with the passage of his binary cards through the sparking mechanisms, by gearing the card-rollers to the axle of the two storage drums; or rather, he geared them to the axle of the base-conversion drum, which was in turn geared to the storage drum axle.

Like the other systems of the computer, then, this aspect of the binary input-output system was timed by the motor that rotated the drum axles. As we have said, this was a synchronous motor powered by sixty-cycle alternating current. In fact, however, the synchronous motor was not required either by the other systems or by this aspect of this one. An induction motor or a direct-current motor would have served as well for the synchronization of all activities timed off the storage drum axle, despite the attendant slight variations in rotation rates, because their *relative* rates would have been kept constant by gearing.

The synchronous motor became essential only when a separate source of power, and so of timing, had to be introduced to achieve the high voltage required in the card-writing procedure. The best way to obtain the desired charring voltage of 5,000 volts was to transform ordinary sixty-cycle alternating current to this range and use the positive peaks of 5,000 volts to fire the controlling thyratrons—which peaks would, of course, occur precisely every sixtieth of a second. But now the rotation of the storage drum axle by an induction motor or a direct-current motor would not do. Only a synchronous motor powered by the same alternating-current source as the thyratrons could ensure that signals from the drum abaci reached the controlling vacuum tubes when the thyratrons were at their peak voltage. (All other activities, of course, including the passing of the binary cards through the sparking mechanisms, would share in this new synchronization, simply because they continued to be timed by the rotation of the storage drum axle.)

Atanasoff gave the following explanation in his 1940 proposal (p. 316):

> The transformer supplies a peak voltage of 5,000 volts. The phase of this voltage is so adjusted that it is near its peak in the indicated direction at the instant when the first brush set is touched by a row of contacts on the abaci. The thyratron . . . is normally held blocked, and is controlled by the corresponding computer element through an intermediate tube . . . . If the reading brush touches a negatively charged contact [the intermediate tube is blocked and the thyratron allowed] to ionize. Almost the entire peak voltage of the transformer immediately appears across the card, which breaks down at once and allows an arc to be formed. . . . The arc automatically extinguishes itself approximately one quarter of a cycle after beginning because the voltage wave passes through zero at this time. . . .

Binary-Card Reading

The reading mechanism was highly similar to this writing mechanism. As we have noted, the arcing between electrodes through a card produced a charred spot that would serve as a conductor of current when the card's equation was read back onto a storage drum for further processing. The char-detection, or reading, mechanism had an arrangement of two sets of electrodes identical to that of the charring, or writing, mechanism, and rollers passed a "charred" card between them. But now the peak voltage used was only 1,600 volts, because its function was to produce an output signal without altering the card. Moreover, it was applied to the electrodes at every bit position of each passing column; and since all thirty pairs of electrodes were to be fired, just one thyratron, under the control of just one vacuum tube, sufficed.

For every row of a card, then, a series of "1's" and "0's" was signaled directly to an add-subtract mechanism and on to a drum abacus, a "1" for every charred spot (where current flowed) and a "0" for every clear spot (where no current flowed, since again no current meant 0 volts at the add-subtract mechanism). Atanasoff's algorithm called for reading an equation from a binary card onto each drum, and his circuit design allowed for a direct reading to either through the add-subtract mechanisms (see fig. 7).

Operating Speed

It is evident now that the operating, or clock, speed of the entire computer, namely, the processing of bits throughout at the rate of sixty per second, was determined by the requirements of the intermediate input-output system. It was, in fact, not only determined, but actually *limited* by those requirements. From a purely mechanical point of view, Atanasoff could hardly have achieved accurate "punching" at a much faster rate, given his particular need: to punch a card—or otherwise mark it so that it could be read later—in thirty parallel rows simultaneously, for fifty successive columns, with the possibility of many consecutive "punches" in a given row or column. But the overriding factor was his need to synchronize this operation with sixty-cycle alternating current, in order to achieve firing (writing) and testing (reading) at voltage peaks.

Thus while Atanasoff invented his electronic spark-recording mechanism in order to have a *faster* method of binary input and output, that mechanism remained the *slowest* link in the most basic operation of his algorithm, the elimination of a variable from a pair of equations. The electronic arithmetic unit and the electrostatic store had to be coordinated with this input-output system, even though, in and of themselves—as Atanasoff recognized—they could have operated faster. The drum memory could have been rotated appreciably faster than it was, without sacrificing reliability, and, of course, the electronic calculations could have been performed many times faster.

The other side of this coin is that in inventing a punching method that was faster and simpler than could have been achieved mechanically for his purposes, he was also able to integrate binary input-output with his memory and his arithmetic unit, just as he integrated decimal-binary and binary-decimal conversion with those units. He thus realized an efficient, balanced, centrally oriented architecture for his computer, and an overall computation speed that was still unprecedented in its day.

Actually, it is doubtful that Atanasoff seriously considered punching cards mechanically, even though it is conceivable he might have done so by *slowing* the drum rotations. Such a method would have introduced a substantial amount of mechanical equipment and the circuitry to operate it, not to mention the time and expense of development. What is more important, he thought he saw a better way to process his binary cards, a way that was much more consonant with his basic conception of electronic computation.

We have now completed our physical description of Atanasoff's computer. Its overall organization is shown in figure 18.

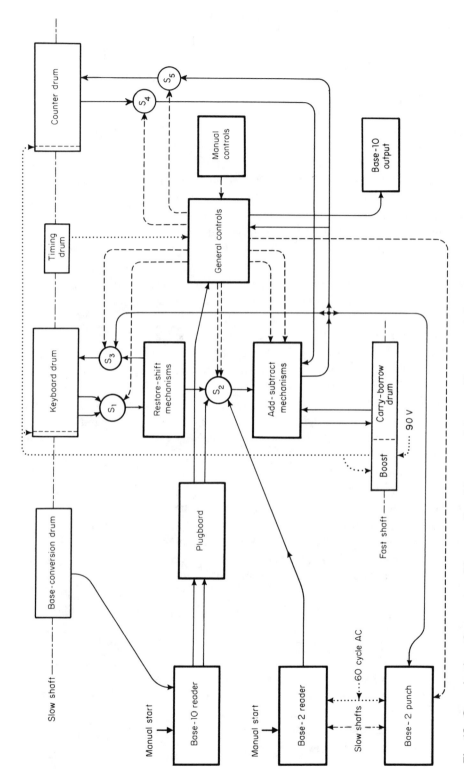

Fig. 18. Organization of computer. The overall architecture of Atanasoff's computer is depicted here. It is, of course, idealized with respect to physical layout. (See also fig. 7.)

Outcome

Despite Atanasoff's considerable ingenuity in its design, his card-charring and char-detection method revealed a low yet significant rate of error at the final testing stage. He and Berry had worked it out carefully and tested it to the extent that was possible prior to actual construction. Atanasoff observed as of August, 1940, "It is believed that the two major difficulties have been surmounted—the perfection of the [add-subtract mechanism] design and the electrical system of recording data" (p. 320). Tests on the finished computer in the spring of 1942, however, revealed an error rate estimated to be in the range of one error in 10^4 or 10^5 bit positions on the cards. This meant that a small set of, say, three equations generally could be solved successfully, but as the sets became larger the results became less reliable.

The difficulty clearly lay in the writing and reading phase of the procedure. The essence of the problem—actually the design problem on which Berry wrote his master's thesis—was to find a satisfactory combination of three elements: peak writing voltage, ratio of peak writing to peak reading voltage, and card material or dielectric. A peak voltage of 5,000 volts for writing and of one-quarter to one-third that magnitude for reading had proved satisfactory in charring tests on several different dielectric materials in 1940. In the spring of 1942, these voltages also worked with several materials in the test runs on small sets of equations, but not on large sets. The card material was suspect, because the one that had worked best in the earlier tests could not now be tried; only a small piece remained, and its source could not be located.

On the other hand, the tiny error rate that arose with the larger sets might have existed all along, showing up only in the more extensive tests. On this possibility, Atanasoff and Berry tried other peak voltages with the dielectrics available to them, hoping to find the desired combination. The demands of war research projects intensified, however, permitting less and less time for this nonmilitary project. Atanasoff, frustrated over an infrequent failure of one aspect of one system of his computer, and an auxiliary system at that, continued to believe he would correct the error, but he never did.

In our opinion, he might well have overcome the difficulty under other circumstances. For example, he might have found a combination of voltages and dielectric that reduced the error frequency, and then supplemented that improvement with checking procedures. In any event, even though the error rate was too high for the computer to be useful without some modification, it was low enough to establish the promise of this novel input-output method, perhaps even to justify a patent on it. We should note here, too, that this failure would not have prevented patenting other novel features of the computer.

Both Atanasoff and Berry left Iowa State for full-time war research positions shortly after the construction was completed, Berry in July of 1942 and Atanasoff in September. Neither man returned to work at Iowa State College after World War II ended. Atanasoff declined an offer of the chairmanship of the physics department

Fig. 19. Clifford E. Berry

and, in fact, never taught again, at Iowa or elsewhere. Berry spent enough time there to complete his graduate studies, writing his dissertation on mass spectrometry and receiving a Ph.D. in 1948. In the fall of 1948, the computer was dismantled, on instructions from the chairman of physics, who stipulated that usable parts be returned to the supply room and the balance discarded. One storage drum was preserved and is today held by the computer science department at Iowa State University, which also has two add-subtract mechanisms reconstructed from vintage parts.

Clifford Berry died in 1963. When the computer he and Atanasoff had built began to figure in litigations, Atanasoff adopted the name Atanasoff-Berry Computer, or ABC, in recognition of Berry's contribution to it (fig. 19).

Computation Time

We conclude our description of what we consider the first electronic computer ever invented by giving our estimate of its computation time, for a full set of twenty-nine equations in twenty-nine variables. In so doing, we will also be providing a brief summary of the overall procedure. The reader can follow the presentation by referring to either figure 18 or figure 26, each of which shows the ABC's organization in a different way.

Computation time is *machine time* plus any *operator time* required beyond machine time. We can estimate the former with greater accuracy than the latter, from the fact that all machine processes were timed off the storage drum axle, which rotated at the rate of once a second. We will give our estimate of this time first, and then our best approximation of the additional operator time.

Machine Time

We sorted out five basic steps and made the following estimates of the machine time for each:

> *a)* convert each thirty-term decimal equation to binary form and enter it on the counter drum, *one and one-half minutes* (at fifteen seconds for each of six decimal cards of five coefficients);
> *b)* write each binary equation onto a card from the counter drum, *one second;*
> *c)* read each binary equation onto the counter or the keyboard drum, *one second;*
> *d)* process a pair of equations through the arithmetic unit to obtain a new equation in one fewer variables on the counter drum, *two minutes;*
> *e)* convert a binary coefficient on the counter drum to a number on the decimal dials, *one and one-quarter minutes.*

There were several possible overall procedures for executing Atanasoff's algorithm on his computer. We chose the following sequence of four stages, and calculated the approximate machine time for each on the basis of the above estimates for the individual steps:

> 1. *Decimal-binary conversion.* Convert each of twenty-nine decimal equations to binary and write each on a binary card. *Forty-four minutes* (29 × 90 + 29 × 1 seconds).
> 2. *Forward elimination.* From these twenty-nine binary equations, read twenty-eight pairs, successively, onto the two drums, solve each pair, and write each reduced equation from the counter drum onto a binary card. From these twenty-eight equations, read twenty-seven pairs, . . . , and write each reduced equation . . . . Repeat for twenty-six pairs, twenty-five pairs, . . . , down to one pair. *Thirteen hours, fifty-two minutes* [(28 + 27 + . . . + 1)(2 + 120 + 1) seconds].
> 3. *Back-elimination.* From these sets of one equation in one variable, two equations in two variables, . . . , up to the original twenty-nine equations in twenty-nine variables, read one of the two equations in two variables onto the counter drum and the equation in one variable onto the keyboard drum, process to eliminate the common variable, and write the new equation in a second single variable onto a binary card; read one of the three equations in three variables onto the counter drum and successively read each of the two equations in a single variable onto the key-

board drum, processing each pair to eliminate the common variable, and write the new equation in a third single variable onto a binary card; repeat for an equation in four variables and three equations in single variables, for an equation in five variables and four in single variables, . . . , for an equation in twenty-nine variables and twenty-eight in single variables. *Thirteen hours, forty minutes* $[(1 + 2 + . . . + 28)(1 + 120) + 28 \times 2$ seconds].

4. *Binary-decimal conversion.* From this set of twenty-nine equations in twenty-nine different single variables, successively read each equation onto the counter drum and convert first one, then the other, of its two coefficients to a number on the decimal dials. *One hour, thirteen minutes* $(29 + 2 \times 29 \times 75$ seconds).

The *total machine time* for these four stages comes to about *twenty-nine and one-half hours.* Notice that stage 2, the forward-elimination procedure to obtain the first equation in a single variable, and stage 3, the back-elimination procedure to obtain all the other equations in a single variable, took equally long, about thirteen and three-quarter hours. Together, of course, they consumed the preponderance of the machine time for the entire solution.

A general formula for the approximate machine time for Atanasoff's computation, based on our estimates of the five individual steps, can be given in terms of N, the number of variables, or of equations, in a set. By stages the times are: (1), $91N$ seconds; (2), $123(N^2 - N)/2$ seconds; (3), $(121N^2 - 117N - 4)/2$ seconds; (4), $151N$ seconds. Thus stages 1 through 4 took about $2N(N + 1)$ *minutes,* or $N(N + 1)/30$ *hours.*

Operator Time

Operator tasks were of two sorts, *on-line* tasks wherein the operator ran portions of the problem on the computer, and *off-line* tasks of some duration which the operator executed apart from the computer. We start with time estimates for the off-line tasks, of which there were four, each separate from but associated with one of the machine stages given above:

1. *Decimal-binary conversion.* In preparation for this stage, punch the original decimal equations onto 174 decimal cards (twenty-nine equations, 6 cards each), eighty instances per card. *Four hours* (at one second per punch).

2. *Forward elimination.* For this stage, prepare twenty-nine stacks of binary cards, the first to consist of the 29 converted equations from stage 1, the second of 28 blank cards for equations to be generated from those twenty-nine original equations, the third of 27 blank cards, and so on down to 1 blank card to hold an equation in one variable. Prepare these 435 cards by marking each with a code indicating both the equation held or to be held

by it (as, "a to k" for its variables) and the number or order of that equation in its stack. Arrange the stacks themselves in the order of their use in the elimination process. *Two and one-half hours* (at twenty seconds per card).

3. *Back-elimination.* For this stage prepare a stack of 29 binary cards, 1 from the original set of 29, and 1 from each of the sets generated in stage 2 down to the single card holding an equation in one variable. Each has already been marked with an indication of the equation now held by it, but prepare further by assigning a new number for its order in this new stack. Also prepare a stack of 28 blank cards, by marking each to indicate the equation in a single variable to be held by it as this elimination proceeds (as, "a" for its variable) and its order in this stack. Place the two stacks (57 cards) in suitable positions for use in stage 3. *Twenty minutes* (at twenty seconds per card).

4. *Binary-decimal conversion.* After this stage, divide the respective pairs of generated decimal numbers on a desk calculator, and record each of the resulting quotients with the correct sign affixed. *One and one-half hours* (at three minutes per division).

The total *operator time* for the *off-line* tasks associated with the four machine stages comes to about *eight and one-half hours*.

Let us look next at the on-line tasks the operator executed while running the computer through those same stages. These included inserting and removing cards, pushing buttons or closing switches, plugging and unplugging cables, and reading dials. Time estimates here are complicated by the fact that portions of the on-line tasks could be done during machine time, whereas we are interested only in operator time beyond machine time.

As with our machine-time estimates, we could sort out certain basic steps the operator would execute repeatedly, assign a time for each, and estimate times for the four stages on the basis of these steps. In this situation, however, it has proven simpler just to list all the steps of each stage, despite a certain amount of repetition. We have also found, from our own simulations of these on-line tasks, that a single time allowance of ten seconds per step yields as sound an approximation overall as we can attain.

Our analysis of the on-line operator time beyond machine time, then, is as follows for the four machine-time stages:

1. *Decimal-binary conversion.* There were eight steps for each of the 174 decimal cards:
   a) plug the five-line cable from the decimal-card reader to the socket of the appropriate set of five add-subtract mechanisms;
   b) place a punched decimal card on the tray of the decimal-card reader;
   c) lower the rack of brushes over the card on the card reader;

d) signal the computer to convert the decimal coefficients on that card to binary form on the counter drum;

e) remove the card from the card reader and lay it aside;

f) take a blank binary card from the appropriate stack and put it in the writing tray;

g) signal the computer to write the equation from the counter drum onto the card in the writing tray;

h) remove the card from the writing tray and place it on the appropriate stack.

Since none of these steps could be done during machine time, our on-line operator time for stage 1 comes to eighty seconds for each of 174 instances, or a total of about *four hours.*

2. *Forward elimination.* There were ten steps for each of the 406 pairs of equations solved to generate a new equation in one fewer variables:

a) take a punched card from the appropriate stack and place it in the reading tray;

b) signal the computer to read the equation from the card in the reading tray onto the counter drum;

c) remove the card from the reading tray and set it aside;

d) take another punched card from the appropriate stack and place it in the reading tray;

e) signal the computer to read the equation on that card onto the keyboard drum;

f) signal the computer to compute the pair of equations on the two drums;

g) remove that card from the reading tray and place it on the appropriate stack for later reuse;

h) take a blank binary card from the appropriate stack and place it in the writing tray;

i) signal the computer to write the equation from the counter drum onto the card in the writing tray;

j) remove the card from the writing tray and place it in the appropriate stack for later use.

Except for the first round of eliminations, steps *a, g, h,* and *j* could all be executed while the computer was performing an elimination (of about two minutes' duration). Our total on-line operator time for stage 2, then, comes to sixty seconds for each of 406 instances, or about *seven hours.*

3. *Back-elimination.* There were again 406 pairs of equations to be solved in this stage, to generate twenty-eight new equations in one variable, but an increasing number of steps for each new equation. We can consolidate the steps as follows:

a) place a punched card in the reading tray, signal the computer to read its equation onto the counter drum, and remove the card from the tray, for a total of 28 instances of three steps each;

b) place a punched card in the reading tray, signal the computer to read its equation onto the keyboard drum, and remove the card from the tray, for a total of 406 instances of three steps each;

c) signal the computer to compute the pair of equations on the two drums, for a total of 406 instances of one step each;

d) place a blank card in the writing tray, signal the computer to write the equation on the counter drum on that card, and remove the card from the writing tray, for a total of 28 instances of three steps each.

The times for these consolidated steps are: (*a*) one-quarter hour (28 × 10 × 2 seconds, with placing the card on the tray done during the preceding machine elimination step); (*b*) two and one-half hours (406 × 10 × 2 seconds, with removing the card from the tray done during the subsequent machine elimination step); (*c*) one and one-quarter hours (406 × 10 seconds); (*d*) one-quarter hour (28 × 10 × 2 seconds, with placing the card on the writing tray done during the preceding machine elimination step). Our total on-line operator time for stage 3, then, is about *four and one-quarter hours.*

4. *Binary-decimal conversion.* There were seven steps in each of the fifty-eight conversions (*after* the initial placement of the specially punched decimal card in the decimal-card reader):

a) place a card in the reading tray;

b) signal the computer to read the equation in a single variable onto the counter drum;

c) signal the computer to convert one of the two coefficients on the drum to a number on the decimal accumulator;

d) record the converted decimal number;

e) signal the computer to convert the other of the two coefficients on the drum to a number on the decimal accumulator;

f) record that converted decimal number;

g) remove the card from the reading tray.

Since none of these steps could be done during machine time, our on-line operator time for stage 4 is about *one and one-quarter hours* (70 × 58 seconds).

The *operator time* for all four *on-line* stages comes to about *sixteen and one-half hours.* And the total *operator time* for on-line and off-line tasks comes to about *twenty-five hours.*

General formulas can be given for the operator times, as was done for the machine time, in terms of the number of equations in a set. These are: $(N^2 + 71N - 2)/360$ hours for off-line tasks; $(9N^2 + 115N - 8)/720$ hours for on-line tasks; and $(11N^2 + 257N - 12)/720$ hours for the combined total. It will be seen that the operator time increased as the square of the number of equations, just as the machine time did, but that the linear term was now a stronger factor. The square term dominated, however, for the larger sets in which Atanasoff was interested.

And it dominated even more so in the formula for combined machine and operator time: $(35N^2 + 281N - 12)/720$ hours, or, more simply, $.05N^2 + .4N$ hours.

Total Computation Time

We have tried to be conservative in our time estimates, that is, to allow more than ample time for each hand or machine operation. Nevertheless, the human operator time remains a very rough approximation, and the machine time, while more accurate than the operator time, is also only an approximation.

Our concluding estimates, then, for the solution of a full set of twenty-nine equations in twenty-nine variables on Atanasoff's computer are:

Machine time, about twenty-nine and one-half hours.
Operator time, about twenty-five hours.
Total computation time, about fifty-five hours.

Comparison with Desk Calculator Time

Atanasoff observed in his 1940 proposal that the solution time for a set of linear equations on a desk calculator increased as the *cube* of the number of equations; he suggested the formula, $N^3/64$, or $.016N^3$, and remarked that for large sets the time would be much greater because of human error (p. 306).

Any such estimates are very rough. Atanasoff's formula was borne out, however, by the work of James Wilkinson, who (with others in England in the 1940s) solved a system of twelve equations in forty operator-hours, more or less, and a system of eighteen equations in seventy to one hundred operator-hours (personal communication). These times are to be compared with twelve hours and twenty-three hours, respectively, for solving such systems on the ABC.

Now Wilkinson's procedure included checking. After a pair of equations was reduced to one equation, the coefficients of each of the three were summed and the results compared by means of the Gaussian factor used to eliminate the designated variable. This extra step increased the total time by only about 10 percent, regardless of the size of the set, so that Atanasoff's formula for hand calculation remains a reasonable working formula.

Let us compare this formula for the solution time on a desk calculator, $.016N^3$ hours, with that for Atanasoff's computer, $.05N^2 + .4N$ hours, which increased only as the *square* of the number of equations because of the parallelism of operations in his vector processing algorithm. Since a human operator can place the decimal point where it should go, the hand method is essentially a floating-point method. The ABC, on the other hand, was necessarily fixed-point, and Atanasoff expected to adjust the coefficients at the outset to prevent numbers from running off scale.

It is hard to say how difficult this procedure would have been, but because the original equations were to have come from physical data it might not have been much of a problem. Wilkinson's experience with his system of eighteen equations is relevant here. He took these from a natural problem and solved them primarily to determine whether substantial round-off error would accumulate, as theory predicted. In fact, he found very little aggregate error.

As for round-off error in the ABC, Atanasoff was actually unable to round off his coefficients during the elimination procedure. His vector algorithm called for shifting the keyboard coefficient to the right each time the sign of the designated coefficient changed, so that the last (least significant) bit was simply dropped away. To avert any substantial loss through this process, he provided fifty bit positions in his coefficients, with the original physical measurements occupying only twenty or so of these.

Our main difficulty in comparing Atanasoff's solution time with that of the desk calculator lies in assessing the time required to catch machine errors in the ABC other than those of round-off or going off scale. As we saw in the preceding section, the ABC's binary-card mechanism failed occasionally. And while this problem might have been overcome either by altering the peak writing and reading voltages or by finding a card material with better electrical properties, such measures would probably not have made the system foolproof.

The natural step to take next, had Atanasoff pursued the problem further, would have been to modify his computer so that the operator could compare results after performing each elimination twice. This would have been an easy change to make on the ABC, by the simple expedient of introducing a control button to cause the keyboard abaci to be subtracted from the counter abaci, with the result punched out on a binary card. Then, if there was no error this card would have no charred spot, a point readily verified visually.

Such a remedy for the ABC's error problem would have increased the computation time by a factor of about two and one-half, yielding a formula of $.12N^2 + N$ hours. Thus, for a full set of twenty-nine equations the total computation time would have been about 130 hours, one-third the desk machine's 390 hours—still an impressive gain for that era.

This procedure of running a problem twice to catch intermittent errors is not such an extreme measure as it may seem. When the ENIAC solved its first problem, the operators ran each subroutine twice and checked the results on an IBM comparator. (Data was entered into the ENIAC initially and written out terminally via IBM punched-card machines.) Subsequently, less checking was needed, and such might also have been the case for the ABC.

Let us close this chapter of the book, devoted to an exposition of Atanasoff's computer as he conceived and built it, by noting that the ABC's real importance to history did not hinge on its overall solution time, or even on whether it could have been made to work successfully. Its importance lay rather in proving the feasibility of automatic electronic computing, through the successful demonstration of electronic switching, logical arithmetic circuits, capacitor storage, regeneration, and other features, all of which are evaluated fully in chapter 5.

CHAPTER 2

Mauchly's Pre-Atanasoff Years

The Evidence

J. Presper Eckert and John W. Mauchly, principal designers of the ENIAC, were for decades regarded not just as inventors of that electronic computer, but as inventors of the *first* electronic computer. Their recognition for this achievement began with the unveiling of the ENIAC in early 1946. It was very well established by the time the ENIAC patent was issued to them in 1964, and grew apace until late 1973, when a federal court found their patent invalid in *Honeywell v. Sperry Rand* (the latter owned the patent at that point). One ground for invalidation, among others of a more technical nature, was that the ENIAC had been derived from the work of a prior inventor, namely, John V. Atanasoff, whose computer we described in the previous chapter.

We are of the firm conviction that the judge in the ENIAC patent case, Judge Earl R. Larson, was correct in this finding of prior invention by Atanasoff. We believe, with him, that Atanasoff's computer was the first electronic computer, and that critical ideas were passed from Atanasoff to Mauchly, in 1940–41, and used by him and Eckert in their design of the ENIAC. (We do believe that the ENIAC was the first *general-purpose* electronic computer, inasmuch as Atanasoff's machine was a *special-purpose* computer.) We discuss Judge Larson's decision at some length in chapter 4.

Despite that finding, however, the recognition Eckert and Mauchly had enjoyed earlier continued to grow, partly because of the efforts of interested parties and of historians apparently unwilling to have the long-accepted version challenged, and partly because of Atanasoff's relative obscurity. His detractors have persisted to the present day in dismissing his role out of hand, often with the argument that a court decision cannot be taken as the true state of events, or, put slightly differently, that the courts cannot settle this sort of controversy.

We find these arguments both premature and unduly cynical, so far as the ENIAC patent suit is concerned, for two reasons. First, a vast reservoir of research data was created by that suit: documents, devices, and other exhibits relevant to the case, together with sworn testimony by material witnesses, all produced under court subpoena and subjected not just to scrutiny by a federal judge, but to examination by attorneys for both sides; arguments presented to the court by those attorneys;

and the final ruling. Very rarely has such a wealth of validated information been available for settling a historical controversy.

Second, a court decision *can* be accepted if critically examined—in light of all the evidence—and found to be correct! (By the same token, it can be rejected, rationally, if examined and found to be incorrect.) Cases on important issues deserve vigorous public debate, in which the facts developed at trial are reported and either acknowledged or countered. Thus far the ENIAC case—on the historic issue of invention of the electronic computer—has undergone very little such debate. Judge Larson's critics, in particular, have not made the effort to address the trial data on which he based his decision.

We will, then, in the next three chapters of this book, present what we consider the relevant evidence from the ENIAC case (and from one other case pitting Mauchly against Atanasoff). In the current chapter, we explore Mauchly's own accomplishments in computing prior to his first meeting of Atanasoff in late December, 1940, together with the research those accomplishments were intended to serve.

In chapter 3, we develop the paths by which Atanasoff's work influenced Mauchly and Eckert in their design of the ENIAC and also of the EDVAC; that is, we document the connections with court testimony and exhibits relating to the transmission of ideas. In chapter 4, we not only present Judge Larson's decision, but also fill out the Atanasoff side with further testimony by him from the ENIAC trial, affording him his day in court, as it were, to balance a preponderance of Mauchly testimony given earlier.

Before we begin, however, we wish to acknowledge that one person has recently come forward to present the Mauchly—or Mauchly-Eckert—side of this controversy, and to argue from at least some of the trial material. That person is Mauchly's widow, Kathleen R. Mauchly, whose article, "John Mauchly's Early Years," was published in the *Annals of the History of Computing* (Mauchly 1984), in response to the stand we took in our *Annals* article, "The ENIAC: First General-Purpose Electronic Computer" (Burks and Burks 1981). As we stated in the introduction to this book, her article is the most thorough effort yet made to reverse Judge Larson's ruling in "the court of public opinion." Moreover, it is, we believe, essentially the case Mauchly himself had been developing in the years before his death in 1980 (see Mauchly 1980). We respond to this "best case" against any significant Atanasoff influence in appendix B.

The Harmonic Analyzer

John Mauchly was born in Cincinnati, Ohio, on August 30, 1907, and moved with his family to Washington, D.C., at a young age—his father was a physicist with the Carnegie Institution of Washington. He graduated from McKinley Technical High School in 1925. After studying electrical engineering at Johns Hopkins University for two years, he switched to the graduate school and earned a Ph.D. in physics in 1932, writing his thesis on a spectroscopic study of carbon monoxide.

Mauchly stayed on at Hopkins as a research assistant for one year, then joined the faculty at Ursinus College near Philadelphia as the sole member of the physics department. He was there from 1933 to 1941, rising to the rank of associate professor, but left to become an instructor at the Moore School of Electrical Engineering, University of Pennsylvania, in the fall of 1941. For this first decade of his career, Mauchly considered himself somewhere between an experimental and a theoretical physicist, with a special interest in applying statistical methods to research data. His chief research at Ursinus was a project he started in 1936, on the effects of solar activity on the earth's weather. He and a dozen students working under the National Youth Administration (NYA), a depression-years relief program, carried out meteorological analyses on the desk calculators of that day.

This procedure was extremely slow, taking about fifteen minutes for each of the thousands of analyses of weather data required by the project—*after* lengthy averagings of the original raw data had been made. Mauchly searched the literature and other sources for some more automatic device to speed up the analyses. In 1939 or 1940, he built a small hand-operated electrical analog computer, a harmonic analyzer, which reduced the time to just three minutes per analysis, a fivefold improvement with which he was greatly pleased. (See his February 24, 1941, letter to Atanasoff in chap. 3.) He and his students continued to use this device at least well into the spring of 1941.

Mauchly's harmonic analyzer actually occasioned his first encounter with Atanasoff on December 28, 1940. He delivered a paper on it and results he had obtained from it to a session of the American Association for the Advancement of Science (AAAS), in Philadelphia, and Atanasoff, a member of the audience, came up afterward to share his own interest in automatic computation.

Mauchly's Testimony

As we shall see, his harmonic analyzer was the only arithmetic instrument Mauchly had built prior to that first meeting, although he had built a digital cryptographic device and was to claim an electronic binary counter at the ENIAC trial. His enthusiasm for the analyzer and for the weather project was evident from his lengthy testimony under examination by attorney H. Francis DeLone of the Sperry Rand defense team. He did, however, run into difficulty over his originality in building it during cross-examination by attorney Henry Halladay of the Honeywell side. (For simplicity, we identify all attorneys as representatives of one side or the other, even though they were not in-house counsel but had been brought in from outside law firms.)

We begin with Mauchly's testimony to DeLone, omitting, however, the detailed physical description Mauchly gave of the device and how it worked. The interested reader can find this description in the ENIAC trial transcript (ENIAC

Trial Records, transcript pp. 11,773–81). We concentrate instead on the rationale behind his harmonic analyzer, as he explained it in court (transcript pp. 11,743–56 and 11,767–82). We first summarize his account, then sample some of the more poignant passages directly.

Mauchly testified that the aim of his weather research was to establish a relationship between activities on the sun and variations in the earth's weather, through the method of harmonic analysis. Physicists were using harmonic analysis to study the influence of solar activities on the earth's magnetic field, and also on the propagation of radio waves. Periodicities had been found in the records of magnetic events, for example, that corresponded to the sun's rotation period of approximately twenty-seven days; those events had then been related to solar flares that recurred— so far as the earth was concerned—with successive rotations of the sun.

Mauchly was very familiar with this magnetic research because of his connections to the Carnegie Institution of Washington. He explained that his father had headed the section of Terrestrial Electricity and Magnetism there. He had visited it often in his youth, and, he explained, he himself had worked in that same department, analyzing such data himself, during the summers of 1936, 1937, and 1938. From this experience, he thought he could find similar, if more elusive, periodicities in weather phenomena. He meant in particular to apply more sophisticated statistical procedures to the results of his harmonic analyses than the physicists were applying to theirs, procedures being developed, he said, by psychologists and mathematicians.

(Harmonic analysis is based on Fourier's theorem that any periodic oscillation [repeating variation, or wave], however complex, can be analyzed into a set of components each of which is a simple regular oscillation [sine or cosine wave]. That is, the original complex wave is the summation of a set of superimposed simple sine and cosine waves, and harmonic analysis consists in computing these basic waves. The harmonic period of a component wave may be either the full period of the related event [the twenty-seven-day period of the sun's rotation, in Mauchly's problem] or an integral division of that period [one-half or one-third or one-fourth of the rotation period, as Mauchly explained in court].)

Mauchly said he gathered his weather data from two sources. He subscribed to the U.S. daily weather map, a composite of information from some 200 stations throughout the North American continent. These were available for just the cost of mailing—about a penny a day! He was also able to obtain earlier U.S. weather maps, free of charge, from the Carnegie Institution, where they had accumulated for some twenty years but were of no use to the magneticians, as Mauchly called them. He testified, in fact, that he was allowed to take away not only the daily weather maps, but even meteorological publications from all over the world, for use in establishing his theory of periodicity in weather variables—a theory rejected by scientists at the U.S. Weather Bureau with whom he talked, as well as by those at Carnegie. Such opposition, he said, only served to challenge him and spur him on.

He began with the variable of precipitation, hoping to establish periodic recurrences of, say, slight increases or decreases in quantity. He set his NYA students to work adding up the daily rainfall data from those 200 stations for those twenty years of maps, and calculating a grand sum or grand average, as he put it, for each day. The next step was the harmonic analysis of these figures. This he and the students did for some time by well-established tabulation procedures—apparently those used at Carnegie to analyze magnetic data—but, as he explained to DeLone, he then thought of a way of simplifying the very tedious and time-consuming calculations such procedures entailed.

Mauchly's simpler way was to have a machine do the job, and the machine was the harmonic analyzer, which he characterized as an analog instrument that worked by analogy to electrical circuits, with results read out as currents on a meter (figs. 20 and 21). With this machine, he said, there was no need to tabulate the daily precipitation totals obtained from the raw data; one just inserted the figures and read off the results. As to the number of positions in which data were to be inserted, he explained that for the sake of a simpler design, he had sacrificed the ideal of twenty-seven (corresponding to the days of the sun's rotation) for just twelve. (He did not mention that he had gone on to design a twenty-seven-ordinate instrument for analyzing these data; see his November 15, 1940, letter to meteorologist H. Helm Clayton, quoted later in this chapter.)

As we have remarked, Mauchly described his harmonic analyzer with obvious pride and in great detail at the ENIAC trial. He concluded by observing that it afforded a tenfold speedup in the analysis of data over the tabulation procedure, and that it achieved the requisite accuracy of about 1 percent. He continued to use the more cumbersome procedure as well, however, until he left Ursinus, because this one machine could not possibly process the mountains of data he had accumulated.

Let us now quote directly some passages of Mauchly's testimony to DeLone (transcript pp. 11,753–54; 11,768–74; and 11,779–82):

> . . . It was a challenge, as I regarded it, to put some of the more recent statistical theories and methods to work in an area where nobody had yet applied these things at all. There was no one at the Institution who was really reducing the data by any kind of a statistical method that would bring out some of the characteristics I thought could be got out. . . . There you meet the positive psychological spur, you might say, of all, and that is that as soon as somebody told me that I had better quit there, stay out of this, . . . that spurred me on.

> And one of the things that I wanted to do, which I thought of as a method for exploring the relationship between the sun and the weather, was a method which had already been employed in magnetism. It is known that the sun rotates on its own axis approximately once every 27 days. . . .

> Could you do the same thing with weather?—was my question. Well, the only way to answer that question that I knew of was to actually try to do it . . . .

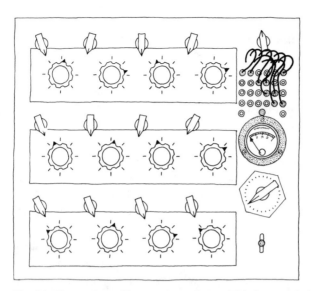

Fig. 20. Front view of harmonic analyzer. This is an artist's sketch of the harmonic analyzer Mauchly built in 1939 to reduce his weather data. Ordinate values were entered on the twelve large potentiometer knobs, their signs set with the twelve smaller associated switches. The resulting Fourier coefficients were read, each in turn, from the meter on the right; the particular coefficient was selected by the switch below the meter. The plugboard above was used to change the scale of the reading on the meter dial.

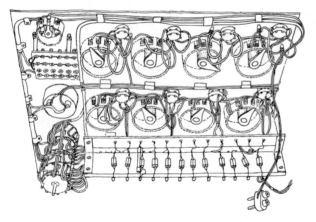

Fig. 21. Rear view of harmonic analyzer. Only the top two rows of potentiometers are visible in this sketch, the bottom row being obscured by the board on which the resistors were mounted. It is evident from the very complex structure to the left of this resistor board that the switch for selecting a given coefficient consisted of many rotary switches mounted on a single shaft.

Now, here is something which the people at the Weather Bureau were sure did not occur. They were very much against the idea that the sun affected the surface weather, and it would take a preponderance of evidence to get them slightly interested in the idea that it might be true.

Now, how old could we go? How far back? Well, we had data back to about 1920. So we had not quite, but close to twenty years of data.

And this was on a daily basis. So we could figure out—if you take 200 stations every day for 20 years, you've got a lot of data to add up even for just one weather variable. We had succeeded in doing most of that, but still those figures in themselves would prove nothing unless they were properly analyzed, and if you wanted to look for 27 day recurrences . . . to at least impute possibility of a solar relationship . . . you had to do more than just submit 20 years of daily data. You had to analyze it to see what the probabilities were that there was such a recurrence . . . every 27 days, . . . or perhaps the phenomena could also exhibit itself as not every 27 days but on half of that period. In other words, every 13½ days there might be something of this sort. When you have a phenomena which can vary up and down with some kind of a period, you quite often have conditions where half that period, or one-third of that period, or one quarter of that period, can also occur in the natural phenomenon.

This is true in musical instruments where you pluck strings, and so on, that you may get vibrations which are due to a string vibrating, so that it acts as a string of half the length, or one-third the length, and so on. It splits into segments in its vibration in such a way that you get harmonics, as they call them, so that in addition to the fundamental musical tone, you get harmonics of this which might have twice the frequency, three times the frequency, and so on.

Well, this kind of thing can occur in the weather phenomena and solar phenomena that I was interested in, too, and this leads to making what is known sometimes as harmonic analysis of the data . . . .

Well, what I devised actually was a tangible physical instrument for helping to do this. The methods of doing it from the mathematical point of view had been worked over considerably, and many people had published what you might call tabulation methods of how you tabulate the data in columns and do certain things with rows, and so forth, so that you come out with a computation, which is simplified by the tabulation method you have used. But the ultimate simplification of all this is—let a machine do it for you. You don't have to write down the figures in columns, or rows, or anything else. You just insert these figures into a suitable machine and read off results.

In this case I thought of a way of simplifying the work by a device which we would call an analog device nowadays. That is, it works by analogy with electrical circuits and you read off approximately by measuring currents on a meter, and what you read on the meter is an approximate result of what you could have obtained by calculating very accurately with a desk calculating machine, or something of that sort.

Very accurately, but much more laboriously . . . . So, at any rate, some-where in the late thirties I thought of this way of simplifying the work we had to do, and it had to be a way which was cheap enough so that I could implement it in my own laboratory, even if I had to buy the parts out of my own pocket, and I found such a way . . . . So we made this compromise, to make a simple device in which you insert 12 pieces of data. Then we devised a way of going from the 27 days of a solar rotation down to 12 values, which were almost the equivalent of, you might say, compressing 27 days into 12.

It was a pleasure to find out after we had gone through building this thing that one of the students could set new data into this and read off the three or four harmonics we wanted in, I think, about two minutes per set of data, where it took us about 20 minutes to do this when we were just tabulating the results and using a desk calculator to aid us in doing multiplications. So it was about 10 to 1 speed up here and we succeeded in getting just sufficient accuracy to be adequate for the purpose we needed. An analog device like this cannot be capable of any accuracy you desire, but here we were about one per cent, which is what we desired . . . .

I don't remember exactly when it was completed, but it got a lot of use in the 1939–40 academic year in order to get the data prepared for a talk which I was not only expected to give but permitted myself to give when I put in a request to be on the program of the AAAS . . . meeting in Philadelphia, in December 1940, I believe.

. . . I used it considerably myself, but I also had one of the students who did nothing but continue to set this up with data and read off the numbers, which I then processed further for the analysis leading to the paper I re-ported at that Christmas meeting.

DeLone did not question the originality of this harmonic analyzer, and Mauchly volunteered no information about earlier electrical versions of it. (Me-chanical versions date back to a tidal harmonic analyzer built by Lord William Kelvin around 1875.) Yet the implication was strong, given the lay setting of a courtroom, that he had devised this analog computing instrument without benefit of prior (electrical) art. Honeywell attorney Halladay, however, broached the issue directly, whereupon Mauchly did indeed claim to have conceived it entirely on his own. There ensued an exchange in which Mauchly was reduced to admitting that he had read accounts of various electrical harmonic analyzers in the literature of the 1930s.

Moreover, under the pressure of the moment, Mauchly made some startling admissions about his inventive practices—defensive admissions not only harmful to his case, but entirely gratuitous in terms of what Halladay was actually asking of him! We quote both sides of this exchange (transcript pp. 12,129–34):

Halladay: Well, had you modeled this [harmonic analyzer] on the work of someone else?

Mauchly: No.

H: Had you read about it somewhere else?

M: I don't believe that I had read about anything electrical of this sort. I read about the differential analyzers and tide calculators which worked by mechanical means—some with pulleys, ropes, and others with friction disks, and things of that sort.

H: Well, do you mean to say, then, that this harmonic analyzer, as depicted on Exhibit 21,379, was built by you entirely of your own conception?

M: So far as I recall, it was. My children weren't old enough to help.

H: Your children weren't old enough to help?

M: No, they weren't at that time.

H: Let me read you something, Dr. Mauchly.

It says: "Possibly you might be able to offer some advice in connection with a research project which we are now undertaking—namely, the construction of a network for the purpose of performing Fourier analyses. The circuit is already worked out (a modification of one used at M.I.T. some years ago) and a 12-ordinate analyzer, of perhaps 1½% accuracy has been built, using G.R. voltage dividers."

Now, does that reading refresh your memory as to a source of information upon which you had drawn in building your 12-part harmonic analyzer?

M: I would like to hear it again, or see it—

H: All right. I will read it again. I am referring to Plaintiff's Exhibit 22,522, part of the file which you produced, Dr. Mauchly, or had produced in response to one or another of the subpoenas that you have referred to before. It's a letter dated November 2, 1940, addressed to General Radio Co., Cambridge, Massachusetts.

And the part I read is as follows. . . .

Now my question was, does that not refresh your memory that the 12-ordinate harmonic analyzer which is depicted in these photographs, Plaintiff's Exhibits 21,379 and 21,378, stem from the work of somebody besides yourself?

M: It refreshes my memory that I apparently wrote down something like that in a letter to General Radio seeking some more information or advice from them with regard to how to use their equipment. It does not really refresh my memory as to what had been done earlier at MIT and how much of modification there was from that. I just don't remember.

H: Would you now agree, however, that the harmonic analyzer which is depicted in the exhibits before you, 21,379 and 21,378—

M: Could I see the letter that we are talking about?

H: —was not entirely of your design, but in fact was a modification of one used at MIT some years prior to 1940 and which had come to your attention in one way or another?

M: I will stand by what I said in my letter. I am not sure that your question applies to the same thing or not.

H: Well, then, I would like you to answer the question, if you would, please.

M: Well—

H: Shall I put it again?

M: Would you show me the material from which I drew this design? I don't recall it.

H: All I am trying to do is to get you to answer the question. If the question isn't plain, tell me it is not plain and I will try to rephrase it. The question is simply this, Dr. Mauchly: Will you not agree that the harmonic analyzer which is depicted in these photographs before you . . . was not entirely of your design, but was a modification of a design used at MIT some years prior to 1940?

M: Well, the answer has to be qualified because I don't know and cannot now recall what the other design was, and I have no reference at the present moment to help me. I can't even see what I wrote in the letter, apparently.

H: Is the—

M: So I have to give a qualified answer.

H: Is the question unclear? Let me try it once more.

M: Yes.

H: If it is not clear, I will try to rephrase it.

Mr. DeLone: I do object to the interruption. The witness went on and endeavored to give a qualified answer and there was an interruption.

The Court: The question has been stated several times, Dr. Mauchly. Please answer it.

M: I stand by my letter which says that I had received some kind of a suggestion from some prior work, which I do not now recall, but which is referred to in the letter and which apparently no one else can find right now. So that this was not a flash from heaven, a full-blown device without any prior suggestion as to how anybody could do anything, but neither was any other calculating machine, as far as I know, and neither were the mechanical type calculators and harmonic analyzers, so far as I know. Each one was building on somebody else's work.

H: Then do you agree, Dr. Mauchly—

M: I agree I did not do this in a vacuum.

The Court (to the witness): Will you wait for the question, please?

H: You agree that you didn't do it in a vacuum, that it was not entirely your own design, and that it was in fact a modification of one used at MIT some years prior to 1940. Will you not agree that that is a suitable and correct statement?

M: I will agree if that's what my letter says.

H: "The circuit is already worked out"—

M: My circuit is.

H: —"a modification of one used at MIT some years ago, and a 12-ordinate analyzer of perhaps one and a half per cent accuracy has been built, using G.R. voltage dividers."

M: Yes. What I am trying to get across in my answer by way of qualification is that the pronoun "one," as I understand the reading now, which occurs in that circuit [sentence] with respect to something that was built at MIT earlier, its antecedent may be circuit. I don't know whether what I did adapt was a different kind of a circuit which came for a different purpose to a harmonic analyzer or whether MIT had already published and I had read about a differential analyzer—I mean, a harmonic analyzer. Excuse me—but, in any case, let me say that I did not build this thing in a vacuum. I have never built anything in a vacuum. I have always used prior art where it seemed proper, and useful, appropriate. You don't do these things without having a few people ahead of you who have worked out something before. I used Ohm's law, for instance. Mr. Ohm published that quite a while ago.

H: Does that end the qualification you wanted—

M: I will say that is the end of the question.

Halladay then introduced Plaintiff's Exhibit (or PX) 123, "Bibliography of Literature on Calculating Machines," a 1938 Moore School document from Mauchly's files. This bibliography, compiled by Irven Travis, the school's computer expert, had apparently been acquired by Mauchly in connection with a course that he sought to take from Travis in 1939, but that was canceled because of low enrollment. It was sixteen single-spaced pages long and included a listing of "Harmonic Analysers" with twenty-five entries; among these latter were two MIT theses that might or might not have borne on Mauchly's 1940 query to General Radio.

Halladay questioned him about the 1930 MIT master's thesis of M. Maxwell Bower, "Periodogram Analysis," but Mauchly could not recall whether he had read it; he could not even be certain he had received the 1938 Travis bibliography before he built his harmonic analyzer (transcript pp. 12,135–36). There was this further exchange (transcript pp. 12,136–37):

M: It's the sort of thing that I would read, but I don't remember now. I don't even remember Mr. Bower's name.

H: As a matter of fact, by the time you had built the harmonic analyzer there had been a substantial volume of literature on the subject, had there not?

M: I would believe so.

H: And you had done research in that literature to educate yourself on how to build such a device, had you not?

M: I had read the literature. I really hadn't done something I would dignify by calling research.

Thus Mauchly was drawn from an initial assertion that he had not modeled his harmonic analyzer on anyone else's work or read about anything electrical of the sort—that he had in fact conceived it entirely on his own—to an admission that he had modified something done earlier at MIT (whether the basic circuitry or the harmonic analyzer itself), and that he had read the considerable literature on the subject beforehand.

As it happened, DeLone was able in redirect examination to question Mauchly on the 1930 Bower thesis, of which he had secured a copy, and to bring out that Bower's machine was very different from Mauchly's (transcript pp. 12,484–86). (No mention was made of the second MIT thesis, "A New Method of Harmonic Analysis," 1927, a Ph.D. dissertation by G. W. Kenrick.) Halladay, however, had made his point, namely, that Mauchly, when pressed, could not be at all sure he had not used another inventor's ideas, even after testifying that he had not. Moreover, Halladay had gained the unexpected bonus of Mauchly's voluntary admissions of his habit of using prior art—and doesn't everyone?—with *no expression of a need to give credit.*

And so it was that a device that could have come off in court as at least a novel and challenging application of the existing analog technology came to be seen primarily as "borrowed" from that technology without acknowledgment. And the real cause for this eventuality was Mauchly's failure to prepare himself as a witness. He allowed himself to be trapped by his own documents! Had he read the material he himself had produced from his files, he would not have made such bold assertions, only to have them discredited. Or, rather, it was his failure to prepare coupled with an arrogant attitude that was his undoing on this and several other occasions.

Mauchly's testimony on his harmonic analyzer does make abundantly clear that this analog device—and the use he was able to make of it—occupied his research time quite exclusively while he prepared to give his weather-analysis paper at the close of 1940. Indeed, as we noted earlier, a letter he wrote on November 15, 1940, to meteorologist H. Helm Clayton, revealed that he was by then deep into the design of a *second* (twenty-seven-ordinate, as against the twelve-ordinate) harmonic analyzer. Moreover, as we shall see in chapter 3, his 1941 notes and letters indicate a sustained enthusiasm for the device up until his visit to Atanasoff in June, with only a gradual abandonment of it as he explored electronic (but not always digital) computing possibilities.

The Cipher Machine

Mauchly's cipher machine was the only digital instrument he built or even planned in any detail while he was at Ursinus College, that is, up until the summer of 1941. It was, of course, not an arithmetic but a cryptographic device.

He built this machine in 1937 and attempted to interest the U.S. Army Security Agency in it, without success. A photograph introduced at the ENIAC trial as PX 21,376 showed three banks of neon indicator bulbs, three columns per bank and nine bulbs per column, with a strip of paper placed along each column to designate a letter of the alphabet for each bulb plus a blank for the space between words. A set of two switches per alphabet bank, one a nine-position and the other a three-position switch, allowed the selection of one letter (or blank) per bank of twenty-seven bulbs, so that three entries could be encoded or decoded with each setting. There were two further small switches for selecting a scrambling code to be used in the operation at hand.

Mauchly's Testimony

Mauchly presented it at trial as a digital device that worked in the base three. But all coding devices are digital, and, as he acknowledged, twenty-seven-position switches would have been preferable to his combinations of nine- and three-position switches. He also said that this was a pilot model, and that a finished product would have had a keyboard on which to enter the message. He made the further, puzzling statement that the neon lamps were not just indicators of letters being processed, but played a role in the coding process; we do not see how this could have been the case.

We quote a few passages from Mauchly's direct testimony on his cipher machine in response to questions by defense attorney DeLone (transcript pp. 11,783–90):

> . . . And this device as I built it was not all to be desired by someone who had this job to do all the time but, rather, as a sort of pilot model to show the principles by which this could be done.
>
> In other words, if someone was doing this all the time, he would probably put the clear text in through something like a typewriter keyboard. There is no typewriter keyboard but, rather, there's some switches on the front panel. If he were doing it all the time, he might want the results to come out as a typed message from a typewriter. Here there is no such but, rather, a bank of lights which indicate the result out.
>
> . . . If I had had a 27 point switch, I probably would have used it, but these were radio surplus parts which were available to me at a low price, and

so the simplest way of getting a selection of any one of 27 values here was to use a nine times three combination of these two switches.

Well, this had a very nice property in this particular case, almost an accidental property. The English alphabet, as we normally use it, has 26 letters in the alphabet, and if we had been dealing with some other language, I would have been stuck. . . . As it was, 27, which is 3 times 3 times 3, was almost ideally suited to what I wanted, because I wanted it to represent at least the 26 letters of the alphabet, and I didn't want to have much more left over as useless, and the one extra thing that I had here, after 26 letters of the alphabet, I thought might be useful to represent a space. . . . You could use it to represent a period, or . . . a dummy . . . .

Well, as I mentioned, 27 happens to be 3 times 3 times 3, and this actually, then, happens to work on a base 3 number system. It's the only device I know of in the digital field which turned out to—the base 3 system was the most appropriate for it and where something actually got built and was working on a base 3 system.

. . . At least I demonstrated it to [a branch of the Government], and the intention was that if they were interested, why I would let them develop it as they wished.

. . . this was the Army Security Agency to which I offered it, because I knew statisticians who worked with them. I was able to contact them easily.

. . . Most of the rest [of the digital devices I developed at Ursinus], I guess, are related more closely to digital computing in the arithmetic sense— counting, and things of that sort.

The cryptographic device I just referred to here is interesting. It's a digital machine. It's interesting because it uses a base 3 number system.

It's also interesting because the read-out lights are not just read-out lights, but in themselves have an electrical property with a kind of matrix function table which signals out a single light to respond to numerous input signals, you see. So there are a lot of things here in the cryptographic device, we will say, which interact and connect with applications to numerical devices and to digital computing devices as generally understood, although this itself did not work with numbers in any arithmetic way.

The Honeywell side did not choose to cross-examine Mauchly on this instrument.

DeLone now questioned Mauchly about his arithmetic digital endeavors while at Ursinus College: a two-neon oscillating device, gas-triode ring counters, and, lastly, a plan to build a vacuum-tube desk calculator with direct application to his weather analyses. We turn to these next.

The Two-Neon Device

Although photographs of Mauchly's harmonic analyzer, his cipher machine, and various pieces of laboratory or experimental equipment were exhibited and discussed in the ENIAC patent case, the only item actually produced and demonstrated was a two-state device whose basic elements were two neon bulbs. This digital device was in fact the only original electronic contrivance created by Mauchly before he met Atanasoff. It deserved to be called electronic because its neons filled an organic function. It was not a computing device, however, or even a "building block" for a computer. We investigate it here because Mauchly insisted in the ENIAC trial that it was a binary counter.

The two bulbs of the device were mounted at the top of a glass rod, through which wires passed to them from a capacitor and three resistors housed in a round cardboard base; wires from this base to the outside could be attached to a battery or other direct-current power source. The two states of the device were one in which the first bulb was on, the second off and, alternatively, one in which the first was off, the second on. When it was connected to a battery of appropriate voltage, the neons switched states continually, blinking back and forth like a railroad crossing signal. When it was connected to a battery of a different (appropriate) voltage, the neons held their (chance) states as long as the connection was maintained (figs. 22 and 23).

Mauchly's Testimony

Mauchly, in direct examination by DeLone, presented his two-neon device only in the latter mode, claiming actually to have built it and used it as a binary counter in that mode. Not until he was cross-examined by Halladay did he acknowledge its possible use as a toy railroad signal, or, as he also termed it in this mode, a relaxation oscillator. Yet it was not even a relaxation oscillator in any practical sense. Whereas earlier relaxation oscillators had been built with vacuum tubes to drive other circuits, neons could produce no useful output signal. We do believe, then, that Mauchly originally developed this device as merely a toy in its oscillating mode, or perhaps a teaching tool embracing both modes in a purely experimental sense, and that he tried to elevate it to the status of counter for trial purposes.

But how could he portray a neon device with no means of entering or reading out signals as a counter? He explained that what it counted were *interruptions in power,* caused by the operator's manually disconnecting and reconnecting it to its power source in a certain way: one disengaged a wire from a terminal momentarily, then retouched it to the terminal, in which process its bulb turned, say, on, while the other bulb turned off; one could do this repeatedly, always disengaging

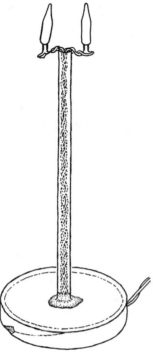
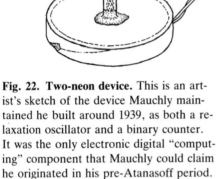

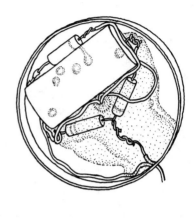

Fig. 22. Two-neon device. This is an artist's sketch of the device Mauchly maintained he built around 1939, as both a relaxation oscillator and a binary counter. It was the only electronic digital "computing" component that Mauchly could claim he originated in his pre-Atanasoff period. The hollow glass rod carried wires from the base to the neon bulbs at the top.

Fig. 23. Underside of two-neon device. This sketch shows a capacitor and three resistors set in wax in the cardboard boxtop that served as a base. The two emerging wires went to the terminals of a battery or other power source. Mauchly explained that this instrument, when connected to a battery of appropriate voltage, could count manually administered power interruptions of very brief duration (about a hundredth of a second). It functioned as a toy railroad crossing light when permanently connected to a different battery, again of appropriate voltage.

the wire only a hundredth of a second or so, then retouching it for the next change of states.

Thus Mauchly's "binary counter" offered the entirely fruitless possibility of counting power breaks, up to "1," through the eye, hand, and brain of the operator, with an impossible requirement that those breaks be no longer than a hundredth of a second. (Even electromechanical relays of that day required a sixtieth of a second to respond.) Mauchly did testify that a switch could have been inserted to enable the device to count more accurately, but he did not claim to have inserted one, and the outcome would still have been fruitless. Indeed, we find the notion of interrupting the power supply and counting the interruptions patently absurd, a desperate ploy incredibly attempted by a man of Mauchly's expertise in 1971, as evidence that he had invented the electronic computer.

(Mauchly also testified that this device could have been connected to others like it in series, to enable it to count power interruptions beyond "1," but not that he had ever equipped it with the circuitry to do so.)

From a historical point of view, it is ironic that Mauchly should have claimed this neon device as a counter. The binary counter used in scaling circuits to count cosmic rays (roughly) had evolved, in the 1930s, from the relaxation oscillator and had constituted an inventive advance over it. What Mauchly had created, by substituting neons for vacuum tubes, was a *degenerate* form of relaxation oscillator, which he nevertheless touted in court as a counter capable of functioning in a computing context!

The precise date of this device was not established in court. Mauchly thought he might have built it a year or two before he met Atanasoff in late 1940. He was unable, however, to locate among the papers he had produced from his files any circuit diagram of it predating their first meeting. Indeed, the *only* diagram of it he could find was a sketch he had drawn for Atanasoff during that first meeting. This sketch included a battery, but not its voltage, so that he had to admit that even this one sketch remained nondescript as to whether it depicted a "counter" or a "relaxation oscillator."

We have no reason to doubt Mauchly's estimate as to when he first built the two-neon device—as a toy. We do doubt that he thought of using neons for counting before he met Atanasoff and discussed the possibility with him. For Mauchly did have papers to show his efforts to design neon counters, with vacuum tubes interspersed for control (and amplification), in 1941, beginning a few days after their meeting. He clearly did have a fascination with these tiny diodes: he had used them in his cipher machine, he had built the toy railroad flasher of them, and he continued to try to design them into counters until mid–August, 1941, only then finally recognizing their deficiencies as computing elements (PX 665; see chap. 3).

We now let Mauchly speak for himself, through excerpts of his testimony before DeLone as he presented his "binary counter" and before Halladay as he defended it. The photographs referred to are PX 21,374 and PX 21,375. The first depicted top and bottom views of the one model Mauchly had preserved, much as we show it in our figures 22 and 23; the second depicted it standing beside a rectifier supply he had built for it after joining the staff of the Moore School, to display it blinking in his office there. A sign, "RR Stop," was posted above the two photographs at the trial, where exhibits being discussed were often mounted on easels. We begin with the testimony before DeLone (transcript pp. 11,792–99):

Actually [PX] 21,374 and 21,375 . . . depict exactly the same piece of apparatus except that in 21,375 a piece of cardboard which was up around the lamps is now down around the base so that you can more clearly see that there are two of these neon diodes at the top of the vertical glass rod . . . .

This I referred to as a binary counter and it's one of very simple construction. All the parts to it are in the base of a cardboard cover for an ice cream half-pint container, or something, and the—a couple of resistors and a condenser in the base and then the two neon lamps which are shown at the top of the glass rod.

At times, it was used with a battery. I had, of course, other power supplies available [at Ursinus]. . . .

. . . one of these alone could be used as a counter providing it had something to count.

The something which it counted in this case would be momentary interruptions in the power supplied to it. In other words, if you had this connected to a battery of steady voltage and that voltage was properly chosen, then one lamp and only one lamp would be on.

Now, if you interrupted that voltage momentarily and restored it almost immediately thereafter, . . . you would find that the lamp that was on was now off and the lamp that was off before was now on.

Now, if you were a little clumsy about this and you just scratched a wire across the terminal, you might have made a dozen interruptions in that process and you might see the same lamp on that you saw before you did this. Now, all that means is that you interrupted the circuit an even number of times.

But if you have a switch which you carefully controlled so that you know that you were interrupting just once, then the lamp will alternate back and forth for each interruption, and you could see it do that if you do it slowly enough.

If you wished to put on other apparatus such as pulse generators, then you could test how fast this thing would respond to such interruptions and how fast it will count.

But in all cases what it is doing is counting in the binary system; that is, it has only two possible indications: either the left lamp is on or the right lamp is on. Never both, but one is on.

. . . the method I was just describing is the way in which you would normally do it if you were using it as a counter; that is, you supply it with a steady voltage, but everytime that you want to switch it, change from one lamp to the other, you interrupt the current for some very short period of time such as, say, a hundredth of a second or less.

. . . [we] should begin with the fact that a gas diode such as these neon elements were can be maintained, continue to carry a current, continue to show a reddish glow . . . , at a voltage which is lower than the voltage necessary to get it started. So you have what is known as . . . a starting potential, and then you have . . . a running potential, thereafter, you see.

. . . What you must do is be sure you have enough resistance in the circuit, that is, you don't have just a battery and a bulb now, but you have a battery, a resistance and a bulb, so that as soon as the 90 volts has started this thing and the, what you call the internal resistance has gone way down because the gas in there has now been electrified by the passage of the current that you have just started,—from that point on you have got to have something which you might say has better sense than that gas has as to what it can stand and what it can do. . . . It is almost a suicide device if left to itself. So you put this resistance in there to prevent it from committing suicide, and then you get, as a result of putting the resistance and the neon lamp together there,—you get something which you can use for testing fuzes. Is there voltage here or isn't there?—providing it's as high as 90 volts.

And what I did in this circuit was to combine two of these lamps with part of a circuit in common and part separate. And if we don't want to go into all the electronics of this thing, what it amounts to is that once one of these bulbs is started, then the voltage available at the other bulb can't be any greater than, we will say, 80 volts, because it only takes 80 volts, we will say, to keep one of these bulbs running. But as soon as you stop the current and the gases in that tube have a chance to cool off and no longer are electrified, then things are apparently equal as far as the two tubes are concerned.

Reapplying the voltage, there will have to be 90 volts somewhere, and whichever bulb starts first, it will be a matter of chance, except that there is another element in this circuit, the condenser [capacitor]. And the way all the things are connected up is that the voltage with the tube that has just been on has so changed the electrification of that condenser that if you restart this circuit there will be less voltage on that tube, it will tend to stay around the 80 volts where it was running, you see, but the other tube will get at least the 90 right away, and it will be the one that starts.

Under cross-examination by Honeywell attorney Henry Halladay, Mauchly explained this same process again (transcript pp. 12,113–21), but this time as much in connection with the drawing he had made during his first meeting with Atanasoff

(PX 22,420) as with the photograph exhibits. At one point, he elaborated on the necessity of keeping the power interruptions brief (transcript p. 12,115):

> I would feel a little better if you explain [the process] as a matter of inter-rupting the circuit periodically rather than making the circuit periodically. In other words, a slight touch, then a long pause, another slight touch would not insure this operation. What we are talking about here is most of the time the battery is connected and you are only periodically and momentarily making a disconnection. Most of the time you are making a connection.

At another point, he spoke of the oscillator mode (pp. 12,116–18):

> . . . Now, if you choose a different battery voltage you cause the other phenomenon to occur. That is, it will flash back and forth without interrup-tion of the battery circuits.

> That's right [the battery shown on PX 22,420 is nondescript].

> That is called a relaxation oscillator, or in the vernacular I used it was a railroad crossing light.
> It flashes, which could be used with a toy railroad train, for instance.

> Yes ["RR Stop" is what the sign at the top of the exhibit says].

> Mauchly and Halladay now had the following exchange over the definition of a counter, and in particular of a binary counter (transcript pp. 12,118–22):

> *Halladay:* When you say "counter," Dr. Mauchly, as of December 1940 what does your term "counter" mean?
> *Mauchly:* Well, broadly it would mean anything that counts, if you are asking how you would specifically apply it to a circuit.
> *H:* No, I am not.
> *M:* —electrical circuit.
> *H:* I am asking for a meaning of a word, and it doesn't give us a meaning to simply repeat the word.
> *M:* All right. A counter is something that counts.
> *H:* Again that simply repeats the word. Do you mean it enumerates succes-sively?
> *M:* It seems that saying a counter is something that counts is too simple, maybe saying it is something that enumerates successively is maybe a little too simple, too.
> *H:* Well, I am trying to find out your use of the term and what it means and to what it applies. In order for a device to be called a counter, would it not

have to enumerate successively and do more than just register on or off or 0 and 1, and particularly to count in a binary mode?

M: If I had a count—well, I shouldn't be asking questions, I guess. I should ask you to clarify your question.

H: The question is unclear?

M: The question is not clear to me, that is right.

H: Would it not be fair to say, Dr. Mauchly, that in order to call a device a counter which counts in the binary mode that it had to do more than show on or off alternately, as was the case with your railroad flasher depicted in [PX] 21,374?

M: We have just got through, of course, with questions and answers which show that this device could operate in two modes, and what was called the railroad flasher mode was not the mode which I called counting. So let's not confuse things by referring to it as a railroad flasher when we are talking about it as a counter.

H: I submit you are not answering the question that was put to you.

M: I am defining my terms, but I will listen to a different question, if you like.

H: No, I think it was a particular question, Your Honor.

M: Well, I have an objection to your question, then. I will put it that way.

The Court: Can you restate it or do you want the reporter to read it?

H: I will try to restate it.

. . . .

M: [It was a binary counter] under the circumstances that you were asking about earlier, that is, where there was sufficient potential in the battery to ignite one lamp but not to cause this spontaneous oscillation between the two. In that case you had to interrupt the circuit to cause [it] to make transfer from one state to another—state here being the state of the left-hand lamp being on and the right-hand off, or another state being the left-hand lamp being off and the right-hand lamp being on. Now, a two-state device which maintains one state until it receives a signal which causes it to change state and always on perceiving that signal changes from one state to the other is to my way of using the word a binary counter.

H: You agree, do you not, that it would be impossible to use that device in that condition with nothing added to count in the binary mode, do you not?

M: I do not agree.

H: Well, what would you do with the device when it got over 2?

M: This device won't count over one.

H: Then under what circumstances, and that is what I am trying to understand, could you say that a device that won't count over 1 is a binary counter?

M: Would you allow a counter that could only count to 8?

H: No, we were talking about one that wouldn't count over 1, which if it won't count over 1 does not permit of successive enumeration. Is that right?

M: I am not answering your question about successive enumeration. I am answering your question about whether this is a binary counter, as I understand it, and I say it is a binary counter.

H: Even though it won't count over 1?

M: That is right.

H: All right.

M: It has two stable states.

H: Well, that then we will take as your definition of a binary counter,—

M: Well, the fact that—

H: —a device which has two stable states and will count to 1? Is that acceptable?

M: Yes.

We would point out here that Mauchly, in the definition he finally provided, simply ignored his own strict admonition as to the need to keep the power interruptions "momentary." For, as we saw earlier, neither of the two on-off states of this device *was* stable when it was awaiting a counting signal, that is, while the power was disconnected. Thus Mauchly had at last provided a good definition of a binary counter. It just did not apply to his device!

Whether or not Halladay caught this subtle evasion, he did not have to reveal it in order to reduce Mauchly to a foolish position before the court, namely, that of defending a "counter" that could not count over "1." Now this is not of itself a foolish position, since binary (ring) counters are not supposed to count over "1." But such counters are supposed to be capable of being connected with others into a counting device that *can* count over "1." Because Mauchly's device could not function in such a series, he could not press his point about "a counter that could only count to 8." For he would then have had to face Halladay's question of what else was required to make this device count "something," and to acknowledge that he had made no attempt to augment it in that fashion before meeting Atanasoff. So it was that he ended up defending a "counter" of no utility at all.

Halladay concluded his cross-examination on the two-neon device with the following exchange concerning the date of its creation. Mauchly had testified earlier not only that it had been "considerably before" he met Atanasoff, but also that he had worked out a circuit for it on paper at the time (transcript p. 12,109). Now Halladay asked him to produce a drawing of such a circuit that predated those of January 1, 1941, and the one he had drawn for Atanasoff four days before. Here are some highlights of that exchange (transcript pp. 12,125–28):

Halladay: Now, where in any of these drawings [PX 665] is there showing of a device to interrupt the current?

Mauchly: Well, right here, but you don't like those because they are later dates.

H: Well, find one right here.

M: I showed it to you and you said that has got vacuum tubes in it. And it does. That is exactly what the vacuum tubes are for. That is a switch to do that interruption.

H: Do you know, Dr. Mauchly, that you have looked for drawings [of a binary counter] that antedate the visit you and Dr. Atanasoff had at the University of Pennsylvania in December 1940 . . . and have not been able to find such?

M: No, I don't know that.

H: Would you not admit that you have tried to find earlier drawings and have failed?

M: Well, the situation is that I responded to several subpoenas which asked me to bring in materials. I tried in each case to find in my files all the things that were required by the subpoena, which included drawings as well as other things, and in each case I tried as best I could. I don't know now—I haven't reviewed it from that point of view to see how successful I was or wasn't in respect to the specific things you are now asking. It never occurred to me to search with exactly that in mind, because I knew at the time that this little thing I had and operated and other people had seen operate long before I met Dr. Atanasoff. So it never occurred to me to look for that sort of thing.

H: Well, you see, Dr. Mauchly, I am not at the moment arguing anything with you.

M: (Interposing) I'm sorry if I sounded argumentative—

H: On the subject of—

The Court: One at a time, please.

M: Yes, I am sorry.

H: On the subject of this device—

M: (Interposing) Yes, okay.

H: —this railroad flasher, as I am more pleased to call it or binary counter as you prefer.

M: Yes.

H: —as reflected on Exhibit 21,374, I am trying to find out if you have been able to turn up any sketches, drawings, or diagrams of such a device that antedates the date upon which you had your conversation with Dr. Atanasoff at the University of Pennsylvania in December of 1940?

M: Shall I answer now?

H: If you can.

M: Well, the answer seems to be in three parts now. First, I do not know whether I could do that or not. Secondly, that I have tried to answer subpoenas which said "Bring in things prior to certain dates," but the subpoena did not say specifically, "You've got to bring in drawings of this circuit," which had a date, a certain date, and so I haven't actually examined that point. I don't know the answer to that one. I cannot respond any further than that. I may be able to later, but I can't now.

Of course, it is a matter of simple logic that if one brought in *all* papers for the relevant period in response to a subpoena, one would not neglect some items because one already knew what they proved, or because one had not been looking particularly for them! It would also seem that, to the extent one might be selective among a huge volume of material, one would try harder to find items to support one's own case. Mauchly did bring in the neon device, did expect to testify about it and demonstrate it, and would surely have wanted to document his creation of it before he met Atanasoff. Evidently no such documentation existed, or exists. He did not produce any for the trial, as he testified he might be able to do.

Ring Counters

We have seen that Mauchly had built a digital neon-bulb device, possibly as early as 1938, but that his own notes placed his earliest attempts to interconnect neons in series in 1941, after his first discussion of computing with Atanasoff in late 1940. Now clearly the interest of the Honeywell attorneys in this timing was not based on any significance in the neon circuits themselves, which were an unproductive effort at electronic digital computing. Nor was it based on any possible derivation from Atanasoff, who was not even working with counters. Their interest, rather, was twofold: they wished to show how little Mauchly had accomplished in digital computing before he met Atanasoff, and they wished to point up the flurry of activity on Mauchly's part that immediately followed his learning of Atanasoff's success in digital computing, as distinct from his own success in analog computing (with the harmonic analyzer he had presented in his paper).

Having discredited the neon device of which Mauchly sought to make so much at trial, attorney Halladay ignored another digital computing endeavor of Mauchly's introduced by DeLone in direct examination, namely, his experimentation with gas-triode ring counters. This exercise, original with Mauchly only in that he substituted gas tubes for vacuum tubes in the counters described in the literature, was, in fact, one for which Mauchly himself had little to say at trial. It was also one that was attributed only to his *Ursinus* period, which included the spring term of the 1940–41 academic year when Mauchly was engaged in that flurry of activity stimulated at least in part by Atanasoff.

We present it here, however, not only because it was examined in the ENIAC trial and does represent the limit to which Mauchly was able to carry electronic digital computing before he met Atanasoff—and even in the months afterward—but also because, like the so-called neon counter, it has been taken by recent writers to constitute original basic computer design (see app. B).

The ring counters of the literature were simply the individual, one-digit-place counters of the physicists' scaling circuits used to count occurrences of physical phenomena. Thus the basic elements of the usual scale-of-two counters were binary ring counters, but ring counters in other bases were also being developed and presented in the literature.

The word "ring" merely signifies the cyclical progression of digits, round and round, as signals enter the counter. Besides this denotive function, it serves to distinguish such a counter in any base from a *blocking* counter in the same base; a blocking counter being one that does not automatically return to "0" from the highest digit in its base, upon receipt of the next signal, but remains there until cleared by a signal from a different input.

The expression "ring counter" is not employed much today, because most counters are binary, and "binary counter" is taken to mean "binary *ring* counter." A different term, "flip-flop," designates a blocking binary counter. (Binary counters and flip-flops are usually distinguished by noting that the former has one input, the latter two.) "Ring counter" has been of use historically, however, in the context of a number of different bases, and it does remain the one simple way of designating just one element in a series of connected elements, as against "counter" taken as the series itself, that is, as the entire counting device. A scale-of-two counter, for example, is a series of binary ring counters connected so as to count a large number of pulses entering in the ones place.

But the essential fact to have in mind, as we review Mauchly's work with ring counters, is that *all* the electronic counters of the 1930s were designed for use in scaling circuits, *for counting only*. In scale-of-two counters, as a stream of pulses entered the ring counter at the ones place, it went from "0" to "1" upon receipt of the first pulse and, upon receipt of the second pulse, returned to "0," whereupon a simple switch signaled the counter in the twos place to go from "0" to "1" to effect a carry. The twos counter, the fours counter, and so on, received pulses and passed along carry signals in similar fashion. The obvious mechanical analogy is to an odometer, except that its dials move continuously; one must imagine an instrument that counts on a similar set of dials, but makes discrete jumps from digit to digit. (A pedometer, which counts the steps of a hiker, may be more apt.)

Ring counters had not yet been developed for use in *computing* circuits, or accumulators, for which a much higher level of sophistication is required, in terms both of reliability and of function. As to reliability, the rough estimates achieved by the scaling circuits of the 1930s—rough because of occasional missed pulses—would not do for computing, where exact results are usually needed and where the slightest error is usually compounded in the arithmetic process.

As to function, the ring counters of accumulators must be capable of receiving many streams of pulses in parallel, one stream to each counter or place. They must also be capable of working with associated switches, analogous to those of the scaling circuits but much more complicated, to perform their arithmetic operations: to perform the simplest operation of addition, they must work with these *arithmetic* switches not only to pass along carry signals, but to pass them on a delayed basis in coordination with the next (higher-place) counters' receipt of their incoming streams of pulses. Finally, they must be capable of receiving orders from control switches as to which arithmetic operation to perform at a given juncture.

Indeed, it was just this development in counters, from *counting* ring counters to *computing* or *accumulating* ring counters operating reliably at 100,000 pulses per second, that made the ENIAC feasible. This advance was accomplished only in 1943, however, and not by Mauchly but by Eckert, working from a design of I. E. Grosdoff's. Mauchly, in fact, while at Ursinus, did not get so far as even to interconnect two of the simple decade ring counters he had adapted from the literature so as to "count something," as he himself put it. That is, he never formed a series in which pulses entering in the ones place passed on a carry to the tens place as the count progressed past "9."

Mauchly apparently chose to use these gas "trigger tubes" instead of vacuum triodes to save money. He testified to that effect to both DeLone and Halladay, in another regard. It is hard to imagine that the difference in cost between gas and vacuum triodes could have been significant, but it may have been in those depression years. Mauchly did, of course, lose the great advantage of speed of operation of the vacuum tube.

Mauchly's Testimony

Mauchly, while he had notes on his 1941 designs of neon counting circuits which he produced at trial, had none on the gas-triode ring counters he testified he had built for experimental purposes. Instead, he was examined about these by Sperry attorney DeLone on the basis of a photograph of still-existing circuit boards with such tubes mounted on them (PX 21,382). Moreover, as with the neon circuits, he made no claim to have gone beyond the physicists' *counting* devices to *computing* devices; he was trying only to duplicate their circuits with lesser elements, a goal he actually failed to reach.

Under examination by DeLone as to digital devices he had built at Ursinus College (other than his cipher machine, which they had just discussed), Mauchly said (transcript pp. 11,791—92):

> Well, the ones that I actually built and experimented with in the Ursinus days were mostly ring counters and counters of—things of that sort, using gas tubes rather than vacuum tubes. I was increasingly aware and very interested in the published literature which showed how vacuum tubes could be used for instrumentation in cosmic ray work. They called them scaling circuits, but from our arithmetical point of view they were counters, and there were lots of examples of those things cropping up in the literature in those days. Both the Journal of the Franklin Institute and the Review of Scientific Instruments were carrying articles of that sort.
>
> [I mean] instrumentation using cosmic rays—using vacuum tube counters in measuring cosmic rays. And if I had had all the means I wanted, all the

money and time, and so on, why, I probably would have been constructing vacuum tube counters of the sort that I saw described in literature as my main extra occupation after teaching. What I did instead was to try to use some of these cheaper components to see whether I could make them do the same job perhaps a little more slowly, but much more cheaply, and that I could not find in the literature. I had to do that on my own. There wasn't anything in the literature, so far as I know, for instance, which showed how to make a binary counter out of these little neon bulbs . . . .

Strangely enough, while he seemed to start out on the subject of the gas *triode* experimentation, he soon shifted to the little gas *diodes* known as neons—or perhaps he meant them all along. He went on then for some time about the neon circuits he designed in the first eight months of 1941.

DeLone did ask him explicitly about the photograph of circuit boards with gas triodes in them, and he replied (transcript pp. 11,801–2):

> Yes. [In PX] 21,382, there are three things we might describe generally as circuit boards, things which I built with my own hands to be suitable for experimental purposes. . . . It's not easy at this time to say exactly what these boards are wired up for but I did use such boards for testing ring counters, primarily with gas trigger tubes of a cold cathode type, and there's two such tubes in this photograph. . . .

He now defined ring counters in terms of their application to "counting or . . . scaling purposes."

Thus it is clear from Mauchly's own testimony that his greatest effort with counters at Ursinus College was to duplicate the ring counters of the physicists' scaling circuits using gas tubes, whether neons or triodes. As we have seen, he did not get so far as to connect two of these ring counters together for use even in this strictly counting sense. And, indeed, to the best of our knowledge, at no time in the future when he had vacuum tubes at his disposal did Mauchly succeed in building an operative counter for use in a computing context.

Plan for an Electronic Desk Calculator

Two late 1940 letters bearing on a plan of Mauchly's to build an electronic desk calculator for use in his weather calculations came to light only after the ENIAC trial. Mauchly did testify at that trial of such a plan, but he did not bring in these letters, which were found by Kathleen Mauchly after his death in 1980. The first, a November 15, 1940, letter to meteorologist H. Helm Clayton, with whom he had worked the previous summer in Canton, Massachusetts, was published in full in the *Annals of the History of Computing* (Mauchly 1982), as one of three previously unpublished items.

We quote the relevant paragraph (p. 248):

> In a week or two my academic work will not be quite so heavy, and I shall begin to give some time to the construction of computing devices. We have further simplified the design of our proposed 27-ordinate analyser. In addition, we are now considering the construction of an electrical computing machine to obtain sums of squares and cross-products as rapidly as the numbers can be punched into the machine. The machine would perform its operations in about 1/200 second, using vacuum tube relays, and yielding mathematically exact, not approximate, results. That is, its accuracy would not be limited to the accuracy with which one can read a meter scale, but could be carried to any number of places if one cared to construct the machine with that many parts. With conventional tubes, it would be rather bulky, but special tubes could be designed to make it very compact.

The second letter, dated December 4, 1940, was to former student John DeWire. An excerpt of it was also published, in a letter to the editor of *Datamation* from Mrs. Mauchly's nephew, James McNulty (McNulty 1980). We reproduce that excerpt here (p. 24):

> For your own private information, I expect to have, in a year or so, when I get the stuff and put it together, an electronic computing machine, which will have the answer as fast as the buttons can be depressed. The secret lies in "scaling circuits," of course. Keep this dark, since I haven't the equipment this year to carry it out and I would like to be the first.

These two letters constitute the earliest documentation of any plan, or even hope, by Mauchly to build an electronic digital computer. And they make perfectly clear that such a project lay in the future; not for "a year or so" did he expect to have such a machine. They also make clear that his more immediate goal was to build the second, twenty-seven-ordinate harmonic analyzer he had already designed. Letters to Atanasoff and Clayton in 1941 confirm that his major research occupation remained the analysis of weather data (see chap. 3). We do believe, though, that his designs of neon circuits in 1941, with vacuum tubes interspersed for switching, were efforts toward this digital calculator. At the same time, we continue to believe, with the Honeywell side, that he was inspired at least in part by his meeting Atanasoff and learning that electronic computing was indeed feasible.

Mauchly's Testimony

We quote the pertinent passages of Mauchly's testimony, beginning with his response to Sperry attorney DeLone as to the purpose of his experimentation with digital devices at Ursinus (transcript p. 11,803):

Well, the purpose was to find out for myself whether I had any hope of designing and building a rather low cost digital device which would make the work that I was doing easier. These weather statistics and things of that sort required so much computation that I had the feeling that I needed some better computing equipment than what was available to me at the time, or even what I might procure if I had had some money just to buy what was on the market.

Under cross-examination by Honeywell attorney Halladay, he was asked specifically what he had intended prior to January, 1941. We reproduce the complete exchange (transcript pp. 12,159–60):

Halladay: By January of nineteen hundred and forty-one, what sort of low cost computer were you in fact thinking of building?

Mauchly: A rather simple one at that time. I was trying to get something within my possible budget. This was why I was very conscious of the cost question here. This was why I was working with neon diode counters, which we discussed earlier. This was why I was working with ring counters, which were gas tubes rather than vacuum tubes, and the various sketches and the various circuits I tried out, and so forth, were all directed toward these low cost computing elements and devices.

H: Were you thinking, as of January, 1941, in terms of something like a desk top sized computer—

M: Something a little more—

H: —although electronic in character?

M: Something a little more as far as its capacity; not necessarily as regards its physical size. I didn't know exactly how its physical size would turn out.

H: Were you thinking in terms of essentially a sort of a simple type of computer as of January, 1941? Is that what you had in mind?

M: Well, my notes are descriptive of that sort of thing, yes.

H: And the notes you referred to are what notes?

M: There are some here in exhibit which describe these things. They are essentially desk calculators capable of storing enough information so that you would not have to repeatedly reenter numbers which were to be used again and again.

H: A keyboard sort of thing?

M: But working from a keyboard, not from a punched card necessarily, or from some other thing. Magnetic tapes hadn't even been thought of then for this purpose.

The notes referred to here are, again, PX 665, a ten-page collection titled "Computing Circuits" by Mauchly and dating from January 1 to August 3, 1941. As it happens, however, not until the two August pages—nearly two months after his visit to Iowa to see Atanasoff's computer—was there any actual depiction or exploration of a *digital* keyboard machine. Six of the other eight pages were devoted to attempts to design basic (digital) circuits from neon diodes controlled by, or switched by, vacuum tubes. And the two remaining pages, both dated January 1, 1941, were devoted, respectively, to an *analog* keyboard machine ("Heterodyne computing machine") and to Atanasoff's computer! We return to these notes in chapter 3.

The Situation as of December, 1940

Before we proceed to the series of Mauchly-Atanasoff contacts that formed the ENIAC "connection," let us sum up the state of Mauchly's accomplishments in computing at the close of his pre-Atanasoff period.

His greatest achievement was an analog computer based to some extent on models presented in the literature of the day. This twelve-ordinate harmonic analyzer was also the achievement that consumed the major portion of his limited research time in the two years before he met Atanasoff. (As the sole member of the physics department of Ursinus College, Mauchly carried a heavy teaching load through the academic year, and he took other jobs in the summer.) The years 1939–40 were occupied first with the design and construction of the hand-operated electrical analyzer, then with its application to weather data; this considerable effort culminated in his presentation of results to a session of the AAAS on December 28, 1940, on which occasion his path first crossed that of Atanasoff.

Late 1940 also saw refinements in the design of a twenty-seven-ordinate harmonic analyzer, with the plan for its construction abandoned sometime in 1941 as digital electronic computing gained the ascendancy in Mauchly's thinking. It is the case, though, as we shall see in chapter 3, that he continued to be deeply committed to this analog approach to weather analysis through the spring of 1941, some five months after he had met Atanasoff, but still before he had visited him in Iowa and seen his computer.

Mauchly's efforts at digital design in his pre-Atanasoff period, on the other hand, were scant, amateurish, and completely ineffectual. His most impressive achievement was not an arithmetic instrument at all, but a cryptographic device that he testified was in a crude state intended for development by Army engineers if they were interested; as it happened, they were not. This cipher machine predated the harmonic analyzer by more than two years, having been built in 1937. It too was electrical, although it used neon diodes as indicators of the activity within. It was definitely a side enterprise; Mauchly had been occupied with weather data analysis since 1936, using desk machines to do the calculations.

At the ENIAC trial, Mauchly presented a second digital device built possibly in 1938 or 1939, claiming it was actually a binary counter he had used in a counting context. This neon device was, basically, a toy relaxation oscillator which, when connected to a battery of appropriate voltage, blinked its two lamps alternately in the manner of a railroad crossing signal. But it could be made to hold one lamp on, the other off, indefinitely if connected to a battery of different (appropriate) voltage, and it was in this latter mode that Mauchly claimed to have used it as a counter. He explained that an operator could interrupt the power momentarily, causing the lamps to reverse states, and that doing so amounted to "counting" the interruptions, up to "1."

Despite this testimony, Mauchly was unable to document any contemplation of neons as counting elements until early 1941, after he had discussed his little flashing device with Atanasoff. Indeed, the only diagram of it in either mode was a nondescript one he drew for Atanasoff during their first meeting. What is more, the drawings of neon counters that did emerge in 1941 all included vacuum tubes to be used in conjunction with the neon bulbs for switching and other purposes.

Thus it is clear, as his testimony at trial ultimately demonstrated, that Mauchly did not develop his pre-Atanasoff neon device as a counter, and that he worked on the basically misguided design of neon counters only after meeting Atanasoff and learning that digital electronic computing really was feasible. Moreover, he claimed in these designs simply to be trying to duplicate the scaling circuits of the physicists of that day, with the cheaper neons replacing their vacuum triodes; that is, he claimed simply to be trying to design counters that could *count*, not counters that could *compute*.

The only other digital devices alluded to at trial as instruments with which Mauchly had experimented were gas-triode rings counters constructed on standard circuit boards. Here, too, however, he testified that he was merely trying to build circuits he had seen described in the literature, for scaling purposes, not for computing. The great challenge in the design of counters for a computer, of course, lay in this very transition from counting one stream of pulses entering at the ones place to computing many streams of pulses entering in parallel at all digit places, and to do so with a reliability that had not been attained in the existing scaling circuits.

So it was that Mauchly did succeed in designing and constructing an operable electrical analog device, though not a completely original one—a noteworthy endeavor, we think, and a successful one so far as its goal of analyzing data was concerned. But Mauchly was not successful in his ultimate goal of proving that patterns in the earth's weather were traceable to solar phenomena. He did not publish the paper he gave in Philadelphia, to the best of our knowledge, or any other paper on his weather project. Nor did he ever get beyond what he termed at trial "indications that there were recurrences which were suggestive of solar effects" (transcript p. 11,809).

As to digital computing, whether electrical or electronic, he conceived of no circuitry that was usable in any counting or computing context in his pre-Atanasoff period, which ended in late 1940, or even in what may be termed his Ursinus period, which effectively ended as Mauchly left Ursinus to drive to Ames that mid-June of 1941. The fact that Mauchly chose to draw so much attention in the ENIAC trial to his two-neon device is in itself persuasive evidence that it was all he had accomplished by way of an electronic digital computing instrument.

Mauchly *was* beginning to be interested in electronic digital computing at the time he met Atanasoff. He had, a month or so before, written to two friends of a desire to construct a desk calculator based on vacuum tubes. He had not, however, worked out any of the circuitry for such a machine, or even drawn block diagrams for it. This was purely an aspiration, one that we believe was given considerable impetus once he learned of Atanasoff's computer.

CHAPTER 3

The ENIAC Connection

The ENIAC

The ENIAC (Electronic Numerical Integrator and Computer) was built at the Moore School of Electrical Engineering, University of Pennsylvania, Philadelphia, under the World War II military title Project PX. Work on the project, sponsored by the United States Army Ordnance Department's Ballistic Research Laboratory, Aberdeen Proving Ground, Maryland, began on May 31, 1943. (The official contract was not signed until June 5.) The ENIAC solved its first problem in December, 1945, and it was formally presented to the public in late February, 1946. It had cost over $400,000.

The proposal for Project PX, submitted at the Aberdeen installation on April 9, 1943, and accepted that same day, was prepared by the two chief designers of the ENIAC, J. Presper Eckert and John W. Mauchly, and by the professor who served as the school's director of war research, John Grist Brainerd. Eckert was a master's student at the school—that April 9, most of which he and Mauchly spent working on the proposal even as it was being presented, was actually his twenty-fourth birthday. Mauchly was an instructor who had earned his Ph.D. in physics at Johns Hopkins University in 1932 and had taught physics at Ursinus College, near Philadelphia, from 1933 to 1941.

After the initial design phase, the project became a team effort. Eckert and Mauchly were in charge of both design and construction, under Brainerd's administration. Three other principal designers, Thomas Kite Sharpless and Robert F. Shaw, both engineers, and present writer Arthur W. Burks, worked closely with them. Some eight or ten others from engineering and mathematics were on this team.

Mathematician Herman H. Goldstine, lieutenant and then captain, was the Ordnance Department's representative for Project PX. He advised and assisted the Moore School group in dealing with the government and with suppliers. Moreover, he had been crucial to the launching of the project in the first place.

We maintain that the end product of Project PX, the world's first general-purpose electronic computer, grew directly out of Mauchly's contacts with Atanasoff and his special-purpose electronic computer. All of these contacts—their first meeting in late December, 1940, their subsequent correspondence, Mauchly's visit to Atanasoff's laboratory in June, 1941, and their further correspondence through October, 1941—were thoroughly aired in the ENIAC patent trial conducted thirty years later. As we indicated in chapter 2, the decision in that case invalidated the ENIAC patent Eckert and Mauchly had been granted in 1964, one of the grounds being derivation of the claimed invention from Atanasoff. Chapter 3 presents and analyzes the evidence and the testimony offered in court by both sides of this dispute over what, if anything, Atanasoff contributed to the ENIAC. Our presentation will be clearer, however, if we first provide some background on the ENIAC itself as it developed at the Moore School.

The April, 1943, proposal to the military had actually been preceded by a much shorter and sketchier memorandum written by Mauchly in August of 1942. This memorandum reflected a year's consideration by him, in consultation with Presper Eckert, of a digital electronic computer that could do the work of the differential analyzer, and do it faster and more accurately. We begin, therefore, with this earliest exposition of the machine that was to be the ENIAC, turn next to the formal proposal, and conclude with a brief description of the finished computer. For a complete account of this period—and thorough documentation—the reader is referred to our article on the ENIAC (Burks and Burks 1981).

Let us repeat at the outset a word of caution given in our introduction. We are arguing for a direct progression of ideas and events in the following stages: (1) ideas Mauchly got from Atanasoff in their 1940–41 contacts; (2) Mauchly's conception, with vital input from Eckert, of an electronic computer as explained in his August, 1942, paper; (3) the Eckert-Mauchly conception contained in the April, 1943, ENIAC proposal; and (4) the completed ENIAC, for which the design was essentially in place by July, 1944. But it was just that, a progression, with substantial change and substantial development from stage to stage. The resulting machine, the ENIAC, was thus a far different computer from Atanasoff's ABC and embodied concepts far beyond those Atanasoff had given to Mauchly.

Mauchly's 1942 Memorandum

Discussions between John Mauchly and Presper Eckert had begun in the early summer of 1941, when their paths happened to cross at the Moore School. Mauchly had enrolled in an accelerated defense training course in engineering and electronics, and Eckert was a graduate laboratory assistant for that course. Then, in the fall, Mauchly resigned his associate professorship at Ursinus College to accept the Moore School's offer of an instructorship, and his talks with Eckert continued.

Mauchly also spent time during that 1941–42 academic year studying the school's differential analyzer, which occupied a large basement room. He had for some time been interested in this device, the crowning achievement of analog computing. Now he had an opportunity to observe the most powerful version of the mechanical differential analyzer on a daily basis (fig. 24).

By August of 1942, when Mauchly wrote his memorandum on electronic computing, this machine was engaged exclusively in calculating artillery trajectories for the Ballistic Research Laboratory (BRL) of the Aberdeen Proving

Fig. 24. The differential analyzer. The differential analyzer shown here was built at Aberdeen Proving Ground by the Moore School of Electrical Engineering, under contract with the U.S. Army Ordnance Department. The Moore School's own slightly more powerful analyzer was built under that same agreement, as a mutual arrangement.

The integrators were the fundamental computing components. They are seen in a row on the left, each with its torque amplifier to drive the appropriate shaft in the center bay. Input and output plotting tables are in a row on the right. Banks of five counter wheels, similar to odometers, are in front, together with a tray for the paper on which they printed.

The center bay had placements for shafts running both crosswise and lengthwise. To set up a problem, the operators interconnected these shafts so as to establish intercommunication among the integrators, the plotting tables, the independent variable motor, and the output dials as required by the problem. Fixed gears, for multiplication and division by constants, and differential gears, for addition and subtraction, were placed in the bay.

Ground. The Army's need for artillery firing tables had become urgent after the United States entered World War II in December, 1941. The preparation of just one table, for a given shell to be fired from a given gun under varying conditions, required the calculation of three or four thousand trajectories from a system of exterior ballistic (differential) equations. (Other ballistic equations, for gun design, could also be solved on differential analyzers, but the trajectory application held wartime priority.)

The BRL had its own slightly less powerful analyzer, both machines having been built by the University of Pennsylvania in the 1930s, under the direction of Irven Travis, as depression "relief" projects of the federal government. And so Army Ordnance established more or less parallel computational programs at the BRL and the Moore School, using their respective differential analyzers to calculate the shell trajectories. But it also recruited and trained human "computers" to calculate trajectories on desk machines, at both installations, by a numerical procedure based on difference equations. Even these combined analog and digital efforts, however, offered little prospect of keeping up with the military demand for firing tables. A much faster method was clearly needed.

Mauchly familiarized himself not only with the analyzer's method of computing trajectories, but also with the very slow hand procedure, and gave some thought to how the latter might be mechanized. Then, bearing the Army's interest in mind, he wrote his memorandum on the merits of an "electronic computor [*sic*]" or "electronic calculator" (PX 1,227 at the ENIAC trial), in which he explained its principles and design in general and cited its advantages over the differential analyzer (Mauchly 1942/1973). He expressed his purpose as follows in the opening paragraph (p. 329):

> . . . It is the purpose of this discussion to consider the speed of calculation and the advantages which may be obtained by the use of electronic circuits which are interconnected in such a way as to perform a number of multiplications, additions, subtractions, or divisions in sequence, and which can therefore be used for the solution of difference equations. Since a sufficiently approximate solution of many differential equations can be had simply by solving an associated difference equation, it is to be expected that one of the chief fields of usefulness for an electronic computor would be found in the solution of differential equations.

He went on to explain that these electronic circuits, the "components" of his proposed computer, were to be "interconnected by cables and switching units," and were to calculate "solely on the principle of counting."

Thus Mauchly intended to solve differential equations numerically, as was done at that time on desk calculators, and he intended to perform the required arithmetic operations with electronic counters, in the manner of a desk calculator's decimal dials. Indeed, as he explained next, he meant the separate components themselves to function individually in the manner of desk calculators (p. 330):

> . . . If one desires to visualize the mechanical analogy, he must conceive of a large number, say twenty or thirty calculating machines, each capable of handling at least ten-digit numbers and all interconnected by mechanical devices which see to it that the numerical result from an operation in one machine is properly transferred to some other machine, which is selected by a suitable program device; and one must further imagine that this program device is capable of arranging a cycle of different transfers and operations of this nature with perhaps fifteen or twenty operations in each cycle. . . .

He then stressed the analogy one more time, stating his intention that it "shall be interpreted rather completely." And, in fact, he and Eckert proceeded to design the ENIAC around a group of identical "calculators," which they called accumulators.

Mauchly noted that a computer made up of discrete interconnected components would have advantages over the differential analyzer with respect to problem setup, machine maintenance, fault diagnosis, and repair. He also noted that a numerical computer would be superior to the analyzer in the ease with which results could be checked.

The greatest advantage, however, of his proposed electronic digital (decimal) computer, would be the increased speed of computation. To indicate the magnitude of this gain, Mauchly worked out time estimates for solving a single artillery trajectory on, respectively, the Moore School analyzer, a desk calculator with a human operator, and his proposed computer. In fact, since his proposed numerical procedure was similar to that used by the human operator, he estimated his machine's solution time by analyzing and comparing the two procedures.

He concluded that his machine could solve a single trajectory in "100 seconds," whereas the desk calculator required "*at least* several hours" and the differential analyzer "15 to 30 minutes." This estimate assumed a "counting rate . . . in the neighborhood of 100,000 pulses per second," based on "the literature on counters, often referred to as scaling circuits."

Before turning from this earlier Moore School "proposal" to the formal proposal of the following April, we should observe that, while both documents primarily addressed the solution of ballistic equations, a much broader capability was envisioned from the start. In fact, we believe that Mauchly meant to solve iterative equations in general, and so more types of equations than could be solved on the differential analyzer.

We should also call special attention to Mauchly's inclusion of a "program device," or "programming device" in the computer he described in that manuscript. For the ENIAC was to be the first programmable electronic computer ever built, and its programmability was an essential element in its being a general-purpose computer. Lastly, we should note Mauchly's choice of counters, contrasting with Atanasoff's choice of logic, as his preferred way of doing arithmetic with vacuum tubes.

From the preceding account of Mauchly's 1942 memorandum, we can abstract three basic convictions on his—and Eckert's—part. The first was that an electronic digital computer could be designed to do the work of the mechanical differential analyzer; as we will show, this arose in large measure from what Mauchly had learned from Atanasoff. The second was that counters were to be preferred over logical operations for electronic arithmetic; this arose from Mauchly's views on electronic desk calculators and the possibility of chaining a large number of them, as fortified by his talks with Eckert. The third was that electronic circuits could be built to compute reliably at a rate of 100,000 pulses per second; this arose, again, from his exchanges with Eckert over the previous year.

Thus, in our view, Mauchly's memorandum represented the point to which he had carried Atanasoff's work and ideas on electronic computation at that time, together with a preference for the counting principle and a goal of extremely fast operation, the latter two highly dependent on the judgment of Eckert. Moreover, these same three ideas, or convictions, were also basic to the formal ENIAC proposal of the following April.

The Eckert-Mauchly Proposal

Mauchly had written his August, 1942, memorandum on the suggestion of Carl C. Chambers, a Moore School professor, with the hope that the school might seek government (Army) support to build the electronic computer he was envisaging. He now gave a copy to Brainerd, the director of war research. Brainerd, however, took no action on it, but apparently after several months returned it to Mauchly along with a note: "John: Read with interest. It is easily conceivable that labor shortage may justify development work on this in the not too distant future. JGB." The note was dated "1/12/43."

But then in March, Goldstine, the mathematician in charge of the BRL's computing activity at the Moore School, became intrigued with Mauchly's talk of a computer that could do the work of the differential analyzer electronically. He asked for a write-up of the device, was given a second copy of the memorandum, and upon reading it requested a formal proposal to submit to the BRL.

Events moved swiftly from that point on. Mauchly and Eckert worked feverishly at writing the technical portion of the proposal, and Brainerd wrote the nontechnical portion. On April 9, 1943, Goldstine drove the three authors down to Aberdeen to present a draft, dated April 8, that lacked several illustrative appendixes; indeed, Eckert and Mauchly worked on it in the back seat of the car as they traveled and for the balance of the day at the laboratory, in a vain effort to complete it. Nevertheless, Goldstine and Brainerd made a successful presentation to the military authorities and their civilian advisors, and the project was approved that same day. The appendixes were finished only some weeks later, but the April 8 date was retained; the completed version became Defendant's Exhibit (DX) 3,361 of the ENIAC trial.

The most striking aspect of this proposal, from today's perspective, is its recurring theme of a link between the new electronic digital computer and the old mechanical analog differential analyzer. It referred to the new device throughout as an "analyzer," an "electronic analyzer," or an "electronic difference analyzer." Its very title, "Report on an Electronic Difference Analyzer," set the stage, and a footnote explained (p. 1):

> The word Difference has been used deliberately. Present differential analyzers operate on the basis of integrating continuously, i.e., by differential increments; the electronic analyzer, although it is believed that it could be both speedier and more accurate, would operate using extremely small but finite differences. It thus is more appropriate to refer to it as a difference analyzer rather than a differential analyzer.

The point made here is, of course, the same one Mauchly had made in his 1942 memorandum, namely, that the proposed electronic machine would constitute a *functional* replacement for the mechanical machine, by solving differential equations more accurately and much more quickly. But the proposal went further, noting a *structural* parallel as well. In a section of the technical portion called "Brief Description of Electronic Analyzer," it drew on Mauchly's earlier analogy to a group of interconnected, programmable electronic circuits—or desk calculators—to observe that the *differential analyzer* could also be viewed as a group of interconnected independently operable units (its integrators, primarily)!

Thus the April, 1943, report not only reinforced Mauchly's original conceptual link between the mechanical and the electronic "analyzers," but went so far as to suggest the actual modeling of the new machine on the old. Now one can argue, reasonably, that the goal was to sell the Army on a machine to do a particular job, solve artillery trajectories, which was currently best done on the differential analyzer, and so it was natural to press the parallelism. But the structural parallelism was as genuine as the functional; Mauchly and Eckert did perceive their electronic machine as a structural replacement of the mechanical analyzer, and they did so pattern it, unit-by-unit, to the extent that was possible. The resulting ENIAC had a distributive architecture (rather than a centralized one, as their EDVAC and later machines had) that can be explained in no other way.

Most of the technical portion of the proposal was devoted, of course, to more specific descriptive material that we need not recount here, since we give an overview of the completed machine in the next subsection. We should point out, though, that while this material was much more detailed than Mauchly's earlier presentation had been, the major design tasks still lay ahead. Whole blocks of the computer were merely sketched, with critical electronic problems skirted. For example, a "counter" was listed and described as one of the basic components, but remained a serious design challenge. Likewise, a "program control unit" was specified, but in terms of purpose only, not of design.

The nontechnical portion of the proposal also had much material that need not concern us here, such as machine advantages and cost and time estimates. It did contain, however, one further rather curious item, which we take to refer to the influence of Atanasoff (and which came up as such a possibility in the ENIAC trial).

A brief introduction had reviewed the existing differential analyzers, chiefly in terms of the number of integrators each had. Special attention was called to the "electronic differential analyzer" then nearing completion at MIT, and to the fact that it was not electronic in the way the proposed difference analyzer was to be; that is, the new MIT machine would not use electronic circuits to calculate, but would retain the usual mechanical integrators and gears for this purpose. The following reference was now made to two earlier electronic *difference* analyzers (p. 2):

> It may also be noted at this point that electronic difference analyzers have been considered for different applications in at least two previous cases, but have not been developed and used.

It seems clear to us that one of these "cases" was a veiled allusion to Atanasoff—to his work and ideas as revealed by him to Mauchly in 1941. We return to this reference later, after our presentation of the Atanasoff-Mauchly contacts and related evidence. Let us just note here that the quoted statement does seem to discount prior invention, with the effect of disclaiming outside influence by Atanasoff or others. And yet its characterization of the two earlier machines as not developed *and used* amounts to a recognition of some device in an advanced state of development—such as the ABC.

We turn now to a brief description of the ENIAC, as it stood upon completion in early 1946. Its official name, Electronic Numerical Integrator and Computer, was determined only after the Army had agreed to build it—in fact, only in late May, just as the team was organized and work was about to commence. While the words "Numerical Integrator" continued to signify the machine's essential character as an electronic differential analyzer, the word "Computer" now signified a wider scope.

Description

The ENIAC was a huge undertaking, both physically and conceptually. It was a digital (decimal), programmable, highly parallel computer that could solve a wide variety of problems electronically. It had a manual setup, however, which meant that its greatest usefulness was for problems of a repetitive nature—problems many instances of which could be solved from a single setup.

The ENIAC was comprised of thirty semiautonomous units, each made up of one, two, or three panels that were eight feet high, three feet deep, and two feet wide. There were forty such panels, lined up side by side in an eighty-foot U along three walls of a room. Within this framework were some 18,000 vacuum tubes (many of them double triodes), 70,000 resistors, 10,000 capacitors, 1,500 relays, and 6,000 manual switches (fig. 25).

Each of the forty panels had a "front panel" of switches to be set and sockets to be plugged into at the time of the initial problem setup. Running above and below these front panels, traversing the machine, were the two parts of a multichannel trunk system of electrical cables or trays. Above was the numerical part, which transmitted digit pulses; below was the programming part, which transmitted program pulses. Communication among all the units of the computer was over this trunk system, segments of which were joined with jumper cables, both to one another and to the front panels, to suit the requirements of the problem to be solved.

Most of these thirty units were, of course, arithmetic units. There were twenty identical accumulators, each holding a ten-digit number and its sign. These operated in pairs, one pair per addition or subtraction, but many pairs could operate in parallel. They accumulated totals as the operations progressed. These twenty units constituted the ENIAC's entire read-write store. There were two other arithmetic units, the high-speed multiplier and the divider square-rooter, which operated in conjunction with accumulators to perform their operations.

The ENIAC also had a read-only store, made up of three function-table

Fig. 25. The ENIAC. The ENIAC consisted of thirty separate units arranged in a U. Transmission of numbers among units was over a plugboard trunk system, somewhat analogous to the bay of shafts in the differential analyzer. To enter a program into the computer, the operators set switches on the front panels and plugged cables for the appropriate sequencing of instructions among the units.

Two people are entering arbitrary functions into read-only memories within the U, while two others are checking the program on the front panels. The IBM card reader and card punch can be seen at the extreme right.

units, which held special numbers or quantities to be used by the accumulators throughout the computation. Each such unit had two parts, a movable matrix (on wheels) located within the eighty-foot U, and two panels of the U to which that matrix was wired. The three matrices had arrays of switches to be set as part of the initial problem setup, the numbers entered in them having been predetermined by experimental or mathematical functions occurring in the equations of the particular problem. In the course of the solution, then, the function-table panels accessed the function-table matrices as required by the accumulators.

The ENIAC's initial input system, by which the numbers constituting the initial conditions of the successive runs of a given problem were entered into the computing units, consisted of an IBM card reader and a unit called the constant transmitter. The previously punched cards were placed in the card reader in batches, by hand, after which they were processed automatically as the machine required their data. Similarly, the ENIAC's final output system consisted of a unit called the printer (a misnomer) and an IBM card punch. In calculating the trajectories of a given shell, for example, punched IBM cards were fed continually into the computer, one trajectory per card, and solutions punched out, many cards per trajectory.

Each of the units mentioned so far had its own local program controls. Instructions were set up on these controls and interconnected by means of program cables to form subroutines, which were then combined at the master programmer into a single program. The timing and coordination of the operations were controlled by a central "clock" called the cycling unit. A circuit in the initiating unit started the computation process.

Thus there were three main tasks in the initial setup of a problem on the ENIAC, all governed by the exact nature of that problem. The first was the setting of switches on the front panels of the units. The second was the plugging of jumper cables to form the electrical connections among units via the machine's trunk system. And the third was the setting of switches on the matrices of the function-table units. Again, when the ENIAC was used to calculate trajectories, the first two tasks constituted a long-term setup, but the third—the setting of a function-table matrix—was different for each shell. Once this recurring task was accomplished for a given shell, however, the computation could proceed automatically through some three or four thousand trajectories (at ten or twenty seconds per trajectory), with only the manual entry and removal of batches of IBM cards.

The December, 1940, Meeting

John V. Atanasoff and John W. Mauchly first met on December 28, 1940, at a session of the American Association for the Advancement of Science in Philadelphia. Mauchly gave a talk attended by Atanasoff, who approached him afterward and introduced himself. The two men found at once that they shared a keen interest in computing methods and devices.

As we explained in chapter 2, Mauchly had for some years been analyzing weather data, hoping to demonstrate a periodicity caused by solar phenomena; his AAAS talk reported his latest results. Atanasoff's interest, however, was in Mauchly's use of a machine to carry out the many calculations required for such a statistical analysis, namely, an electrical harmonic analyzer he had built in 1939. Atanasoff had already studied the feasibility of analog versus digital devices for large-scale computation, and also of mechanical or electrical versus electronic elements for either mode, and had now designed and was building his own electronic digital computer. Mauchly, too, had set a goal of building, in the next year or so, a keyboard-operated digital calculator based on electronic counters, again for detecting meteorological patterns; but he was also planning a second, more powerful, electrical harmonic analyzer.

The two physicists were quickly caught up in their mutual interest. After a brief discussion of Mauchly's analog machine, Atanasoff told of his digital one, and Mauchly told of his two-neon relaxation oscillator that he would later testify was also a binary counter. At some point, Atanasoff invited Mauchly to come to Iowa to see and learn about his computer. The immediate documentation of this conversation is two sheets of Mauchly lecture notes on the backs of which both men wrote or drew as they spoke. We present this material, as it was exhibited and examined at trial, in the current section.

We also present the pertinent trial testimony of both men concerning their initial meeting. Throughout this chapter, however, we will mainly paraphrase, rather than quote full passages; we have quoted Mauchly sufficiently in chapter 2 to convey the tenor of his testimony, and we do so for Atanasoff in chapter 4. Material is drawn from both pretrial (deposition) and courtroom testimony.

Later notes and letters, of course, contain further evidence of what transpired during the December, 1940, meeting. These we reserve for their appropriate time frames, in the sections that follow, with the exception of one item actually queried in the context of that meeting.

The Evidence

Mauchly's lecture notes on the reverse sides of which he and Atanasoff made further notations were assigned number PX 22,420 at the ENIAC patent trial of 1971–72. Portions of their entries were no longer legible, at least in xeroxed form—attorneys did have access to the originals—but a number of equations, curves, and diagrams could be made out by the two witnesses. Significant among these were, in Mauchly's hand, a circuit diagram of his two-neon device, nondescript in mode as we explained earlier, and also an equation for a Fourier series; in either Atanasoff's or Mauchly's hand, a square containing elements of a matrix of some sort; and, in Atanasoff's hand, his name and "ISC," for Iowa State College.

Atanasoff's Testimony

Under direct examination as a Honeywell witness, Atanasoff testified in his pretrial deposition (pp. 747–57) that he had a "stock in trade" of what he told people about his computer at that time: how many tubes it used, that it worked on "vacuum tube principles," and that it was a "proper" computing machine—his expression then for *digital*—"as contrasted with an analog machine." He recalled that Mauchly was very interested in this description. Asked whether Mauchly had in turn described any work of his own toward a digital or an electronic computing machine, such as work with ring counters or other vacuum-tube devices, he said he had not. He estimated that they talked for "perhaps a half hour."

When shown the lecture notes, Atanasoff testified that the drawing of a matrix, with its lettered elements, seemed to be his; he thought it was a "very sketchy way" of illustrating his discussion of systems of equations in relation to his machine. He also identified Mauchly's diagram as an "unstable" circuit using neon lamps, with significance for counting, and tied it in with his recollection that Mauchly had on some occasion explained such a circuit to him.

Atanasoff was then asked about references in later letters (Mauchly's inquiry of January 19, 1941, and Atanasoff's response of January 23, 1941) to a cost estimate of two dollars per digit for his computer, an amount that by implication had been discussed during the December, 1940, meeting. He said it referred to the fact that each of his memory drums held 1,500 binary digits, and the cost of the machine could be expressed in terms of those bits as, on average, two dollars each. He observed that this had seemed to him a "very modest" production figure.

During cross-examination in this same deposition, a Sperry attorney asked Atanasoff if he would agree that Mauchly's circuit drawing depicted a binary counter (pp. 1,083–86). Atanasoff, however, agreed only that the circuit would count if "equipped with components of the proper characteristics," and, further, that even if so equipped it would *count* only.

Atanasoff's actual courtroom examination on his first meeting with Mauchly (transcript pp. 1,971–74) was very brief. (The deposition remained, of course, part of the trial record for purposes of judgment.) He added to the earlier testimony only his recollection that he had informed Mauchly of his computer's "capacity" and its "speed," and his characterization of the two-dollar-per-bit figure as a "marginal cost" to be expected if one wished to expand the machine's capacity. The Sperry defense did not cross-examine Atanasoff in court concerning this meeting.

Mauchly's Testimony

Mauchly's direct examination at trial, as a Sperry witness, was used primarily to identify the items recorded by the two men during their conversation (transcript pp. 11,808–18). He did testify that once they had exchanged remarks on their respective work—Atanasoff on his "electronic calculator," he on his "two-neon-diode circuit"—Atanasoff declined to reveal much about his machine, but expressed a willingness to show it to him and tell him "all about it" if he were to visit Iowa. He thought "the most" Atanasoff did was to make him "very curious," because he represented the computer to be a "very low cost device" such as he was seeking. Once more the two-dollar-per-digit matter from the January, 1941, correspondence was injected, with Mauchly remarking that except for those letters he "might never have remembered this . . . real point of curiosity" on his part.

Under Honeywell cross-examination (transcript pp. 12,139–60), Mauchly could not recall telling Atanasoff of any plan of his own to perform harmonic analysis other than with his current machine. As for Atanasoff's computer, he remembered learning that it used vacuum tubes, but not how they functioned; in fact, he said, this was "exactly the point" where Atanasoff was saying he would have to go to Iowa to learn more. He also acknowledged learning that it was digital, though under different terminology. On the other hand, he could not recall whether he was told that the purpose of the machine was to solve large sets of simultaneous linear equations, or whether that was part of what he "had to go to Iowa to learn."

When asked if he would have recognized, as of 1940, that a computer capable of solving such equations with vacuum tubes represented a considerable advance in the art, he replied that it depended on the economics: the question of cost could not be eliminated, because there were lots of desk calculators with this capability that were cheaper than an electronic machine. To the question of whether the estimated cost of two dollars per digit would not have satisfied this economic condition, Mauchly responded that indeed it had, that he went to Iowa on this basis—assuming, however, "that the speed of vacuum tubes would be fully utilized in this device," for "why else would he use vacuum tubes?"

Finally, asked if Atanasoff had spoken to him in Philadelphia of his computer's regenerative memory, he thought not: this, in fact, was "the secret teaser which he was reserving to lure me to Iowa." Mauchly said they talked in the "order of magnitude" of "twenty minutes or a half an hour." The Honeywell cross-examination on that initial meeting ended with the following exchange between Halladay and Mauchly (transcript pp. 12,160–61):

> *Halladay:* And then when Dr. Atanasoff spoke to you in December, 1940 of a computer which would provide a discrete variable attack on these vast problems of physics and such, it excited your imagination, did it not?
> *Mauchly:* Yes, yes, it did.

In Sum

Several significant points of agreement emerge from the testimony of Mauchly and Atanasoff as to what transpired during their first encounter: Atanasoff revealed that he was building a computer, that it calculated with vacuum tubes, that it was digital, that it was very cheap on a per-digit basis, and that it was of very large capacity. Mauchly revealed nothing of any digital device except his two-neon circuit. Atanasoff invited Mauchly to Iowa, promising to show him his computer and tell him all about it. Mauchly was greatly interested in what he had learned thus far and was eager to learn more.

There are also several significant disparities in their testimony: Atanasoff said he informed Mauchly of his machine's speed; Mauchly said he did not learn this, but expected the computer to take full advantage of electronic speed. Atanasoff said he told Mauchly that his machine was designed to solve large sets of linear equations; Mauchly said he could not recall learning this, though he later acknowledged learning that it was to solve vast scientific problems. Atanasoff said he told Mauchly how many vacuum tubes his computer required; Mauchly did not address this point. (All of these items were clarified in the documents of early 1941, and *all* in Atanasoff's favor, as we shall see in the next section.)

Finally, there were four significant features of Atanasoff's machine *not* included in what he termed his "stock in trade" of freely given information at that time, features we presume he purposely withheld from Mauchly as from others. The first was that his computer had a memory based on capacitors. The second was that it provided a means of regenerating that capacitor memory. The third was that it performed its arithmetic functions on a logical rather than a counting basis. And the fourth was that it calculated in the binary rather than the decimal system. Atanasoff was not asked about any of these. Mauchly was asked only about the regenerative memory; he thought this was a deliberately reserved feature—reserved, he said, to lure him to Iowa.

Now, this list of facts conveyed and facts withheld seems to us an entirely reasonable one, on the occasion of meeting a fellow scientist for the first time and discussing one's unpatented invention. In short, Atanasoff revealed the revolutionary nature of his computer, together with some impressive specifics, but held back the essence. As Mauchly himself put the matter in court, Atanasoff told him *that* his machine used vacuum tubes to compute, but not *how* it used them.

But Mauchly went on to interpret Atanasoff's invitation to visit Iowa and learn all about his computer as some sort of deceptive gesture, and this we find unreasonable to the point of being preposterous. The evidence indicates quite simply that Atanasoff behaved cautiously toward a stranger, at the same time that he felt and expressed a strong desire to share his ideas with him in the near future. Any prospect of gain on his part, through the withholding of critical information, seems especially absurd, considering that Mauchly himself was offering nothing in return beyond being an apt learner.

We do have to wonder, too, at Mauchly's identification of *three* different aspects of Atanasoff's computer as *the* particular inducement for him to visit Iowa: how such a machine could be so inexpensive, how it used vacuum tubes to compute, and how its (regenerative) memory worked.

The Interim Period

In this third section of our discussion of the Atanasoff-Mauchly influence, we present ENIAC trial documents and testimony for the period between their first meeting, in Philadelphia on December 28, 1940, and their second, in Ames, Iowa, from June 13 to June 18, 1941. These documents fall into two categories: a sheaf of Mauchly notes (PX 665), to which the Honeywell side referred in cross-examining Mauchly on his *pre*-Atanasoff achievements (see chap. 2); and ten Atanasoff-Mauchly letters, five by each to the other.

It will be recalled that Honeywell used PX 665 in a negative fashion, to show what little progress Mauchly had made in electronic computing *before* he became acquainted with Atanasoff. We now look at that file again, very briefly, as introduced by the Sperry Rand side, and show what little progress Mauchly was able to make in the spring of 1941, *after* he had learned that computing with vacuum tubes was indeed possible, but *before* he had seen Atanasoff's computer! We will take a closer look at just one page, dated January 1, 1941—four days after the Philadelphia meeting—a page of drawings that we believe reflects Mauchly's impression of Atanasoff's computer in a highly significant way.

Most of this section is concerned with the Atanasoff-Mauchly correspondence of this interim period. As before, we quote only those passages that pertain to our inquiry, and we mainly paraphrase the trial testimony.

Mauchly's Notes

As explained in chapter 2, PX 665 is a ten-page collection of notes, titled "Computing Circuits" and dating from January 1 through August 3, 1941, that explores various possibilities for electronic computing devices. A January 1 note (p. 9 in the file) depicts a keyboard-operated "Heterodyne computing machine," concerning which Mauchly testified that he was thinking of adding or subtracting "frequencies corresponding to the decimal digits" of two numbers to produce a third frequency representing their sum or difference (transcript p. 11,857). Based on a standard method of designing AM receivers, this idea indicates that Mauchly, as of January 1, 1941, was entertaining the use of analog methods to add and subtract decimal digits as a step toward an economical desk calculator.

The two August 3 pages (file pp. 10 and 11), to which we return later in discussing the period after Mauchly's visit to Iowa, are significant here only in that one was a further analog attempt and the other was his earliest written effort to design a *digital* keyboard calculator.

All but one of the remaining seven pages in PX 665 depict neon gas diode circuits with associated vacuum triodes for switching or control. Two are dated January 1 (file pp. 2 and 8), one is dated April 6 (p. 5), and three are dated April 8 (pp. 3, 4, and 6). Of these Mauchly testified variously (transcript pp. 11,849–57) that he was exploring ways of connecting a number of his "neon diode counters" in sequence to get carry effect, of building a neon-diode circuit with more than one stable state so as to "store more than a binary number," or of getting "one vacuum tube to control . . . more than one binary digit."

Now these latter notes tend only to confirm that Mauchly was inspired by his earlier conversation with Atanasoff to try to design digital counting circuits utilizing vacuum tubes in conjunction with neons. The one remaining note, however, reveals a further, very critical piece of information passed by Atanasoff to Mauchly during that conversation. This is a fourth January 1, 1941, note (p. 7 in the file) containing two somewhat similar circuit diagrams. Of it Mauchly testified to Sperry attorney DeLone as follows (transcript pp. 11,855–56):

> . . . the nature of both those diagrams seems to be that I was expecting one control tube to actuate some sort of solenoid actuator, which in turn would rotate stepwise some kind of a switch. . . . But you also have the question of how do you read out the switch.
>
> And I would say in the lower diagram, why, there seemed to be a number of resistances connected around switch points, whereas there is a tube connected to the one at the top.

DeLone then called Mauchly's attention to a notation in the lower left-hand corner of the page. Mauchly read this as "At. machin" and "150 tubes" and confirmed that it was in his hand. He added:

. . . I think that meant that at that time just after I had met Atanasoff that he must have communicated to me that he expected around 150 tubes in this machine.

No explanation was given as to why this particular citation of Atanasoff's machine was made on this particular page of drawings; nor was attention called to a question mark set off to the right of them. It is clear to us, however, on inspection of these drawings, that Mauchly did not just happen to write himself an unrelated note as to how many tubes he understood Atanasoff to say were in his computer— to us the number looks more like "450" than "150"—but that Mauchly was here trying to duplicate Atanasoff's machine or some part of it on the basis of what he had learned a few days earlier.

The lower diagram does indeed show a stepping switch driven by a vacuum tube, with resistances for reading out (digits, presumably); it includes a device marked "rotator." We take this circuit to be an electromechanical counter driven by a vacuum tube.

The upper diagram also has a rotating component, in the form of a circle with spokes to the center; without resistances, however, but rather with a reading brush on the end of a wire connected to its second vacuum tube. This latter is definitely *not* a stepping switch, and with its mechanism for reading out (digits), it does appear to be a disk or drum that is also written on by its first vacuum tube.

Our interpretation of this page of diagrams, then, is that it depicts, first (top), a rotating storage device written on and read from by vacuum tubes; and second (bottom), an electromechanical counter driven by a vacuum tube. We have no way of knowing how the two are related, of course, or even of being certain of our interpretation. One critical fact does emerge with certainty, though: Mauchly *knew,* from his very first discussion with Atanasoff, that the speed with which the ABC could compute was limited by the mechanical rotation of some sort of drum or other turning component associated with its memory function!

There is even evidence that Atanasoff was publicly announcing, at this time, the fact that his computer had a *separate* memory. The January 15, 1941, issue of the *Des Moines Tribune* carried an article on the "Computing Device" being constructed at Iowa State College, with a photograph captioned "Machine Remembers," and a reference to the "memory" of the "giant computing machine" in the accompanying legend (PX 670). The picture shows Clifford Berry holding a module containing all ninety tubes (forty-five envelopes) of the ABC's restore-shift mechanisms, to which Atanasoff referred in court as the "regenerative tubes" of his "memory regenerating circuit" (transcript pp. 1,856–57; see fig. 15).

A lesser, but still significant, fact that emerges from this January 1 note is that the two men did discuss the ABC in some detail, at least to the point where Atanasoff revealed the number of tubes it would use (in some connection). (The *Des Moines Tribune* article, incidentally, specified that the machine would "contain more than 300 vacuum tubes.")

Atanasoff-Mauchly Correspondence

Letter 1

Mauchly was the first to write after the AAAS meeting, in a letter dated January 19, 1941 (PX 675):

> I am wondering how your plans with regard to computing devices are working out. Need I say again that I await with some suspense the time when you will be able to let me have more information? How the recording-end functions is the biggest puzzle, I guess—450 digits at less than $2 per digit sounds next to impossible, and yet that is what I understood you to say, approximately.
>
> Your suggestion about visiting Iowa seemed rather fantastic when first made, but the idea grows on me. I've gone so far as to note that our Spring Recess is March 21 to 31, whereas the meetings in Washington are about May 1. . . .
>
> If you aren't too busy, perhaps you can drop a few hints as to your progress.

This letter confirms the trial testimony of both men that Mauchly, during their Philadelphia meeting, was excited by Atanasoff's description of his computer and was anxious to learn more. It also confirms the testimony of both, in effect, that Atanasoff's figure of "$2 per digit" represented a very low cost estimate for a machine of very great capacity. And now Mauchly's mention of "450 digits," not queried in regard to that meeting (or later), confirms that Atanasoff *had* told him of his computer's purpose; indeed, had told him the exact size of the sets of linear equations it was designed to solve, though he had *not* told him it was to compute in the binary system!

For Mauchly's "450 digits" could only represent a *decimal* equation of thirty coefficients in fifteen places, or one equation from Atanasoff's original decimal set; whereas of course the 1,500 digits per drum of Atanasoff's testimony (at two dollars each) represented the equivalent *binary* equation of thirty coefficients in fifty places. (Notice, incidentally, that the implied cost estimate of $6,000 for the entire computer turned out to be quite accurate.)

We must acknowledge that Mauchly was laboring under an illusion, by a factor of about one-third, as to the projected cost of Atanasoff's machine, in that he took "digit" to be *decimal* digit. We submit, however, that Mauchly would have thought two dollars per *binary* digit "next to impossible," as well, had Atanasoff been willing to divulge his use of the base two; for that low figure was attained only by his introduction of the inexpensive capacitators as (regenerated) memory elements, another withheld feature.

Letter 2

Atanasoff replied on January 23 (PX 676):

> Your letter came as a pleasant surprise. . . .
> As you know we have been in active construction on the computing machine for more than a year, and now our plans have assumed rather definite form, and progress is good without too much attention on my part. I am expecting shortly to hear whether or not I will receive a grant-in-aid from an outside source to help in completing the machine. All in all progress is very satisfactory to me, and I expect that our plans will commence to mature in about a year.
>
> I paid Dr. Urry a visit at the Geophysical Laboratory in Washington, and he took care that I received a complete picture of the activities of his organization. . . . Just after my return we had a visit from Dr. S. H. Caldwell of M. I. T. who gave me a rather complete picture of calculating machine activities in the country. His visit gave me the urge to attempt the construction of a differential analyser on a dime store basis. I believe this attempt could be rather successful, and it is something I have laid away for future activity when my work gives out.
>
> By all means arrange to pay us a visit in Ames during your spring recess if this is possible. . . . I can think of many things that I would like to talk to you about. This list includes statistical Fourier analysis, resistance harmonic analysers, computing machines of all kinds, and I suspect there are plenty of other things. I will be glad to have you as my guest while you are in Ames. As an additional inducement I will explain the two dollar per digit business.

This letter is noteworthy for its confirmation, in writing, of Atanasoff's earlier promise to tell Mauchly all about his computer; for he could not explain the cost factor without revealing the nature of the memory and of the arithmetic unit with which it interacted—in short, without revealing the two most fundamental features of the machine. The letter's chief significance, however, lies in its reference to Caldwell's visit and the "urge" that visit gave Atanasoff to try to construct a very cheap "differential analyser" when he had the time.

Samuel Caldwell actually paid Atanasoff several visits, on behalf of the U.S. government, in order to confer about the defense work Atanasoff was directing at Iowa State College. But this first visit, on January 6, 1941, had the further purpose of evaluating the computer project and advising Research Corporation as to its funding. Caldwell did very shortly recommend that it be supported, on the bases both of this inspection and of an earlier reading of the August, 1940, description. Atanasoff was notified in early March that it had been approved.

Atanasoff explained in his pretrial direct examination (deposition pp. 713–20) exactly how he demonstrated his computer, in its early 1941 state, to Caldwell and others who visited him in some official capacity. He and Clifford Berry put it through "a routine," using "test sets" they had developed from the start for testing major components as they were constructed, and also for "illustrating the operation of the machine in further detail." He said that this routine was continually extended as they went along, to embrace more components and more operations. We have reserved our presentation of it for a more pertinent time frame, namely Mauchly's June, 1941, visit to Iowa; let us just note here that it was apparently impressive enough by January of that year to secure Samuel Caldwell's recommendation.

Now, while the Caldwell visit mentioned in Atanasoff's January 23, 1941, letter was explored at trial, the precise nature of his idea for "a differential analyser on a dime store basis" was not developed in connection with *that* letter, but in connection with later letters and papers, commencing with Atanasoff's May 31, 1941, letter to Mauchly. This postponement was for the better because, as we shall see, "dime store" seems to have connoted to Mauchly a cheap demonstration model of the *mechanical* analyzer, rather than the cheap *electronic* digital replacement for the analyzer that Atanasoff described in his May 31 letter.

Letter 3

Mauchly wrote on February 24 (PX 699):

Your invitation and the promised explanation are indeed powerful inducements, and I hope that I shall be able to take advantage of them.

Somewhere—was it in Nature?—I saw an article on differential analyzers which included a picture of an analyzer constructed largely from Mecanno parts. I think the "dime-store" analyzer ought to be successful. If it did no more, it would justify itself in merely aiding students to understand the process of mechanical solution of differential equations. Incidentally, do you consider the usual d.analyzer an "analogue" machine? If so, how about a polar planimeter?

My crew of N.Y.A. people has been augmented, but perhaps not for the better. The new members are fit only for adding machine work . . . .

We recently had to take the back off of the harmonic analyzer for a few minor adjustments. Some of the IRC 50,000 ohm resistors had aged to the point where they were not within the one percent. that we desire. Someday we'll put wire-wounds in and forget them, I hope.

So far we have turned out about 1400 harmonic analyses—each one for eight Fourier coefficients. At 3 minutes per, this has taken just 70 hours. The tabular method that we would otherwise use takes about 5 times as long (but does yield more accurate values, which we do not need for this work). Hence we figure that we have saved 280 hours . . . [ellipsis in letter] which is about the amount of time it took to design and build the device. It wouldn't take that long to build another, of course, especially if one purchased precision wire-wound resistors to assemble.

Here's hoping that you have made progress in Iowa, and that I'll get out to see it all.

This letter confirms once more Mauchly's deep involvement in his meteorological research, together with his dependence on and enthusiasm for an analog device to conduct it. At the same time, it also confirms his hope of accepting Atanasoff's offer to reveal his own digital computer.

It responds positively to Atanasoff's suggestion of a "dime-store" analyzer, but, as we have said, it does not see this as a real computer at this point, but as a model. The fact that Mauchly had just recently read about someone else's construction of such a model no doubt influenced his interpretation of Atanasoff's remark. (Mecanno sets in England were analogous to Erector sets in America, and it was Douglas R. Hartree who had built a differential analyzer largely of Mecanno parts.)

Finally, this February 24 letter of Mauchly's delves into the analog-digital distinction as apparently discussed in Philadelphia. He was cross-examined at length by Honeywell on this point (transcript pp. 12,147–58). (The reader will recall that Atanasoff was the first to use "analogue" to distinguish one of two computing modes, and did so in his 1940 proposal, though he used "proper" for digital devices.) Mauchly first ridiculed the examining attorney's suggestion that he had learned of the distinction from Atanasoff. He even went so far as to claim only to be doing "what lawyers sometimes do," in asking whether Atanasoff considered the differential analyzer and the polar planimeter "analogue" machines. "I was asking him what he thought," he explained, "I was not saying what I thought."

In the end, however, as in a previous courtroom exchange we quoted in chapter 2, Mauchly was forced to back down. He had to admit, partly on the basis of a later document of his own (PX 847, dated August, 1941, discussed later), that he had got the "classification" and the "terminology" from Atanasoff; that, in fact, he had learned the expression "analogue device" as applied, for example, to his own harmonic analyzer, from Atanasoff.

Thus it was true, Mauchly *was* asking, not telling. But he was doing so in this particular instance for information, not just as a lawyer's ploy. In either case, though, this bit of testimony makes clearer than ever that Atanasoff was being a good deal more trusting than Mauchly, as the two sought to form a professional relationship, and that Mauchly was capable of a ruse to take what he could without giving in return.

Letter 4

Atanasoff wrote on March 7 (PX 712):

By all means pay us a visit . . . . At present I am planning to attend the Washington meetings at the end of April.

Several of the projects which I told you about are progressing satisfactorily. Pieces for the computing machine are coming off the production line, and I have developed a theory of how graininess in photographic material should be described, and have also devised and constructed a machine which directly makes estimates of graininess according to these principles. We will try to have something here to interest you when you arrive, if nothing more than a speech which you make.

Here is revelation of still further content in the Atanasoff-Mauchly exchange during their first meeting. Just what projects Atanasoff had told of, other than the photographic one cited in his letter, is not clear, but he did testify under direct examination as to his major project of that spring, one for which Caldwell visited him periodically (transcript pp. 2,111–16). This was the development and use of an analog device for analyzing factors in, and devising aids to, antiaircraft fire. Atanasoff would very probably not have done more than mention this project to Mauchly, if that, because it was classified.

Letter 5

Mauchly's plans for a visit to Ames during the spring recess fell through. The two then hoped to see each other during professional meetings in Washington at the end of April, and possibly in Ames in June. Mauchly wrote on March 31 (PX 744):

> . . . I hope to see you in Washington when you come to the meetings, and you can let me know then how things would shape up for a June visit.
>
> I haven't been able to do more than just paper-work for a computing machine here, so I haven't yet found out how practical my ideas are. We've been very busy running barometric pressures through the harmonic analyzer in order to get some more information on the 12-hour tidal oscillation.
>
> It's good to hear that your "production line" is producing. Is there any chance that you can now disclose more information?

This letter confirms not only that Mauchly had not made any progress on building a computer before he met Atanasoff, as we saw earlier, but also that he had not made any progress toward building it (or even learning whether it was practical) by the end of March 1941! It also, once more, confirms his overriding commitment to his weather project at the same time that it presses Atanasoff for "more information" on his computer.

Letter 6

Atanasoff wrote on April 22 (PX 756):

> Please excuse my delay in writing. I am undertaking some of that activity that is taking the country by storm; as a result everything else that I should do suffers.

I shall be at the Washington meeting and am looking forward to seeing you.

Atanasoff's reference here is undoubtedly to his antiaircraft project.

Letter 7

Atanasoff wrote again on May 21 (PX 778), expressing disappointment at not having seen Mauchly either in Washington, at Dr. Urry's residence, or at Collegeville en route to New York, and adding:

. . . Several matters connected with the defense project which I have undertaken made my visit in the East a very strenuous one. I am looking forward to seeing you in some way or another in the very near future.

Letter 8

Mauchly wrote on May 27 (PX 789):

It was a disappointment to me, too, not to get in touch with you while in Washington. . . .

Well, anyway, there is more than a little prospect of my making the trip, starting from here about the tenth of June. I have a passenger who will very likely pay for the gas, and that will help.

From your letters I have gathered that your national defense work is unconnected with the computing machine. This puzzles me, for as I understand it, rapid computation devices are involved in N.D. [National Defense]. In a recent talk with [Professor Irven] Travis, of the E.E. School at U. of Pa. I asked him about this, and the matter seemed the same way to him. But if Caldwell has looked over your plans (I think you said that he was out there) and hasn't seen any N.D. possibilities, I suppose that that means your computer is not considered adaptable to fire control devices, or that they have something even better. Travis (who goes into active duty with Navy this week) pointed out the advantages of lightness and mass-production for electronic computing methods, but said that when he was consulting with General Electric over plans for the G.E. differential integraph they figured it would take about one-half million dollars to do the job electronically, and they would only spend 1/5 of that, so they built the mechanical type with polaroid torque-amplifiers. . . .

There is no way of knowing how Mauchly came to mention "fire control devices" in this letter. It is highly unlikely that Atanasoff would have divulged even the subject matter of his defense project at their first meeting. It is possible, on the other hand, that Mauchly had surmised its nature from the involvement of Caldwell, whom he mentions in the same context, and from his own connection to Travis, who also worked in antiaircraft fire control for the Navy. When cross-examined on this issue at trial (transcript pp. 12,182–84), Mauchly acknowledged learning before he went to Iowa that both Samuel Caldwell and Thornton Fry had reviewed Atanasoff's plans for a computer (though he may not have known on whose behalf). He did not, however, recall learning that Atanasoff's national defense work was in antiaircraft fire control.

Mauchly's May 27 letter does reveal the extent to which he had developed contacts at the Moore School even before he took the defense training course there, in the summer of 1941, and in particular the extent to which Travis was willing to talk computers with him.

Letter 9

On May 31, 1941, Atanasoff wrote what turned out to be a letter of tremendous import for his—and Honeywell's—later claim that he was directly involved in Mauchly's inspiration for the ENIAC (PX 795):

> I think that it is an excellent idea for you to come west during the month of June . . . . We have plenty of room . . . .
>
> As you may surmise, I am somewhat out of the beaten track of computing machine gossip, and so I am always interested in any details you can give me. The figures on the electronic differential integraph seem absolutely startling. During Dr. Caldwell's last visit here, I suddenly obtained an idea as to how the computing machine which we are building can be converted into an integraph. Its action would be analogous to numerical integration and not like that of the Bush Integraph which is, of course, an analogue machine, but it would be very rapid, and the steps in the numerical integration could be made arbitrarily small. It should therefore equal the Bush machine in speed and excell [sic] it in accuracy.
>
> Progress on the construction of this machine is excellent in spite of the amount of time that defense work is taking, and I am still in a high state of enthusiasm about its ultimate success. I hope to see you within two or three weeks.

Atanasoff, for the second time, has referred to an idea he got during a visit by Caldwell, but now he has spelled it out to leave no doubt that he was speaking of a real computer, electronic and digital like the ABC, and not at all a model of the mechanical "analogue" differential analyzer. This Caldwell visit seems to have been the one mentioned in the earlier letter of January 23, since there was apparently no intervening visit: Mauchly's comments on the "absolutely startling" one-half-million-dollar estimate for an "electronic differential integraph" would have brought it to mind again, and in that context Atanasoff disclosed it more fully. For he knew that he could build an equivalent machine for much less money by using his electronic arithmetic and storage techniques.

Atanasoff's reference to the differential analyzer as a "differential integraph," or simply an "integraph," is of interest. This usage was fairly common at the time and was, in fact, the more accurate designation, because the basic computing unit of the analyzer was the *integrator*. To solve a set of differential equations on the analyzer, one first converted them into a set of *integral* equations and then integrated the new set on the machine. What Atanasoff was proposing also required an initial conversion, into a set of *difference* equations, for numerical solution on the digital machine. This numerical solution, the discrete equivalent of integration, entailed many small addition steps, that is, the accumulation of a large number of arbitrarily small increments. (Recall that among the problems the ABC was originally conceived to solve were sets of *partial differential* equations, which required a preliminary conversion into sets of *linear algebraic* equations.)

The reader will have recognized this proposal of Atanasoff's (which he had said in his January 23 letter was something *he* meant to pursue when he had the time) as basically the proposal Mauchly was to set forth in his August, 1942, manuscript, except that Mauchly stipulated counters for both storage and arithmetic. In other words, this disclosure by Atanasoff in May, 1941, just before Mauchly's visit to Iowa to see the ABC, is a prior written expression of the basic idea behind the ENIAC, conceived as it was for the electronic solution of differential equations by just such a numerical procedure.

And so, with Atanasoff's letter of May 31, 1941, we have an added vital point of contact—an added *ENIAC connection*—beyond his computing principles themselves as embodied in his digital electronic computer: we have his original idea for applying digital electronic principles to a more advanced computer, one that would solve sets of differential equations by solving corresponding sets of difference equations.

Atanasoff was asked at trial, under direct examination by Honeywell, to explain this idea for "converting" his machine into an "integraph"; because of the critical nature of this issue, we quote his response directly (transcript pp. 2,128–29):

> . . . the Bush analog integraph consisted of a series of integrating devices which could be . . . arranged in different ways so that you can bring it into analog with differential equations. Now it struck me—and may I say, while it was stated at the time that it could be used for solving partial differential equations, it was not a handy device for handling [these] and it was better adapted to dealing with ordinary differential equations . . . .
>
> Now, it suddenly came to me that, of course, the computing elements which we had constructed could be used for this integration. The signals could either come out of a stored table of values and these values could be stored in abaci of the type which we [had] been using for storage . . . , or the data could come from a rotation of a mechanical shaft, in which case it would need an analog visual transformation before it entered the machine.

The steps could be made as small as possible; and while . . . it would be necessary [to divide it] into steps, since the steps could be made as small as you please, this disadvantage rapidly vanishes. The thing that you would have as a great advantage of this would be the increase of order of accuracy from one part in a thousand to one part in 10,000, which was characteristic of the Bush integraph, . . . to a much higher value, almost arbitrary higher value, depending upon the design characteristics which you built into the other form of integraph [in] which data was presented and added in little increments . . . .

Mauchly testified, under cross-examination by Honeywell (transcript pp. 12,188–89), that he was uncertain as to how he had taken Atanasoff's reference to "action analogous to numerical integration, but not like that of the Bush integraph"; he added, however, that he could "conceive of a way in which this would seem to mean something" to him. There ensued the following exchange between Mauchly and Halladay:

> *Halladay:* At that time it was meaningful to you, was it not, employment of numerical integration to produce a numerical result to differential equations which were attacked in a different way by the Bush Integraph? Is that not a correct way of saying it, without regard to the specifics?
> *Mauchly:* Yes.
> *H:* And is it not correct to say also, Dr. Mauchly, that this excited your interest at that time?
> *M:* To some extent, I am sure.
> *H:* And that you were interested in a machine which would take arbitrary small steps very rapidly in accomplishing numerical integration?
> *M:* Yes, I was.
> *H:* And you, therefore, went to Ames, Iowa, to see whether or not Dr. Atanasoff had a device under construction directed toward that, among other ends, did you not?
> *M:* I went there to learn as much as I could. This could be included in that, I guess.

Letter 10

The last letter in this correspondence prior to Mauchly's visit to Atanasoff's laboratory was one from Mauchly dated June 7, 1941 (PX 808):

> At present I can't say when I shall arrive . . . .
>   . . . you might expect to hear from me that Friday or Saturday [June 13 or 14], or perhaps find me on your doorstep at some late or early hour.
>   Enclosed is an announcement that you might find interesting as an example of what goes on around here. I suppose there are similar enterprises

all over the country. Pennsylvania is going in for a defense training course for high school graduates, too. I thought I might be teaching in that program, but they haven't notified me of any need for my abilities—and they have hired one of our chemistry men. One hears, all around, that physicists are in so much demand, but it doesn't seem so in the neighborhood of Collegeville.

I'll finish this letter when I see you.

The Situation as of Early June, 1941

At the end of chapter 2, we paused to assess the state of Mauchly's computing accomplishments just prior to his first encounter with Atanasoff in December, 1940. Let us now make a similar assessment of the state of those accomplishments just prior to his June, 1941, visit to Atanasoff's laboratory, and also the state of his knowledge of Atanasoff's computer at that time.

The State of Mauchly's Accomplishments

We saw in chapter 2 that there was neither evidence nor testimony to demonstrate any progress by Mauchly toward an electronic digital computer in his pre-Atanasoff period. The only original electronic device he could point to was a two-neon relaxation oscillator that could have been at most a part of a binary counter, and the earliest evidence for *it* was actually a sketch drawn during the first Atanasoff-Mauchly meeting. He probably had built such a device prior to that meeting, but he certainly had not equipped it to function as a counter or any other computing component. Mauchly did write to Clayton and DeWire, in November and December, 1940, respectively, of an *intention* to build an electronic calculator in another year or so, one that would be of use in the weather project that was his primary research focus while he was at Ursinus College.

Mauchly *had,* in 1939, developed and built an electrical analog computing machine, a twelve-ordinate harmonic analyzer, which he used intensively to analyze his weather data. He also mentioned, in his late 1940 letter to Clayton, a design improvement in a twenty-seven-ordinate harmonic analyzer he was planning to build.

We believe Mauchly was stimulated, in his quest for an electronic calculator, by his initial discussion with Atanasoff. But again he made no progress at all through the spring of 1941 in designing even a simple counting device. He continued to concentrate on neons as the basic counting element, to be controlled now by interspersed vacuum tubes, and he was never able to hit on a design or to build any counting device consisting in a series of interconnected circuits. All of his testimony on these efforts, as evidenced in his circuit drawings, certifies only that he was attempting such and such a design or experimenting with such and such an idea, *not* that he achieved any success whatever.

Indeed, Mauchly's letters for this period between his first and his second meeting with Atanasoff indicate that their initial discussion served more to excite his curiosity about *Atanasoff's* accomplishments in electronic computing than to spur him toward accomplishments of his own. He repeatedly asked Atanasoff for more information and expressed his anxiety to get out to Iowa to see the computer and learn all about it. Moreover, his testimony on these letters (and those Atanasoff wrote him) confirmed his excitement about the computer, his curiosity as to how it worked, and, finally, his added anticipation upon learning that it could be "converted" into a digital electronic "integraph."

The State of Mauchly's Knowledge
of Atanasoff's Computer

Atanasoff testified in the ENIAC patent case that he had a stock-in-trade of things he told people about his computer at the time he met Mauchly in late 1940: it computed with vacuum tubes, it was digital, it was designed to solve very large systems of simultaneous linear equations, it was inexpensive on a per-digit basis, it used roughly so many vacuum tubes, and it operated at a certain speed. He felt sure he had conveyed all of these points to Mauchly during their first discussion in Philadelphia.

Mauchly's testimony confirmed that he had learned most of these features of Atanasoff's computer in that original conversation; the letters he and Atanasoff exchanged over the next several months confirmed all of the rest, with one exception. This was the machine's operating speed. Of it, Mauchly maintained stoutly that he drove to Iowa fully expecting to find a machine designed to operate at electronic speed throughout. A note of his dated January 1, 1941, however, clearly shows that in fact Atanasoff *had* informed him of a mechanical limitation to his machine's overall rate of computation.

Thus, whether or not he knew the exact rate at which the ABC was to eliminate a variable from a pair of simultaneous equations, Mauchly *did know*, from the very beginning, that each such computing cycle was subject to the rate at which a drum or other rotating mechanism could perform its function.

It is quite clear, on the other hand, that Atanasoff withheld certain essential features of his computer, both during their December, 1940, discussion and in his letters over the next five months. He did not claim to have told Mauchly, and we believe he did not tell him, that: it had a memory based on capacitors, it had a means of regenerating its capacitor memory, it calculated on a logical rather than a counting basis, and it calculated in the binary rather than the decimal system. (As the *Des Moines Tribune* article of early 1941 tends to suggest, Atanasoff could have mentioned a separate memory, but in all probability he would not have linked it to his use of capacitors as memory elements.)

Now, in addition to what Mauchly learned of Atanasoff's computer in their original discussion, there is one further critical fact, mentioned above, that he learned from one of Atanasoff's letters just before he traveled to Iowa. On May 31,

1941, Atanasoff wrote to Mauchly that his machine could be converted into a computer to solve differential equations by numerical integration, that is, digitally. This piece of information, together with the electronic computing principles embodied in Atanasoff's computer, would turn out to be of vital importance for the later conception of the ENIAC.

Further Mauchly Testimony

We have seen that Mauchly acknowledged, during cross-examination by Honeywell attorney Halladay, a degree of excitement over the prospect of learning firsthand about Atanasoff's computer. But he also testified on the very subject of his state of mind as he drove to Iowa—his hopes and expectations. And, just as he had previously volunteered information detrimental to his own cause, he did so again, in the same precipitous fashion as on that earlier occasion, unasked and apparently out of pique over Halladay's line of questioning.

The earlier testimony, at least, did not involve either Atanasoff's work or any other digital enterprise; it was addressed to Mauchly's customary use of the ideas of others, as in the design of his harmonic analyzer. The damage then was to his character and did not necessarily encompass the issue of invention of the ENIAC. Now, however, in a single stroke, he managed to destroy his own underlying thesis, namely, that he had made progress toward the design of an electronic computer before he met Atanasoff, or certainly before he went to Iowa to see Atanasoff's machine. He had, in fact, not even tried to invent a computer before he went to Iowa!

Halladay had come to the subject of that visit and was pressing Mauchly as to how carefully he had read the copy of Atanasoff's 1940 manuscript provided him in Iowa (transcript pp. 12,163–64):

> *Halladay:* Do you want the Court to understand, Dr. Mauchly, that you spent from Friday evening, June 13th, 1941, until Wednesday morning of the following week and did not read the Atanasoff manuscript with professional care?
>
> *Mauchly:* It's that phrase "with professional care," I guess, that is the hooker here. I read whatever I read then with sufficient care to satisfy myself as to what it was that seemed to be described there, but not—I wasn't interested then in all the details of what was described there, so why should I read details? Now, what you call "professional care," of course, I haven't learned yet.
>
> *H:* Well, I was assuming that you had your own standard of professional care.
>
> *M:* I did indeed, yes.
>
> *H:* And it was asked in that sense. Now, if you didn't use your usual standard of professional care in reading the Atanasoff manuscript, then tell us that you didn't.

Mr. DeLone: Well, I object to the form of the question. It's a play on words. The witness wants to know what Mr. Halladay means by professional care, and the term is undefined.

The Court: (To the witness) You may answer.

M: It may be a little hard for anyone to understand at this late date just what my attitude was and why I did what I did in those days, but I was searching for ideas which might be useful to help me in computing. I wasn't even thinking about inventing computing machines. I was just thinking about: Could I build something, for instance, which would help me. But I was looking in various places for better ways of implementing computational jobs, many of which I had in front of me if I wanted to do research work, on weather, and such things.

I came to Iowa with much the same attitude that I went to the World's Fair and other places. Is there something here which would be useful to aid my computations or anyone else's computations? . . .

It might seem at first that Mauchly was contradicting his prior testimony about the calculator he had been thinking of building as of January, 1941. He wasn't, though; he was just elaborating. He was saying that he had not been thinking about *inventing* a computer. He did want to *build* a computer—if he couldn't find one—but a computer based on the ideas of others, not one that he would originate. The *research* the machine was to facilitate remained his primary focus.

What, then, of his two letters of late 1940 (to Clayton and DeWire) announcing an intention to build an original calculator in the next year? We would respond that future plans one sets forth in letters to friends may reflect no more than the most tentative hopes or dreams; that Mauchly's own evidence and testimony, in a courtroom setting, proved only the merest "toying" with ideas of a very rudimentary nature; and that here it is Mauchly himself declaring that he was *not* trying to invent a computer on his own.

We proceed now to Mauchly's visit to Iowa, as elucidated in the ENIAC trial, and also as witnessed in an earlier case. Let us just note, at this juncture, that we are leaving the issue of what Mauchly had achieved in electronic computing independently of any Atanasoff influence and are taking up the issue of what he learned from actually seeing Atanasoff's computer and discussing computing with him in person over a number of days. We believe we have demonstrated, thus far, that Mauchly achieved nothing of any significance on his own prior to that visit, and we expect now to demonstrate that he learned a great deal of value for the design of both the ENIAC and the EDVAC from that visit.

Mauchly's June, 1941, Visit to Iowa

Mauchly reached Atanasoff's home at dusk on Friday, June 13, and left on Wednesday morning, June 18. In that period, he gave one lecture, on his meteorological research and his harmonic analyzer, to a small audience gathered by Atanasoff in the physics building of Iowa State College; visited the statistics department with Atanasoff; spent the weekend discussing computing in general and Atanasoff's computer in particular with Atanasoff, either at his home or in his laboratory; spent the daytime hours of Monday and Tuesday on campus, for the most part talking with Atanasoff and/or Berry and watching demonstrations of the computer; and spent Monday and Tuesday evenings at Atanasoff's home in further discussion. He brought along his son, Jimmy, then six years old, giving him over to the care of Mrs. (Lura) Atanasoff and the three Atanasoff children.

The entire case for what Mauchly learned about electronic computing in Ames in 1941—what he was shown, what he was told, what he was given to read, and how he responded—rests on the testimony of the two principals (Berry had died) and of observers on the scene. No notes or drawings of the sort scientists make as they discuss physical phenomena appear to have been preserved. And although Atanasoff has asserted that Mauchly took notes on his 1940 description of the ABC (Atanasoff 1984, p. 255), no such notes were produced at trial.

The case falls into four convenient categories: the state of Atanasoff's computer in mid-June, 1941, together with the availability of demonstration routines; Mauchly's experience of the computer; Mauchly's reading of the 1940 manuscript; and Mauchly's interest or involvement in what he was learning. We will look at each of these in turn, as testified to at trial or in deposition by the various parties.

State of the Computer

As will be apparent in our discussion of Mauchly's experience of the ABC, he neither confirmed nor disputed Atanasoff's testimony concerning its state at the time of his visit, or, for that matter, the tests Atanasoff claimed to have conducted on his behalf. Rather, he pleaded a lack of detailed recall, based in part on a lack of interest. On the other hand, his acknowledgment of certain of these demonstrations does confirm Atanasoff's version in large measure. Accordingly, we confine ourselves here to that version.

Atanasoff testified, under direct examination, that his computer was being worked on very actively in June of 1941 (deposition pp. 805–6), with modular parts removed and the machine generally "all torn up" (trial transcript p. 2,133). Indeed, a reference point frequently used at trial was the July 8, 1941, due date of Clifford Berry's master's thesis, on the binary-card writing and reading system of the computer, a date toward which he was working somewhat feverishly. In preparation for Mauchly's visit, however, Atanasoff had Berry remount on the main frame all units and mechanisms that were in completed form.

Atanasoff gave the following details on the condition of his computer as of Saturday, June 14, when he first took Mauchly to see it (direct, deposition pp. 775–82 and transcript pp. 1,932–33; cross, transcript pp. 2,424–40 and 2,461–62). Components in place were:

a) substantially all the modular add-subtract mechanisms;

b) the entire bank of restore-shift mechanisms;

c) the low-speed axle, the two memory drums, and the base-conversion drum;

d) the high-speed axle, the boost commutators, and two bands of the carry-borrow drum;

e) several sets of writing/reading brushes;

f) the decimal-card reader;

g) much of the wiring among the above devices;

h) the binary-card writing (punching) and reading units, plus one unit of their electric punching circuit;

i) the synchronous motor drive;

j) the power supply and regulator and the filament transformer.

Not in place were:

a) the major portion of the carry-borrow drum to be mounted on the high-speed axle;

b) most of the writing/reading brushes;

c) most of the transformers for punching and reading the binary cards;

d) much of the wiring among the major components;

e) the timing controls to be mounted on the low-speed axle;

f) the general controls for the overall operation of the computer.

It is clear from these two lists that whereas the carry-borrow drum remained to be built, its *design* was complete: two bands were in place and operative on the machine. Likewise, the *design* of the writing and reading circuits for the binary-card system was complete; a unit of it was also installed on the machine. (While this latter system was the one with which Atanasoff ultimately had serious difficulty, the fact remains that he and Berry proceeded with the full complement of these punching units on the basis of their design as of June, 1941.) Thus we can conclude that the only units not *represented* in final design form on the computer at the time of Mauchly's visit were its automatic controls.

Demonstration Routines as of June, 1941

For demonstration and testing purposes throughout its construction period, the ABC was equipped with a number of temporary control devices (direct, deposition, pp. 716, 718, and 804; transcript p. 1,943). The chief of these at the time of Mauchly's visit were:

a) a special one-cycle switch that could be activated to produce a single revolution or any number of successive revolutions;

b) a cycle-counting device that kept track of the number of successive revolutions;

c) means of connecting a cathode-ray oscilloscope to a particular terminal of a drum abacus, in order to display the string of voltages (binary digits) held by that abacus;

d) contacts positioned on the low-speed axle for arithmetic control purposes, and also for synchronizing the activity of the oscilloscope with that of this axle.

Now the demonstrations took two forms. One entailed the *static* testing of components on bench equipment, with the results of operations read out on voltmeters (one binary digit per meter). The other entailed the *dynamic* testing of the computer itself, with results displayed on the oscilloscope (one binary number at a time, since there was just one oscilloscope). As construction progressed, of course, more and more testing could be conducted on the machine, rather than on the bench.

Atanasoff mentioned, in the course of extended direct examination (deposition pp. 782–83 and 805–26; transcript pp. 2,054–85), only two setups for testing components off the machine, one for testing the binary-card punching system, the other for testing the add-subtract mechanism. He was not asked to describe the former, but was examined about the latter in great detail. He had, in fact, produced for exhibition an almost exact model of his add-subtract mechanism, together with a model of the testing apparatus, which was likewise exact except that it read out its results on display lights instead of on meters. He was thus able to give live demonstrations, in deposition and in court, of both the add-subtract mechanism itself, as he put it through the binary operations of addition and subtraction, and the test equipment with which he had routinely demonstrated it to select visitors.

We quote his most concise description of this static test set, given in deposition (p. 783):

> . . . [It] included a transformer for supplying the filament current, a power supply, including a transformer, rectifiers to produce a plus 120 volts and minus 120 volts which were utilized in the circuits in question, and also produced the input signals necessary to actuate the add-subtract mechanisms; and then also available were the additional, the deflection instruments [meters] necessary for describing the output of the . . . mechanism.

As we have indicated, he went on in his deposition to a demonstration of the operation of the add-subtract mechanism, just as he would do later in court. We need note here only that the courtroom procedure was the longer of the two, including an explanation of the binary system, a verification of several additions and subtractions in conjunction with the (truth) table from his 1940 manuscript, and a tracing through of the electronics in conjunction with the circuit diagram from that same manuscript.

He also stated that his demonstration routine in 1941 consisted (p. 782):

first, of a test of the [add-subtract] unit on the test set; second, of a test of the regenerative principles applied to the memory circuits of the machine; and third, tests of the computing power of the machine to add and subtract.

We turn now to those machine tests.

Atanasoff gave his best running account of the dynamic testing of the computer proper while under direct examination in pretrial deposition. We have paraphrased the highlights of this presentation of demonstrations he could conduct on his computer at the time Mauchly visited—or, rather, at the time Caldwell visited, since that earlier event of January 6, 1941, was actually the subject of this examination (deposition pp. 713–20).

He explained that the condensers in the memory drums were such that they could be charged by a touch of voltage (from a battery) to their exterior terminals. Starting with all terminals at zero potential, and working with corresponding circles of condensers on the two drums, he (or more likely, Berry) touched those terminals requiring "1's" with a negative voltage and left those requiring "0's" uncharged.

The oscilloscope was now attached to a keyboard abacus, the motor turned on, the one-cycle switch activated, and the test of the restoring (or regenerating) circuit for this keyboard abacus conducted. It was observed that the pattern on the oscilloscope, a series of hills and valleys representing the digits of the binary number on the abacus, remained unchanged, however many times the drums were rotated and the abacus terminals swept by the brush from its regenerating circuit. A similar test was then run on the regenerative feature of the add-subtract mechanism associated with the corresponding abacus of the counter drum.

Next, Atanasoff said, various other demonstrations were conducted, with the oscilloscope still attached to the counter abacus. The addition and subtraction of two numbers were checked against results worked out on paper. The "borrow" step was tested separately by leaving the counter abacus empty (all "0's") placing a "1" in the least significant position on the keyboard abacus, then subtracting and observing all "1's" on the oscilloscope. The shifting operation was illustrated by having the counter abacus empty and some random number on the keyboard abacus, then performing successive additions and observing that the original pattern appeared on the oscilloscope after the first, the second, the fourth additions, and so on, but in each instance had shifted one position.

Following this description, Atanasoff testified that the decimal-card reader had been installed sometime during the first half of 1941 but was in the instrument shop at the time of Mauchly's visit; in fact, he thought he recalled taking Mauchly to the shop to examine it (deposition pp. 721 and 766–67). At trial, however, he revised this earlier testimony, saying he now felt certain that the decimal-card reader was in place on the machine by the time Mauchly arrived (direct, transcript pp. 2,430–34). We are inclined toward the courtroom version, inasmuch as it was a conscious emendation, but in any event the discrepancy is minor, since on either version Mauchly was exposed to the device.

Its being in place, of course, would mean that it was possible to demonstrate the decimal-card reader and the base-conversion drum dynamically, and also to enter numbers on the drums from their original decimal form instead of directly on the condenser terminals from a battery. That is, a much longer test sequence could be conducted than was previously possible. (As we shall see, Atanasoff in court did recall running this particular sequence for Mauchly.)

Atanasoff also testified in his deposition (p. 804) that prior to Mauchly's arrival on the scene Berry had installed the device for counting drum rotations. This, too, allowed a much longer sequence of operations, since Berry could now hold the one-cycle switch closed for, say, 100 rotations (addition times) and then verify that the outcome was in accord with his hand calculations.

Finally, Atanasoff mentioned both the static and the dynamic punching units of the binary-card system that had been built prior to Mauchly's visit (deposition pp. 805–6; transcript p. 2,136). In fact, he was queried about these only in regard to that visit, and, as we shall see below, he claimed to have shown Mauchly this process as a separate operation, both on the bench and on the computer, rather than as part of the overall elimination procedure.

Mauchly's Experience of the Computer

In this subsection, we present the ENIAC trial testimony of both Atanasoff and Mauchly. We also present the testimony of Dr. Sam Legvold, a doctoral student of Atanasoff's in 1941, as taken in deposition for the *Sperry Rand v. Control Data Corp.* (CDC) case, in which Sperry charged infringement of the Eckert-Mauchly regenerative memory patent (owned by Sperry) and CDC claimed derivation of that memory from Atanasoff.

Atanasoff's Testimony

We have condensed and consolidated Atanasoff's courtroom and deposition testimony on Mauchly's experience of the ABC into a single account (direct, transcript pp. 2,132–37 and 2,163–68; direct, deposition pp. 760—65, 773, 802–5, 820–26, and 844–52; cross, transcript pp. 2,430–34 and 2,707–10; redirect, transcript p. 2,883).

Atanasoff testified that he took Mauchly to see the ABC twice over the weekend, once alone and once with their family members, and that although he did not turn the machine on, they examined it together in some detail, placing their hands on individual components in the process. He said that each of these sessions lasted two or three hours. Then, on Monday morning, he had Berry give Mauchly a demonstration, as he (and perhaps others in the laboratory) looked on. He said that Mauchly and Berry "immediately went into discussion of the various details."

Thereafter, there were "frequent occasions" when Mauchly was "around the machine" with Berry, Atanasoff, and/or Legvold, in the course of which it was demonstrated to him "in detail." Mauchly even participated with Berry in "manipulation of the parts of the machine," helping him "in various minor assembly jobs." Atanasoff said that there were a few occasions when he was called off to other duties—it was summer and he was not meeting classes—and left Mauchly alone or with Berry, but that he himself spent a total of ten to twelve hours with Mauchly on the campus, most of it in the presence of the computer. He added that he answered all of Mauchly's questions freely, and that Berry too was under no constraint to withhold information.

As to the particular tests performed, Atanasoff said these were exactly the tests he had described earlier. Addition and subtraction were demonstrated on the test set, so that Mauchly "saw the logic system perform" as the input buttons were pushed in accord with a chart and the output signals verified on meters. He recalled specifically that both he and Berry were present during a detailed discussion of "the logic element." He also recalled that they spent some preliminary time on the binary number system.

Atanasoff testified to at least two sessions in which they demonstrated addition and subtraction on the computer, checking the oscilloscope results against calculations made on paper. He said that numbers were put in "either by reading them through from punched [decimal] cards or by placing them on with a charged probe," and that Mauchly was shown the "operation of converting a base 10 card to base 2 numbers" through the "card reader and the base-conversion drum onto the counter abacus." He was also shown the "process of punching [binary] cards" on the single punching unit of the computer, but not the reading in of numbers *from* a binary card. Lastly, they operated the computer in a continuous series of additions, using Berry's "revolution counter" to tally up the desired number of cycles.

As for the permanent controls not yet on the machine, Atanasoff testified that they were "discussed with him in a general way," but that "exact structural details" were uncertain and were not given. He commented that these controls were required for the machine to solve sets of simultaneous equations, and that this process was not demonstrated. It was explained to Mauchly, however; "every one" of the steps in eliminating a variable from a pair of equations was spelled out, and the method of "non-restorative division" which Atanasoff considered "enormously important" for shifting in the binary system explicated.

Atanasoff estimated that, of the "25 to 30 hours" he was with Mauchly altogether, two-thirds to three-quarters were spent discussing "computing machines, either this exact computing machine or the theory of computing machines or the future of computing machines." He said that these discussions began almost with Mauchly's arrival, at dinner Friday evening, and continued through the weekend, on the campus, and in the evenings upon their return to his house. They had discussed the ABC "in some detail," he testified, before they made their first visit to it.

In answer to questions on topics other than his own computer, Atanasoff recalled discussing the term "analog," as contrasted to "computing machines proper," his "clumsy term" for "digital" at the time. He remembered little discussion of Mauchly's harmonic analyzer outside of his lecture, and none of any digital concept of Mauchly's other than his experimentation with "neon glow lamps."

He remembered having promised (in his May 31 letter to Mauchly) to discuss the subject of converting the ABC into a digital integraph, and he believed he had done so. He could not recall the actual event, however, because the plan that had occurred to him (during Caldwell's visit) was "theoretical"; it had not been "formulated into terms of exact machinery," and "the exact interrelation of the parts in order to do a structure analogous to Bush's had not been worked out."

Sam Legvold's Testimony

Sam Legvold pursued his Ph.D. in physics, serving as a graduate assistant under Atanasoff from 1939 to 1942, and he worked for Atanasoff in Washington during the war. He later became a professor of physics at Iowa State, having received his doctorate in 1946, and was in fact instrumental in persuading Clifford Berry to return and complete his doctorate (in 1948). He testified in both the ENIAC case of *Honeywell v. Sperry Rand* and the regenerative memory case of *Sperry Rand v. Control Data Corp.* We summarize his testimony on the Mauchly visit to Ames, as given in deposition for the memory case in 1967, and then quote a few passages (Regenerative Memory Trial Records). Legvold appeared for defendant CDC, represented by Allen Kirkpatrick; plaintiff Sperry Rand's attorney was Laurence B. Dodds.

Legvold explained that he did not work on the computer project, but on Atanasoff's antiaircraft project, whereas Berry worked on both, though mainly on the former. He said that he was very close to Berry, and to Atanasoff for that matter, and was interested in their computer; that he spent considerable time with all the people on that project; moreover, that he had frequent need to consult with Berry in the computer room because Berry was "our electronics expert" on the antiaircraft project.

Legvold's testimony verifies Atanasoff's own account, as given in the ENIAC case, of the state of the ABC and the availability of test routines in mid-1941, though it does so in less detail and with one or two slight discrepancies. It confirms that Berry was working hard on the binary-card punching system, which Legvold characterized sympathetically as "a whole new realm of trying to get information in and out of the machine." Finally, it verifies—even enhances—Atanasoff's recollection of the amount of time Mauchly spent with the computer and the degree of his involvement in the activity surrounding it during his visit.

Legvold said he had personally witnessed the running of tests and demonstrations on numerous occasions, and in particular by May or June of 1941. He described these in terms of the rotating drums, the brushes for writing on and reading from the drum contacts, the memory regeneration or "jogging" function, and the "built-in logic for doing addition [and] subtraction." He attested to the existence of the base-ten card reader and the fact of conversion to the base two on the drums, but could not recall the details and could not recall how information was taken out at the end. And he insisted, under challenge by Dodds, that "they had the bugs out" of the computer and that what it could do despite its lack of binary input and output was "sophisticated."

As to his recollection of the Mauchly visit, he told Dodds that while he could not cite specific conversations, he was sure Mauchly, Berry, and Atanasoff had discussed computers "to the hilt." "I recall the tenor of the encounter," he said. "I do recall this right."

That "tenor" had been conveyed to Kirkpatrick, in direct examination, as follows (Regenerative Memory Trial Records, Legvold deposition pp. 23–24):

> My most vivid recollection . . . is of Mauchly in shirt-sleeves helping Clifford and the other people working on the computer, doing things with the computer and on the computer. This, I think, indicates something of the spirit of the meeting that he had with these people. They were deeply interested in computers and were exchanging freely all ideas that they had about computers, and Clifford would detail for him all of the operations, as I recall it now.
>
> I would have a tough time picking any particular part of the machine and saying that I know he saw this and that. All I know is that the spirit of the encounter was one of free exchange of ideas, and I think everything involved here—Clifford Berry and J. V. Atanasoff and Mauchly all together were mutually stimulating people relative to work on computers, . . . and the whole spirit of this meeting of Mauchly with Berry and Atanasoff was one where ideas were freely exchanged and where all principles involved in the computer were there . . . .

Asked specifically about his recollection of seeing Mauchly "in his shirt-sleeves" in the computer room, he elaborated that Mauchly was "in and out from there" to the shop, which was only fifteen feet away. He then said (p. 25):

> . . . The computer was there and they were working on it, and they could demonstrate how it operated and all aspects of what went into this computer were there for him to see and for him to criticize, or whatever he wanted to do, and I would say that it was typical free exchange of scientists discussing their research project . . . .

Kirkpatrick quoted Mauchly's testimony, in deposition for this same case, that he did not recall spending "more than perhaps a half hour in the presence of this incomplete device," revised later to possibly "an hour and a half," and asked Legvold for his estimate of Mauchly's time with the computer. Legvold responded (p. 27):

. . . it was my recollection that he spent at least two working days there, which would be essentially more like 12 to 16 hours with the machine, and then I am sure he spent more than this discussing this with Atanasoff in periods outside the room where the computer itself was.

Legvold also testified to the presence of items in the room other than the physical hardware (p. 28):

. . . there was a desk in there on the east wall . . . and all the sketches and all of these things related to the machine were there at the desk. . . .

Dodds, in his cross-examination, induced Legvold to acknowledge not only that he could testify to no specific topic of conversation among Mauchly, Berry, and/or Atanasoff, but also that his declarations as to the tests he had seen run, and the success thereof, depended on his trust in the statements of Berry or Atanasoff. There ensued the following exchange between Kirkpatrick and Legvold, in redirect examination (pp. 47–48):

Kirkpatrick: Dr. Legvold, referring back to the occasions when you saw Dr. Mauchly in the computer room in the physics building with Berry and/or Atanasoff, were they talking about the computer or were they just passing the time of day about something else?
Legvold: Their conversations were about computers.
K: About the computer in the room?
L: About the computer in the room and how to make computers in general.
K: Would you say that that predominated over small talk or other talk?
L: Oh, yes, yes. . . .
K: Now, when Berry or Atanasoff, around the time of Mauchly's visit or before, were telling you that the machine worked, do you believe that it really worked or they were kidding you?
L: Oh, the machine would perform certain operations . . . .
K: Did you know them well enough to feel each of them were enough of a scientist to tell the facts and not try to balloon up your impression?
L: Oh, no.
K: What is your answer?
L: I say certainly the machine would perform these functions, and I don't know any reason in the world to say that it wouldn't, you see, because it was far enough along so that it would do certain of these things at the time of his visit.

Kirkpatrick then asked Legvold his impression of Mauchly's ability to absorb and understand what he was told about technical matters. Legvold responded (p. 52):

Anyone who knows Dr. Mauchly knows he is a sharp cookie, bright fellow, and that he has a high understanding of electronics and computer technology, and he had, I think, a high interest in it at the time he paid the visit here . . . .

Finally, under recross-examination by Dodds, Legvold recalled that Mauchly had given a seminar on his work with an analog computer, and he acknowledged that this machine could have been the topic of "one of the many conversations" about computers that he had overheard.

We return now to the ENIAC patent case and Mauchly's own testimony as to his experience of the ABC in 1941.

Mauchly's Testimony

In his direct courtroom testimony in the ENIAC case (transcript pp. 11,825–27), Mauchly said that his clearest visual recollection was of an angle-iron frame and a long axle with "a rotating drum which was studded with contacts, which presumably were connected to condensers inside." He also remembered "other pieces . . . hung on this framework," but no details of these beyond the fact that they were "intended for working in cooperation with the drum to produce the [expected] results." He was uncertain as to whether vacuum tubes were mounted on the machine in any way, but recalled clearly that Atanasoff had shown him "things with tubes in them" that he was proposing to use.

In the Honeywell cross-examination, Mauchly commented (transcript pp. 12,190–99) that there "really wasn't much to look at as far as looking at a machine was concerned." He did recall that it was possible to put things on and take them off, and that chassis holding vacuum tubes were of this modular sort. He said he did not see all thirty such mechanisms, but was aware that thirty would be required. He could not recall seeing any input-output devices for receiving punched cards, or a high-speed shaft in addition to the low-speed one. In these last two instances, however, he did not deny their presence; he simply had no recollection of them.

We quote Mauchly's direct testimony on any demonstrations he was given in Atanasoff's laboratory (transcript p. 11,828):

Well, I referred to the fact that demonstrations were made. Now, the motor was turned on, the drum rotated—that I believe I saw. How much further a demonstration could have been made with what I saw there I can't really recall now, because of a, you might say, slight confusion in this respect. My memory tells me that there was another table in that room, I believe and, so, some of the parts of this machine could be demonstrated by tests made when the equipment was off the machine. So what I can't remember now is how much was demonstrated off the machine and how much was demonstrated on the machine. But whatever was demonstrated was demonstrated in a very non-automatic way, you might say, by specific operations of buttons or with some way not—maybe the simplest way to say this is it did not have any way of putting in numbers or taking out numbers in an automatic way. If you wanted to demonstrate anything, you had to apply electric charges to wiring elements or something in the machine in order to even get a small set of digits in there so that you could demonstrate that anything happened.

This statement does seem to preclude the availability of the decimal-card reader for the automatic input of coefficients, which Atanasoff had testified was in place and was demonstrated (although charges were also applied directly). But, as we noted earlier, Mauchly did not deny seeing such a device when cross-examined about it.

Mauchly also said, in direct examination (transcript p. 11,832), that he understood—perhaps had been told in Philadelphia—that the machine's purpose was to solve sets of linear simultaneous equations. He acknowledged further, in cross-examination (transcript pp. 12,191–96), that he "was perfectly aware of what the machine was supposed to do" as it was demonstrated to him, and that a (basic) function of the machine was to store data on one drum and transfer it to the other and to do arithmetic operations in between. He acknowledged, too, at this time, that he may well have taken his coat off during the demonstrations, and may have helped Berry place components in or remove components from the computer.

Other specifics elicited later in cross-examination (transcript pp. 12,205–8) were, first, that Mauchly had "no doubt whatsoever" that the drum with the brass studs—the extant one was exhibited in court—was a *regenerative storage device,* and, second, that to the best of his recollection he had seen the "add-subtract mechanisms demonstrated in one way or another" (on the machine or on a test set). He also acknowledged seeing an oscilloscope and believed he had looked at it "and would have asked what did these things represent" ("0's" and "1's"). This particular encounter ended with the following exchange between Mauchly and Honeywell attorney Halladay:

> *Mauchly:* . . . I was perfectly convinced by what Dr. Atanasoff said and what I saw at that time that his add-subtract mechanism did in fact work.
> *Halladay:* And did in fact accomplish addition electronically?
> *M:* Work in this case means that it did produce a result which was the result of adding two binary digits and getting an answer, or subtracting two binary digits and getting an answer.
> *H:* And doing so electronically, correct?
> *M:* It used vacuum tubes, yes.

Halladay then pursued the issue of the speed of operation of the add-subtract mechanisms, whether or not this was *electronic* speed (transcript pp. 12,208–9). Mauchly did not recall investigating their speed at all, other than to ascertain that it was adequate for the rate at which numbers were taken from or put on the drums. He remembered this latter rate as "sixty revolutions per minute," but still had no memory of a fast axle, much less of its rate of 900 revolutions per minute.

Sperry attorney DeLone followed up on Mauchly's failure to remember the fast axle by citing Atanasoff's testimony that his "boost and carry drum" (what we have termed the carry-borrow drum) was not in place in June of 1941 (redirect examination, transcript pp. 12,487–89). It was not, of course, and Halladay had not indicated that it was, but in fact the *boost* commutators were in place on the fast axle, and so were two temporary carry-borrow bands. Indeed, Mauchly could not have seen addition and subtraction executed on the computer—as he seems to have—without the carrying and borrowing operations!

(The reader may recall that the fact of two shafts of different speeds is theoretically irrelevant, since the functions of the fast one could have been performed on the slow one. It was just a matter of simplifying the construction of the *pulse-time* commutators by having a separate, faster, shaft for them. The obvious rate for that shaft would have been sixty rotations per second, as compared to the slow shaft's one per second, or per *addition time,* but Atanasoff found it simpler to use a rate of fifteen per second and to incorporate four complete "cycles" in each rotation.)

As to the amount of time he spent with the computer, Mauchly testified that he "was in and out of that computer room probably more than once" (direct, transcript p. 11,827). When queried in cross-examination (transcript pp. 12,189–90), he first said:

I don't really know, don't remember how much time. Perhaps too much.

And then:

Well, I am perfectly prepared to say that I spent lots of time, that I spent hours in the basement there. My own impression of what was happening was not that I was looking at something which was just standing there and which was no particular joy to look at but, rather, that I was talking and discussing with Dr. Atanasoff and Clifford Berry during these periods. It was more a communication between person and person. . . .

As to what topics were discussed, aside from this computer, Mauchly mentioned his own weather work and (possibly) some of Atanasoff's other research. "But primarily," he said, "we were talking about what can you do to make better computing machines" (direct, transcript p. 11,824). Asked by DeLone for details of these latter conversations, he could recall none. Finally, during cross-examination on the particular topic of Atanasoff's idea for an electronic digital computer to solve differential equations, Mauchly acknowledged, very cautiously, that Atanasoff did tell him *that* this could be done (transcript p. 12,203). As for hearing *how* it could be done, he again pleaded lack of recall:

> He may have told me something about this. I don't know to what extent
> he did. From my frame of reference, why, as I said before, I can't recall now
> how many different things Dr. Atanasoff may have talked to me about and
> how many things he told me . . . .

He now added, however,

> . . . but I do recall that my knowledge about all these things was not zero
> when I came to him.

With that, Halladay returned to just what Mauchly's computing ideas and ventures
had (or *had not*) been prior to his June, 1941, trip to Iowa.

Mauchly's Reading of the 1940 Manuscript

We now present the ENIAC trial testimony of the two principals on the issue
of Atanasoff's August, 1940, description of his computer and its availability to
Mauchly during his visit. It will be recalled that this document was written in the
early months of construction, both as a proposal to raise funds and as the foundation for a patent application. It embraced all aspects of the envisioned computer in
considerable detail; Atanasoff had, after all, spent nearly three years (1935–37)
investigating the possibilities for a computer of the ABC's capacity, and another
two (1938–39) developing this particular one.

Atanasoff's Testimony

In court (direct, transcript pp. 2,133–37), Atanasoff testified that he handed
Mauchly an original (typed) copy of his manuscript (PX 455)—there were two such
copies—on Monday morning, June 16, at his laboratory. He said Mauchly then had
that copy in his possession for the balance of his visit, reading and discussing it in
the evenings, and in fact asked but was denied permission to take a copy back east.
We quote, in part:

> I remember an occasion in my home—he had the copy in his possession,
> he took it to my home and read it evenings and I remember us in my own
> study at home, in my home, reading and discussing the manuscript. . . .

On this same subject in deposition, Atanasoff had stated (direct, p. 765) that
Mauchly had "full access" to the manuscript, and:

> . . . We discussed it in the evening. We discussed it during the day. We
> discussed it, as far as I could see, in detail. I know of no part of it that was
> not covered in discussion at this time. . . .

As to precise topics of the manuscript he remembered discussing, he cited the carry mechanism, the logic of the add-subtract mechanism, and binary addition and subtraction (transcript pp. 2,137–38). Regarding the logic, he said:

> . . . I remember discussing with him the symbol, the 3,2 symbol . . . and the other symbols, and the meaning of this, and I told him he could follow through for himself how the thing worked, and I don't positively know that he followed it through in complete detail and understood every step but he certainly understood the general features of [the mechanism].

Regarding binary addition, he said they "certainly discussed" the table in his manuscript that showed addition and subtraction in terms of high and low voltages (what we have called a truth table), but he recalled no details of this discussion. He then commented:

> I remember his expressing his surprise that the base 2 number system was advantageous for computing. I am not sure that he was entirely convinced as to this.

Asked if he had shown Mauchly how binary addition and subtraction were accomplished, he replied:

> I did so show him. As a matter of fact, he seemed—I believe from my memory that he was not proficient in base 2 calculations at the time he came to visit me.

Mauchly's Testimony

Mauchly's entire direct testimony on his exposure to the August, 1940, manuscript (transcript pp. 11,833–35) was devoted to his memory, first, that he was not permitted to take a copy away, and, second, that it contained few details of the proposed computer. Both of these "protestations" amount to a denial of any real attention to the actual content, a defensive posture the Honeywell side was quick to focus on in cross-examination. The passages are brief and seem worth quoting in full:

> I don't remember much about it. My main memory on this is, you might say, a little—or the other way around, in that what impressed itself on my memory was that I was not allowed or given anything that I could take home to read. In other words, anything that I was given to read, I had to read there. He did not want any information in written form to leave his laboratory offices, and so he made it very clear to me. I mean, it wasn't any—no bashfulness, you know, about this. It was a clear understanding that "We do not want this material circulated, and so, to prevent this from happening, why we will not let any written material be taken away."

Well, my general memory is that while I was there, he gave me to read some material, which I think was prepared originally to describe what he was proposing to do to any organization that might be available to give him a grant to help him do it. In other words, it was something like a proposal.

[As to a recollection of its contents,] not specific, no. I mean, the general idea was just describing the utility of wanting to solve certain classes of problems, and here was a way of doing it, and then some detail—but not very much—as to how you would do it.

The cross-examination about the first point became bogged down on the issue of whether Mauchly had actually asked permission to take a copy of this document home (transcript pp. 12,167–69). While he maintained he did not remember how the matter arose, he said he would have liked to take a copy, indeed, that he usually did try to take home "sales literature from a store, or something," so as to refer to it later. He also acknowledged that he understood the reason for Atanasoff's denial of permission to be a concern over patent problems.

The cross-examination about the second point became bogged down, too— and much more seriously—on the issue of how carefully Mauchly had read the manuscript (transcript pp. 12,161–67). When first shown PX 488, a photocopy of the original PX 455, he refused to concede that it was the document he had read in Iowa (even though he had had an opportunity to examine it prior to his testimony) at the same time that he declared he had probably read "the one" he saw in Ames "very carefully." There followed the argument over his degree of "professional" care that we touched on earlier, during the course of which he suddenly volunteered that he had not even been thinking of inventing a computer before he went to Iowa, that he was merely looking for help with his computations.

Mauchly then pleaded a loss of interest in the details, once he had seen the machine and learned in general how it was to work:

. . . there wasn't any great reward for me at that point to try to memorize the details that were in this book. There was no sense to me in doing that, and that apparently is the difficult thing to get across, that why should I use my professional time in reading a description of a device which I wasn't particularly interested in thereafter when I could be talking to Atanasoff about things which he and I were both interested in. . . .

He did recall their discussing ferromagnetic storage elements, which were among possibilities listed in Atanasoff's manuscript. And he acknowledged an interest in "the way he described the various applications" for his computer:

. . . It's always interesting to see how somebody else puts in words things that you already know . . . .

The Halladay-Mauchly exchange on this issue ended as follows:

Mauchly: . . . So I was more interested, as I say, in general discussions with Dr. Atanasoff than I was in trying to photographically memorize a book, which I did not do.

Halladay: I didn't suggest that and I still don't know whether you read the manuscript . . . with any kind of care, professional or otherwise.

M: I read it with some kind of care, what there was of it and as much as I cared to read.

H: Now you say as much of it as you cared to read.

M: Yes.

H: Is there any question but what you read from the first page to the last page?

M: Well, I will have to admit that my memory again does not extend to saying exactly how many pages there were or whether I read the last page or not. I know that in the last year or so when I have learned a little more about what was intended in that machine I have been a little surprised because there were things that I had never heard of.

We have seen, throughout Mauchly's testimony thus far, an interspersing of this plea of a lack of interest—or loss of interest—in Atanasoff's computer with that of a lack of recall. We turn next to a closer look at this stance as he was examined about it in the ENIAC trial.

Mauchly: Was He Interested?

Atanasoff was not examined in court on the degree of interest, per se, that Mauchly displayed in his computer, and he commented on it in just one instance that we have found. He was speaking in general of the demonstrations and discussions among Mauchly and himself and Berry (direct, transcript p. 2,136):

> . . . He seemed to follow in detail our explanations and expressed joy at the results, at the fact that these vacuum tubes would actually compute. . . .

Of course, it is implicit in the testimony already cited, of both Atanasoff and Sam Legvold, that Mauchly *was* greatly interested. Extended explanations and demonstrations of the sort they described do not occur, after all, without real encouragement from the listener/observer. And since Mauchly himself admitted his exposure, at least, to a sizable bank of information at the hands of Atanasoff and Berry, it would seem indisputable that he was interested to a significant degree.

Yet he strongly disavowed any but the most superficial interest, even as he acknowledged the various things he had learned. Moreover, unlike Atanasoff, he was examined at trial explicitly on this issue, by the Sperry side, and he had much more to say about it than came out in his account of the experience itself. This subsection, then, presents Mauchly's further testimony on this matter of his interest, or lack thereof, in what he learned about electronic computing while he was Atanasoff's guest in Iowa.

But there was another witness with strong views on Mauchly's involvement with Atanasoff and his computer. This was Lura Atanasoff, to whom Atanasoff was married from 1926 until 1948. (They were effectively separated when he moved to Washington in 1942.) We present her testimony as taken in deposition, in 1967, for the *Sperry Rand v. Control Data Corp.* suit over the regenerative memory patent.

Mauchly's Testimony

Mauchly first expressed his disappointment in Atanasoff's computer when asked what he had learned about the two-dollar-per-digit cost estimate which had, in part, drawn him to Iowa (transcript p. 11,829):

> I didn't learn a great deal about that, as I didn't really try to do the cost accounting . . . , but what I did learn, was you might say, the essence of what I wanted to know. It was a disappointment because there had been nothing said, so far as I can remember, in my meeting with him [in Philadelphia] which indicated that this machine was not a fully electronic machine. That's why my great curiosity, my great wonderment, how in the world did he do this for $2.00 a digit, because I was thinking in terms of all electronic things, which would make use of electronic speeds . . . .

It seems strange that Mauchly should have *assumed* the machine was entirely electronic, just on the basis of not having been told that it was not! In fact, he knew perfectly well before he went to Iowa that the speed with which Atanasoff's machine would compute was limited by a mechanically rotating component. As we saw earlier, his own notes reveal his understanding, from that first meeting in December of 1940, that there was a drum or disk of some sort working in conjunction with separate electronic components. Nevertheless, in court he proceeded to express his disillusionment at this "discovery" (transcript pp. 11,829–30):

> . . . almost immediately, of course, when I got out there I began to learn . . . you could see the drum idea and you could begin to put together the picture that this is a mechanical gadget which uses some electronic tubes in operation, but it's still restricted in speed and was not what I was interested in from the point of view of electronic speed gadgets. . . .

This explanation, that he himself had been interested in a completely electronic computer, is another curious twist. He had been aspiring—and was still only *aspiring*—to an electronic desk calculator, with a keyboard input that would have severely limited his overall rate of computation. As he himself had expressed it, this machine would perform its arithmetic operations as rapidly as the numbers could be punched in. (A mechanical output would also have affected its overall rate.) But, in any event, there was no reason for concern over Atanasoff's drum memory, because his calculator was to use counters, which incorporate their own storage function along with their arithmetic function! He could very well have utilized the new electronic switching technology of Atanasoff's add-subtract mechanism in designing a calculator based on counters.

He and Eckert *did* utilize that technology in the ENIAC, which was a (decimal) "counting" computer, as opposed to a "logical" computer, in its basic arithmetic design. What is more, they utilized Atanasoff's *logical* principle, as well, in the EDVAC; for it they abandoned counters, which combine arithmetic and storage, and turned to separate arithmetic unit and memory. And in 1952, they received a patent on a serial binary adder whose claims included the basic ideas of Atanasoff's add-subtract mechanism.

Thus Mauchly *was* interested in the add-subtract mechanism. In fact, he was interested in the *drum memory,* as well. He and Eckert used the principle of regeneration it exemplified in the EDVAC, in the form of a mercury-delay-line memory, and in 1953 they obtained the memory patent to which we have referred on this and other forms of regenerative memory, including Atanasoff's.

In his ENIAC trial testimony, Mauchly went on to declare a certain concern over the fact that the ABC was a special-purpose computer, but then generously allowed that he could have overcome that limitation and moved on to the general-purpose computer *he* envisioned had it not been for the speed limitation (transcript pp. 11,830–31):

> . . . The fact that it was also being built specifically to solve a special class of problems, rather than a general class, was quite obvious, of course, but that wasn't the thing that really worried me because what I had been looking for were ways of getting high speed electronic things without paying too much for them and—sure, this was electronic in part, but . . . .

> What I [mean is] that his machine was specifically, deliberately constructed with the idea that when finished it would solve simultaneous linear algebraic equations, which is a very special class of problems, class of problems that occurs over and over again in many places: statistics, physics, engineering, whatnot, but there are many other problems which cannot be handled in this way and my interest was in solving some problems which couldn't be handled this way. And I could see that other people would be interested in solving still other problems which couldn't be handled this way. So my general interest was in trying to get computers which would be versatile and not restricted to some one class of problems. That did not worry me about his machine, the fact that he was designing that. That was his privilege, if that's what he wanted to do. . . .

And yet Mauchly himself had aspired only to an electronic desk calculator, and had at this point succeeded in inventing only a neon-diode relaxation oscillator!

Finally, in this direct testimony, Mauchly went so far as to claim that because of his disappointment in Atanasoff's "gadget," he devoted much of their discussion time to "going over this point of 'Why don't you do this and why didn't you do that?' " (transcript p. 11,830). As in his earlier remarks of this nature, however, he offered no specifics as to how he would have improved the ABC.

Lura Atanasoff's Testimony

Lura Atanasoff held a bachelor's degree in home economics and the arts from Iowa State College. In her deposition for the memory case (Regenerative Memory Trial Records), she gave her occupation as "artist"—she was a painter. She appeared for defendant CDC, represented by attorney Allen Kirkpatrick; plaintiff Sperry Rand was represented by Laurence B. Dodds.

Mrs. Atanasoff testified that, whereas her husband had had complete trust in Mauchly, she had been "suspicious" of him. For one thing, she felt "imposed on" and "completely ignored" by him, so that she became "kind of critical." For another, she was alarmed that Mauchly was "so prying" and her husband so responsive to the many questions about his computer. She was worried because the computer was not patented, and she was still more worried when she learned that Mauchly too was interested in building one. She therefore listened to the conversations held in her presence, she said, and had no doubt as to what was being discussed.

She attributed Atanasoff's openness about his computer to his being "proud" and "full of enthusiasm . . . because this was his brainchild." When she cautioned him to be careful about what he told Mauchly, he assured her that this was " 'a fine, honorable man,' " and that she need not worry because " 'I don't think his [machine] will work.' "

Most of her direct observations were made in the Atanasoff home, especially during meals. She said that all their mealtimes were devoted to "entertaining Mr. Mauchly," despite the presence of four children, that there was "no polite conversation," and that "he kept asking, he asked, always asking questions."

She said she did have occasion to go into her husband's study, where he and Mauchly were talking in the evening, to call one or the other to the telephone. There she observed that they were "deep in conversation," looking at "some drawing" from one of their briefcases, and "comparing notes," with Atanasoff saying " 'Now, this can't work that way.' "

Shown Mauchly's deposition testimony that he had been in the presence of the computer for half an hour, all told, or perhaps an hour and a half, Mrs. Atanasoff recalled the visit the two families had made to see the computer on Sunday afternoon. She felt sure she had been in the presence of the machine "an hour" before she took the children outside to give the men some time alone.

She could also infer that Atanasoff and Mauchly had spent time with the computer on the weekdays, because "they always appeared together talking," they would mention that they "had just left the physics building," and Mauchly would comment at table on having seen the machine.

Mrs. Atanasoff characterized her husband as "very kind," "always helping, doing good for people, and he was very kind to Mauchly." She observed that even when he asked her to testify he still "had some loyalty to the man" and felt apologetic to her over having been "deceived" by him, saying that he did not think Mauchly had intended to use his ideas at first, " 'but maybe knowing these he just, you know, used some of them.' " She thought this was "very kindhearted, because he could just hate the man."

As to any preparation for her deposition, she said that Atanasoff at some point had told her the suit involved possible appropriation of features of his computer, and that he had asked her just to tell what she remembered.

We now present some passages from Lura Atanasoff's testimony, beginning with the direct examination (all page references are to Regenerative Memory Trial Records, Lura Atanasoff deposition).

> . . . [My husband] was always working on [the computer], and then when he got actively producing it . . . , then he put more and more time, everytime he had a minute, his weekends, and everything. Well, of course, that affects a family, you know, you know it. (Pp. 6–7)

> . . . And sometimes the children and I would go over to see his progress, what he was doing, because that was just his life. (P. 7)

> Well, when I was really aware [of John W. Mauchly] is when he appeared at my home late one hot afternoon and was to—I learned he was to spend a few days with us. (P. 9)

> Yes, with his little boy. I was surprised. I didn't have the guest room ready, and I had to fly around . . . . And he was kind—immediately I thought he was kind of an odd, quiet fellow, didn't say a word. The only thing he said, "I thought my wife needed a vacation," that was his only apology for bringing the little boy along, and I thought, "Oh." (P. 10)

> . . . And the morning came and breakfast was over, so here they go with their briefcases, they go over to the college and no one says, "Will you babysit with my little boy," who was six or seven. . . . (Pp. 10–11)

> . . . even once [at the table] he took a note—he took a little card or something out of his pocket and jotted down, "Now, that's interesting," and so on. He is one of those superior professors, you know. (P. 11)

> . . . I said [to my husband], "You must be careful until this is patented." (P. 12)

Because he talked so much, you know, like a little boy with something grand that he has built . . . . And [my husband] thought I was [being] suspicious. . . . So I felt kind of mean . . . and I kind of resented [Mauchly], I was ashamed of it. (P. 12)

When my son called me . . . , maybe two or three months ago, he said, "There is some litigation . . . about daddy's computing machine." And I said, "Who? That old Mauchly?" Immediately I thought of that man, because I thought of his questions . . . . (P. 12)

Well, it seemed that Mauchly had kind of struck a stump, you know . . . , and at this meeting in the East, . . . these learned men get together . . . and talk. So Mauchly was trying to do something that my husband was doing, and this had gone along, so he came—here he drives clear out to Iowa to see about it, not my husband going out to where he was. And then he spends all his time with us and he really spent it, he never patted a child, he never paid any attention to the children, he came in and left his little boy and he went on about his business. (P. 13)

Well, it seemed like longer than it was. It was probably four or five days. I'm sure it was at least four. I'm surprised—well, you know, you remember disagreeable things, and maybe that's why I remember him so well. (P. 14)

Never did we go out. I always cooked the meals, and here we had the four children and three grown-ups, we had a big table, we always had the meals together, it was just a homey affair. He came to our home and shared our home. (P. 14)

Jimmy, this little boy, [my children] made him fit in pretty well, he was a little bit of a problem, but he—Mr. Mauchly was daddy's guest and it was our business to make things as pleasant as we could. That's the way we always did. But you have your kind of secret feelings about it when you are being imposed on. (P. 16)

And I was always proud of [the computer] because I thought my husband would be famous some day, and I would be very proud of it for my children's sake, too. So that's why I was really suspicious and watching. (P. 18)

Mrs. Atanasoff seems to have grown somewhat resentful—if not suspicious—of Sperry attorney Dodds, as well, as he cross-examined her on her knowledge of: technical aspects of her husband's computer; its state when he left Iowa; his 1940 manuscript; the present lawsuit and the Eckert-Mauchly patent in question; and precisely how long Mauchly had spent with the computer. CDC attorney Kirkpatrick intervened several times on her behalf, and the session ended with the following exchange among the three parties (p. 38):

> *Atanasoff:* . . . But if he is—he's a doctor of physics and mathematics, it might not take such a long time for him to gain knowledge, have you thought of that?
>
> *Dodds:* Can you tell me, Mrs. Atanasoff, are you being compensated for coming and appearing for taking your testimony this morning?
>
> *A:* Slightly.
>
> *Kirkpatrick:* Let the record show Mrs. Atanasoff has been subpoenaed, and I believe the fee has been tendered to Mrs. Atanasoff.
>
> *D:* I am not speaking about the statutory subpoena fee, I am talking about compensation for her time, and she said she had been compensated.
>
> *A:* Well, just that $9.47 [for bus fare], or something like that, for my total compensation.
>
> *D:* Have you any arrangement with Mr. Kirkpatrick or his client for compensation for your time later?
>
> *A:* No.
>
> *D:* I think that's all.
>
> *A:* I didn't expect any compensation.

The Post-Iowa Period

We have now reached a watershed in Mauchly's computing career. On the one hand, he had learned from Atanasoff how to compute with vacuum tubes, and also *that*—if not exactly *how*—this knowledge could be applied to the solution of differential equations. On the other hand, he was about to meet Eckert and begin to profit from his electronic expertise and ingenuity; for it was at this point that Mauchly left Ursinus College, effectively, and joined the Moore School of Electrical Engineering of the University of Pennsylvania.

Mauchly spent the summer of 1941 as a student at the Moore School, taking an intensive defense training course in electrical engineering and electronics for which master's candidate Eckert was the laboratory supervisor. At the end of that course, he was made an instructor, and he stayed until early 1946, both teaching—he became an assistant professor in 1943—and participating (with others) in government-sponsored research projects that included the design and construction of the ENIAC and the initial design of the EDVAC.

The evidence from this final period of the Atanasoff-to-Mauchly ENIAC connection consists in letters between Mauchly and Atanasoff; an exchange of letters by Mauchly and meteorologist Clayton; and certain Mauchly notes which show, once more, the very rudimentary state of his own work in digital computing *and* his conscious dependence on the ideas of Atanasoff as he groped his way toward a machine he himself might build. We present all of these documents in a single chronological sequence, as the best way to portray the development of Mauchly's thought at this critical juncture. That is, rather than treating the correspondence and the notes in separate sequences as we did for the interim period, we present each document as it occurred. (Most of the notes actually fall conveniently between two letters.)

Mauchly's June 22, 1941, Letter to Atanasoff

However long Mauchly may have meant to remain in Ames, he left rather abruptly upon receipt of a telegram from his wife informing him of a job prospect at the American Optical Company (ENIAC Trial Records, transcript pp. 12,210–11). He wrote to Atanasoff the Sunday after the Wednesday on which he had left Ames, as follows (PX 816):

> The trip back here was uneventful, except for the fact that I was carrying on a mental debate with myself on the question of whether to teach [a defense course for high-school graduates] at Hazleton [Pa.], or to learn something at U. of Pa. My natural avarice for knowledge vied with that for money, and won out, so after obtaining assurance . . . that some one else [could] take the Hazleton work, I dropped that and prepared to become a student again.
>
> I drove to Southbridge, Mass., Friday evening [June 20], and looked through the American Optical plant on Saturday morning. They seem quite serious in their intentions toward me, but no decision was to be made for several weeks.
>
> On the way back east a lot of ideas came barging into my consciousness, but I haven't had time to sift them or organize them. They were on the subject of computing devices, of course. If any look promising, you may hear more later.
>
> I do hope that your amplifier problem has been licked by some adequate design. The tubes that I ordered two weeks ago aren't here yet, so I couldn't try anything here even if I had time.
>
> I forgot to ask what happens to Clif Berry after he gets a master's degree—does he stay on for Ph.D. work?
>
> Please give the enclosed note to your wife. We enjoyed our trip very much, and hope you can stop here some time.

This letter confirms a number of facts about the state of Mauchly's aspirations in late June, 1941, as he "prepared to become a student again": (1) he was actively seeking a new permanent position, (2) he felt a need to learn more electronics, (3) he was excited and stimulated by his visit to Atanasoff—almost overwhelmed, it seems, by the new possibilities for computing devices that were occurring to him as he drove home, (4) he did not yet know whether any of these ideas would prove "promising," and (5) at this point he still had neither the time nor the equipment to "try anything."

In court, the Honeywell side asked Atanasoff about the "amplifier problem" mentioned by Mauchly, but he could not recall what that might have been (transcript, p. 2,170). The Sperry defense used this letter only to establish when Mauchly left Ames (transcript pp. 2,701–5 and 11,833–36).

Mauchly's June 28, 1941, Letter to Clayton

The next letter, written by Mauchly to meteorologist H. Helm Clayton on June 28, now confirms the specifics of what had excited and stimulated him in Iowa: (1) he had seen a computer (Atanasoff's) *that he regarded as electronic*, (2) he considered this machine very fast overall and was not bothered by the limitation imposed by its mechanically rotated memory, (3) he had no quarrel with its specific purpose of solving systems of linear equations, indeed was aiming at even *less* generality himself, (4) he regarded the ABC as nearly completed, and (5) he had both the understanding and the conviction that it could be "adapted" (his term— Atanasoff had said "converted" in his May 31 letter) to solve systems of differential equations more rapidly and much less expensively than the mechanical differential analyzer could. We quote (PX 822):

> Up to a few days ago I was in hope of making a trip to Massachusetts this June with the possibility of returning the Sundstrand machine which you so kindly lent us. Now it appears that I can't do that, and within the week I shall properly pack the machine and forward it by express.
>
> I know this must have inconvenienced you already, and I feel that we owe you a great deal for the loan.
>
> Immediately after commencement here, I went out to Iowa State University [*sic*] to see the computing device which a friend of mine is constructing there. His machine, now nearing completion, is electronic in operation, and will solve within a very few minutes any system of linear equations involving no more than thirty variables. It can be adapted to do the job of the Bush differential analyzer more rapidly than the Bush machine does, and it costs a lot less.
>
> My own computing devices use a different principle, more likely to fit small computing jobs.
>
> While at Iowa, I talked on the construction of harmonic analysers. But I haven't done anything about the 27-ordinate cost-estimate as yet.
>
> All of my time since coming back from Iowa has been taken up with an Emergency Defense Training Course at the Univ. of Pa. I had a chance to teach for the summer in a defense course given to high school graduates, but turned that down in order to become a student myself. I am working in electrical engineering and electronics. Whether or not I am given a defense

job involving electronics later on, the training will be helpful in connection with electronic computing devices.

I haven't had any chance to work on weather problems recently. I did hear Rossby talk at Iowa City—concerning the training of students for meteorology, etc. Let's hope your own work is getting along well.

This clear statement by Mauchly himself of what he had learned from Atanasoff was, of course, very damaging to the defense. Plaintiff attorney Halladay, however, had the task of seeing that it received its due exposure in court. He could, and did, use it to refresh Mauchly's memory as he cross-examined him on his conversations with Atanasoff in Iowa, especially on their discussion of an electronic digital integraph (transcript pp. 12,202–3). But he had also, quite shrewdly, succeeded in introducing it during his examination of Atanasoff, five months earlier, on this same topic of Mauchly's Iowa experience.

Atanasoff appeared in June, 1971, Mauchly not until November. On the earlier occasion, Sperry attorney Ferrill argued against the admission of PX 822, since as a letter neither by nor to Atanasoff it was not a document on which he could be examined (transcript pp. 2,020–26). Halladay countered that this letter was the first "recordation" by Mauchly of his trip to Ames, on which Atanasoff would be testifying, and that he wished only to "read it into the record as of the [applicable] time frame"; moreover, that it had been an exhibit in Mauchly's deposition and that on that basis he could lay a foundation for its introduction during Atanasoff's testimony if necessary. Judge Larson then read the letter and received it into evidence, and Halladay did subsequently read it into the record in its sequential order as he examined Atanasoff (transcript pp. 2,170–71).

Thus Halladay, by exhibiting immediate support for Atanasoff's testimony from the very hand of Mauchly, gave the judge cause to form a positive opinion of Atanasoff's credibility almost at the outset of the trial. The contrast between the two adversaries would surely loom larger, then, than it otherwise would have, when the letter was produced a second time to challenge Mauchly's protestations of disappointment in Atanasoff's computer. And, as we shall see in chapter 4, Judge Larson did address both this document and the issue of the relative credibility of these two witnesses in his final decision.

We should note in passing that Mauchly was also pressed about his statement, in this same letter to Clayton, that his own devices used "a different principle, more likely to fit small computing jobs." Halladay did not, however, ask him to identify this "different principle," or the Atanasoff principle from which it differed, but focused instead on the fact that it was better suited to "small computing jobs" (transcript p. 12,204). In short, the Honeywell side wished to stress that Mauchly had aspired only to a digital desk calculator before seeing the more powerful ABC, rather than that a "different principle" was also involved.

Clearly, though, Mauchly's allusion was to his own continued preference for the *counting* principle (the use of counters to do the arithmetic operations of a computer) as opposed to the *logical* principle Atanasoff had just revealed to him (the use of logical mechanisms to do those operations). The only possible alternative is the analog principle (of his harmonic analyzer) as against Atanasoff's digital principle, and the context seems to rule that out. It is curious that Mauchly employed the present tense, "My own computing devices use," inasmuch as he had just informed Atanasoff that he had not been able to "try anything" yet. This implication that he was further along than he actually was does tend to confirm our earlier conjecture that Mauchly may have exaggerated his progress when he wrote to friends somewhat removed from his activities. (This is the second letter to Clayton that appears to stretch the truth.)

Clayton's July 20, 1941, Letter to Mauchly

Clayton's reply of July 20 (PX 841) is of interest on two scores. First, it expressed the high regard of a fellow scientist for Mauchly's work in harmonic analysis, and also his disappointment in Mauchly's gradual shifting from analog to digital computing devices as evidenced in his letters of the past eight months. Second, it set forth what had to be one of the earliest proposals for a computing center!

Clayton wrote that he regretted "very much that you are considering the relinquishment of constructing the 27-key harmonic analyser," since "it would certainly be a very useful machine for the kind of work I am doing." He then remarked on "the only drawback . . . with all machines of that type," the expense of having "a skilled man to operate [it] and keep it in order," and went on to suggest the establishment of a "central point where different interests could send data to be analyzed so that one machine could serve several parties each paying for the work as it was finished."

(Atanasoff had alluded to such a center earlier, in his August, 1940, manuscript [p. 306]: "It is the hope of this writer that eventually some sort of computational service can be provided to solve systems of equations accurately and at low cost for technical and research purposes.")

Keeping an open mind, however, Clayton asked: "What was the address of the man you wrote had developed a machine involving a new principle? I would like to know more about it."

Mauchly's Notes of August and September, 1941

As we explained earlier, two pages of Mauchly notes in file PX 665 postdated his Iowa visit. On one of these, dated August 3, 1941 (file p. 11), he again considered the analog calculator, as he had done on January 1, except that then he was thinking of adding frequencies and now he was thinking of adding phases or time delays. He made clear in his direct examination that these were "not necessarily all novel ideas," but were things he was "continually going over and considering" (transcript p. 11,858).

The other page, also dated August 3 (file p. 10), was Mauchly's earliest attempt to set down the requirements for a digital (presumably electronic) calculator. This hand-written sheet, headed "Some Requirements of '*ideal*' computing devices (keyboard type)," is significant not only as his first written effort, but as a step beyond the calculator he was considering in late 1940 "to obtain sums of squares and cross-products" (see chap. 2). That is, it envisioned an extension of the desk calculator to a more "general purpose" version of that calculator, a *multiregister* version which would include the "ability to transfer numbers from one register to another *at will* [italics added]."

It should be noted, however, that while this brief listing of the requisites of computing devices, in August of 1941, constituted an advance over his late 1940 plan for the next year's accomplishment, it also demonstrates that Mauchly, some eight months later, remained far from his goal of actually building a working machine. It was the barest outline of desiderata, with no drawings of circuits to execute the arithmetic functions he postulated. As Mauchly himself testified, in direct examination on this page, he was "just trying to set down what I thought were some very highly desirable characteristics for any computing machine that I would build" (transcript p. 11,858).

A third August 3, 1941, page, introduced in court as separate file PX 851, envisioned modifying circuits for recording coincident pulses that Mauchly had found in the literature. Its chief interest lies in its reference to an article in the July, 1938, *Review of Scientific Instruments*, "Circuits for the Control of Geiger-Mueller Counters and for Scaling and Recording Their Impulses," by (cosmic-ray physicist) T. H. Johnson, in which vacuum-tube circuits were depicted that incorporated *neon diodes* as integral elements.

The incipient, exploratory nature of Mauchly's thinking about digital electronic computers well after his visit to Atanasoff is borne out by yet another August, 1941, file, PX 847, even as the influence of Atanasoff on that thinking is also reflected. Two of its four pages were typed and appear to have been the beginning of an abandoned paper or proposal. This portion, titled "Notes on Electrical Calculating Devices," had an opening section, "Analog versus impulse types," which contrasted the obviously limited accuracy of analog devices with the theoretically unlimited accuracy of impulse or digital devices. It concluded that "it is in the field of impulse machines that major improvements in speed and accuracy are to be expected and sought for."

This first section acknowledged in a footnote: "I am indebted to Dr. J. V. Atanasoff of Iowa State College for the classification and terminology here explained." We referred to it earlier, with regard to a letter in which Mauchly asked Atanasoff if he considered "the usual d.analyzer an 'analogue' machine." We wish only to add now that Mauchly, as a result of his experience in Iowa and after one more fleeting consideration of an analog desk calculator (see above), seems at last to have come down on the side of digital devices. (The analog approach did die hard, for in September he twice more toyed with the possibility of using frequencies to perform addition and subtraction [PX 869].)

In a second section of this typed part of PX 847, Mauchly considered the economy of a digital desk calculator as against that of a larger machine. We quote this passage in full:

> For speedy (and noiseless) operation, vacuum tubes and associated circuits are the obvious answer. There are no essential difficulties in designing V.T. apparatus to do the job of ordinary mechanical calculators—but after taking care of stability, freedom from error, ease of servicing, etc., one might conceivably wind up with a design too costly to build. But economically feasible designs are possible. At present it may not be possible to build a commercial competitor for the desk-type mechanical computer, but larger machines for more involved, more lengthy, or more specialized jobs are practical. In some cases it is possible to materially descrease [sic] the number of tubes and circuits required in a large machine by having many similar operations performed by one and the same unit, which is switched around in sequence, rather than by having many duplicate units operating simultaneously. This is possible because the time required for a single elementary operation may be very short compared to the time allowable for completing one "step" in the calculation.

The influence of Atanasoff here seems as obvious as what Mauchly now finds "obvious": there would be "no essential difficulties" in replacing mechanical apparatus with electronic apparatus, and "larger machines" might be more "economically feasible." In cross-examination, he characterized as "prophetic" his thinking that it might prove more successful, as of that period, to build larger machines than to try to compete electronically with the desk machine (transcript p. 12,234).

The third page of this file PX 847 makes even clearer the Atanasoff influence, this time on the specific topic of how to design a multi-register electronic desk calculator! The two-page typewritten portion had had a third section outlining the "basic units of any computing device," namely, "input registers," an "adding (or add-subtract) mechanism," and an "output register," and explaining how these units functioned in the "ordinary mechanical computing machine." It concluded:

> In a flexible computing machine, it would be very desirable to have all registers in which numbers may be stored immediately usable as either [input or output registers]. One might imagine a computing control panel such as the following:

The third, hand-written page then depicted a keyboard panel that provided for putting numbers in (a machine) and combining them algebraically in a succession of arithmetic operations (as, for example, $xy + w - s/t$).

The tie-in with Atanasoff is the notation "JVA" in the upper left-hand corner of this page, together with the notation "Wilks" (presumably mathematical statistician Samuel S. Wilks), from which it seems fair to infer that Atanasoff had discussed with Mauchly ways in which his own vacuum-tube technology could be adapted to keyboard calculators of some complexity. Or perhaps Mauchly was referring to a section of Atanasoff's 1940 manuscript, "1. General Principles of Computing Machines" (pp. 309–10) that is a fuller account but has some striking similarities to Mauchly's.

Curiously, Mauchly was not examined by either side about this particular page of PX 847, but singled it out himself for comment (transcript p. 12,233):

> . . . You did not question [this] page, but . . . there's a diagram here which shows a layout for something which could almost be the present day electronic desk calculators that are being manufactured and sold all over the country now. That's a small, simple keyboard in which you enter numbers, enter operations, enter other numbers, and get results thrown into registers which you can then use again.
>
> This design here, you see, antedates the present desk calculators of the electronic type by some years.

No mention was made of his reference to "JVA."

Mauchly was cross-examined about the fourth and last page of this August file, which seems not to have been part of the more formal paper (transcript pp. 12,218–21). He testified that its three diagrams were circuits having "to do with counting," and that the neon bulbs which alternated with vacuum triodes functioned as "organic part[s]" of those circuits. It is noteworthy that Mauchly, while still trying to use neons to count, was also trying now to use vacuum tubes to count, rather than just to control or switch the neons as he had done in the early months of 1941. (This page, dated August 15, 1941, is not related to the August 3 file PX 851, in which Mauchly sought to modify circuits involving neons; for Johnson's use of a neon was to discharge a capacitor at a certain point in the counting.)

Mauchly also testified that he never got back to writing more of the research paper contained in PX 847, but claimed that it reflected only a small fraction of his thoughts on electronic digital computing at the time, perhaps "five per cent or 10 per cent of what I might have written had I got back to it" (transcript p. 12,232). Yet his next set of notes, begun on September 21, takes up more or less where this one left off, and is of the same tentative character as this and previous sets.

In that next file, PX 869, Mauchly explored condenser memory, various electronic decimal counters, a mechanical switch for subtraction by complement, a mechanical clock for timing papers at meetings, a cost estimate for some undesignated device that included neons and batteries, and, finally, layouts for a keyboard-operated calculator to do cross-products and squares. It is noteworthy that the primary focus of this file was still a desk machine of the type anticipated back in late 1940, *not* a digital electronic integraph. It is also noteworthy that its first page addressed a method of adapting "J.V.A.'s binary condenser system" to the "decimal system," in a way that was "like Stibitz's use of relays for decimal calculation" (binary-coded decimal representation), and ended with the question, "What sort of add-subtr. circuit can be devised?"

Under direct examination (transcript pp. 12,228–32), Mauchly testified that he might have created certain undated pages of PX 869 prior to the September 21 date entered on some of them. Two undated pages, block diagrams of the special-purpose desk calculator, were the sort of thing he had done at Ursinus College, he said, but also might "well have been [done] during the Moore School 1941–'43 period." He added that he could be certain only that they were "unrelated to the August '42 memorandum," but were "the kind of thing that I was hoping to do with small statistical [keyboard] machines." For our purpose, the essential point confirmed by this testimony is that Mauchly had made no progress at all toward the ENIAC as of September, 1941.

Mauchly's September 30, 1941, Letter to Atanasoff

With an opportunity now to study the Moore School's differential analyzer, in his new position as an instructor of electrical engineering, Mauchly began to think seriously of the possibility of building an electronic digital replacement for that mechanical analog device. And once again he turned to Atanasoff's suggestion for "converting" his own computer into an integraph, weighing the Atanasoff concept of logical add-subtract mechanisms coacting with a separate (regenerative) memory against that of counters to perform both the arithmetic and the memory functions in the manner of the desk calculator.

He wrote to Atanasoff on September 30—Atanasoff had not replied to his June 22 letter—as follows (PX 870):

> This is to let you know that I still have the same living quarters. but a different job. During the summer I looked around a bit while sounding out the Ursinus people as to promotions and assistance; I finally gave up the idea of taking an industrial job (or a navy job) and stayed in the ranks of teaching.
>
> The Moore School of Electrical Engineering is what I have joined up with, and they have me teaching circuit theory and measurements and machinery—but only 11 hours a week instead of the 33 that Ursinus had developed into.
>
> As time goes on, I expect to get a first-hand knowledge of the operation of the differential analyzer—I have already spent a bit of time watching the process of setting up and operating the thing—and with this background I hope I can outdo the analyzer electronically.
>
> A number of different ideas have come to me recently anent computing circuits—some of which are more or less hybrids, combining your methods with other things, and some of which are nothing like your machine. The question in my mind is this: is there any objection, from your point of view, to my building some sort of computer which incorporates some of the features of your machine? For the time being, of course, I shall be lucky to find

time and material to do more than merely make exploratory tests of some of my different ideas, with the hope of getting something very speedy, not too costly, etc.

Ultimately a second question might come up, of course, and that is, in the event that your present design were to hold the field against all challengers, and I got the Moore School interested in having something of the sort, would the way be open for us to build an "*Atanasoff Calculator*" (à la *Bush* analyzer) here [Mauchy's italics]?

I am occupying the office of Travis, the man who designed the analyzer here (duplicated at Aberdeen); I think I told you that he is now in the Navy, so I have no opportunity of benefiting by his rich experience.

I hope your defense efforts have been successful, but not so time-consuming as to stop progress on the computer. When you are East, arrange to see us. Perhaps you would like to look over the diff. analyzer, etc.

Convey my best regards to your family, and Clif Berry and all the gang.

Several points are clear from this letter: (1) Mauchly wanted to outdo the differential analyzer with an electronic digital version; (2) he was aware that for any electronic digital computer he might build, he would have to use some features of Atanasoff's computer (most basically, his electronic switching); (3) he was aware that the "counter" principle, on which he had been working before he learned of Atanasoff's "logical" principle, might not be feasible for either a relatively small desk calculator or the larger analyzer; (4) he wanted Atanasoff's permission to use some of his ideas along with counters if that principle proved the better choice; (5) he understood very well, from the explanations given him in Iowa, Atanasoff's suggestion for converting the ABC into an electronic digital integraph; (6) if he could persuade the Moore School to construct an electronic digital integraph, and if the version based on Atanasoff's machine "were to hold the field" against that based on counters (or any other), he wanted Atanasoff's permission to build such an "*Atanasoff Calculator*" at the Moore School; (7) he held the ABC (and the "progress" being made on it) in high regard.

Because this September 30, 1941, letter, like the June 28 letter to Clayton, figured strongly in Judge Larson's decision in the ENIAC case, we will quote directly from Mauchly's own defense of it at trial. Under examination by Sperry Rand, he could not "detail explicitly" the "different ideas" mentioned in this letter, but asserted once more that they concerned "the possibility of using entirely electronic circuits" for solving "any kind of a scientific or engineering job" (transcript pp. 11,840–41). He did say, however:

> . . . I was still, of course, very much interested in the problems of weather and statistical ways of dealing with them, but as evident just before this in the letter, I was closely looking at the operation [*sic*] analyzer and hoping that there would be a way of doing what the analyzer did with electronic computing circuits too.

As to what he meant in asking, "would the way be open for us to build an *'Atanasoff Calculator'* (à la *Bush* analyzer) here," Mauchly referred only to the "previous correspondence" where "Atanasoff had told me that he had some ideas for . . . building something which would give results similar to those obtained with the Bush analyzer . . . at MIT" (transcript pp. 11,841–42). Mauchly asserted that he "didn't know exactly what his ideas were," and that Atanasoff "had merely mentioned to me, I believe, in some letter that he thought" he had such ideas. He concluded, in what was becoming a typical roundabout (and gratuitous) fashion even in direct testimony (p. 11,842):

> . . . My contact with him had indicated that he was a pretty intelligent fellow who had a lot of good ideas and also that he wasn't always saying everything he knew; that if he didn't tell you exactly how he was going to do it, why, it didn't mean he didn't know. On the other hand, it didn't mean that he was going to tell you if he did know. So this was sort of an open proposition, you might say, that if you got something which you think might be useful to build perhaps the Moore School could be interested in doing this. In other words, I was looking for any kind of project which would assist calculation. It looked to me as if the Moore School was a good fertile field in which to get such things supported and, after all, they had built the differential analyzer.

The Honeywell side did not cross-examine Mauchly specifically on this letter, but confined itself to getting him to admit, as we saw earlier, that part of his purpose in going to Iowa was to learn what he could of the idea Atanasoff had suggested in his letters (transcript p. 12,189); that Atanasoff may well have told him in Iowa how he meant to adapt the ABC to do the job of the analyzer (transcript p. 12,203); and that the computing devices he himself had been thinking of building used a different principle and were intended for smaller computing jobs (transcript p. 12,204).

Atanasoff's October 7, 1941, Letter to Mauchly

Atanasoff replied on October 7 (PX 899):

> I am delighted to hear that you are teaching in the Department of Electrical Engineering at the University of Pennsylvania, and I will be sure to get in touch with you the next time I come east which should be in the very near future. At that time we can discuss our mutual interest in calculators.
> Our attorney has emphasized the need of being careful about the dissemination of information about our device until a patent application is filed. This should not require too long, and, of course, I have no qualms about having informed you about our device, but it does require that we refrain from making public any details for the time being. It is, as a matter of fact,

preventing me from making an invited address to the American Statistical Association.

We greatly enjoyed your visit . . . .

Neither Atanasoff nor Mauchly was examined on this letter, which makes clear, in writing, that Atanasoff had indeed explained his computer to Mauchly, that he intended to take out a patent on it, and that, by implication, he was *not* prepared to grant Mauchly free use of his ideas. This is the last of the correspondence between these two men that we will quote. (There was a further exchange concerning the possibility of an Atanasoff visit to Mauchly in Philadelphia; a final letter from Atanasoff dated October 30, 1941 [PX 924], expressed regret that this visit, too, had fallen through.)

Mauchly's Notes of October, 1941

File PX 903, the last Mauchly exhibit predating his August, 1942, proposal for an electronic computer, marks his earliest attempt to investigate on paper the problem toward which that computer was aimed, namely, the solution of sets of differential equations by the numerical integration of corresponding sets of difference equations. In a four-page note dated October 8, 1941, and several undated but obviously related pages of this file, Mauchly considered the incidence of error for such a method. He also estimated the time needed for the execution of 1,000 steps of integration at fifty seconds.

It is of interest that for this latter calculation Mauchly assumed a counting rate of 10,000 pulses per second, not the 100,000 pulses per second that he would assume in his memorandum. Clearly this and many other refinements had yet to be worked out, and just as clearly Mauchly profited in this regard from his Moore School association with Eckert, even as he had profited from and continued to profit from his earlier association with Atanasoff. He himself testified to his ongoing exchanges with Eckert that first summer, almost from the time of his arrival for the defense course (transcript pp. 11,808, 11,864–65, and 12,218), and to Eckert's contribution to the memorandum a year later (transcript pp. 11,875–77 and 12,234–35).

As we leave this last of the Mauchly 1941 notes on possible computing instruments, we should emphasize that none of those notes contained a circuit design of any promise for use in the ENIAC. In fact, Mauchly's proposal some ten months later included no electronic design, and neither did the final April, 1943, ENIAC proposal itself. Rather, the nontechnical portion of that formal proposal gave assurances that the current work on counters in the United States would be helpful, in a general way, as specific design was undertaken; and the technical portion discussed, very briefly, some simple circuits known at that time (DX 3,361). Moreover, as we observed in chapter 2, it was not Mauchly, but Eckert, who ultimately designed the decimal counter used in the ENIAC.

Further Developments

Although Atanasoff and Mauchly ceased to correspond after October, 1941, they were to have further personal contact at the Naval Ordnance Laboratory (NOL) in Washington, D.C., where Atanasoff began working in 1942. It is unclear from their testimony exactly when Mauchly first appeared at Atanasoff's desk in the NOL, whether in early 1943 as Atanasoff thought or somewhat later as Mauchly seemed to think (transcript pp. 2,228–30 and 2,754–56, and pp. 11,912–13 and 12,282–83, respectively). They did agree, however, that Atanasoff helped Mauchly secure employment as a consultant within the division of acoustical mine testing, of which he was chief, and that Mauchly was there, for a day or so a week, at least from September of 1944 until early 1946. (Mauchly's teaching load at the Moore School had been all but eliminated in the spring of 1944, and he needed to supplement his greatly reduced income.)

Atanasoff and Mauchly also agreed that although Mauchly was hired to consult in statistics at the NOL, he later became involved to some degree in a short-lived computer project headed by Atanasoff.

It does seem very probable to us, from the vividness of Atanasoff's recollection (transcript pp. 2,228–30; deposition pp. 855–65) and the lack of countertestimony by Mauchly, that Mauchly did pay Atanasoff several informal visits prior to September, 1944; that he learned in that period that Atanasoff was no longer working in computing; and that he informed Atanasoff of the ENIAC project, but gave no details. In any event, they agreed on one particular visit to Atanasoff's office at the NOL in late August of 1944, one in which Eckert was also present.

Mauchly and Eckert were in Washington on August 30, 1944—Mauchly's thirty-seventh birthday, as it happened—to consult with Army Ordnance attorneys about both the ENIAC and the EDVAC patents. They stopped in at Atanasoff's office afterward to ask his help with the mercury-delay-line memory they were developing for the EDVAC (Electronic Discrete Variable Computer). Indeed, Mauchly, Eckert, and Atanasoff all agreed in court on the precise purpose of their visit, namely, to benefit from Atanasoff's expertise in piezoelectric crystals (transcript pp. 12,281–82, 17,629–30, and 2,233, respectively).

Pertinent to our inquiry at this stage is a note, PX 2,409, written by Mauchly shortly after that August 30, 1944, trip to Washington, a note in which he specifically addressed his impression of Atanasoff's computer upon seeing it in 1941. Titled "Situation as of September 10, 1944," it was a lengthy assessment of Mauchly's employment situation at that time and his prospects for the future, including possible full-time employment at the NOL when the ENIAC project was completed. We confine ourselves here to the portion on the ABC.

As read aloud in the Sperry examination (p. 11,916), Mauchly had written:

> My present anxiety as to the "next move" can be explained by giving some history.
> While still at Ursinus College, I began thinking about electronic means

of rapid computation. I wanted rapidity and flexibility of electrical circuits to do more complex calculations with less error and greater speed than can be obtained by use of ordinary commercial mechanical computing machines. Just before coming to the Moore School, in 1941, I visited Atanasoff, in Ames, Iowa, to see his electronic calculator and talk with him about the general problem. I thought his machine very ingenious, but since it was in part mechanical (involving rotating commutators for switching) it was not by any means what I had in mind. Eckert remembers that when I first came to the Moore School, I had computing machines as my favorite topic of conversation. . . .

Mauchly was to testify later that he "did not normally keep a diary," but sometimes wrote things out "in times of stress or perplexity" in order to get a perspective from which to deal with them (transcript p. 11,997). Later still, Sperry attorney DeLone took this September, 1944, note to be an instance of that intermittent practice (transcript p. 12,492). This part of it, however, appears to us to be more a rewriting of history, for the record, than a statement of facts on which to get a perspective. Mauchly's portrayal of his reaction to seeing Atanasoff's machine, some three years after the event, is clearly at odds with his own notes and letters at the time, as we have seen.

But the timing of this note is even more suspicious. Mauchly and Eckert were actively pursuing a patent on the ENIAC, and Mauchly, at least, was aware of Atanasoff's contribution. They were also beginning to consider a patent on the EDVAC, with its regenerative memory and its serial adder, and again Mauchly, at least, was aware of Atanasoff's contribution. Did it seem necessary to acknowledge for the record some contact with Atanasoff and to discount his influence? In any case, Mauchly's doing so in terms of "[mechanically] rotating commutators for switching" did effectively deny Atanasoff's most fundamental contribution to the ENIAC, electronic switching circuits, at the same time that it avoided any recognition of a memory of the sort to be used in the EDVAC.

The visit by both Mauchly and Eckert to Atanasoff on the very day they had consulted the Army Ordnance patent attorneys is puzzling and, in our view, reprehensible in light of later events. They went, after all, to seek advice on *their* regenerative memory from the man who had originated the concept of such a memory and passed it on to one of them—indeed, from the man who had given that one a job as well—and then proceeded to patent *his* concept as their own! We return to this patent issue in chapter 4.

Mauchly's Testimony in an Earlier Suit

Mauchly was deposed for the *Sperry Rand v. Control Data Corp.* suit in October of 1967, some four years prior to his appearance in the ENIAC trial from which we have been quoting. As we have indicated, Sperry, in this earlier suit filed in 1964, charged infringement of the Regenerative Memory patent granted to Eckert and Mauchly in 1953, and CDC countercharged derivation from Atanasoff. The case was in fact pursued in court for just a few days in 1972 and was finally settled in 1981, upon the insistence of the court, through an undisclosed agreement between the two parties. Oddly enough, a judgment *was* rendered on this patent, by Judge Larson in his ENIAC trial decision; he found derivation from Atanasoff, among other "infirmities."

We present this finding in chapter 4. Our interest here is in Mauchly's stance in this prior case involving Atanasoff, both as it reveals new information and as it contrasts with his stance in the ENIAC case. As with the depositions of Lura Atanasoff and Sam Legvold in this same memory case, Mauchly had been subpoenaed at the request of defendant CDC, even though he was strongly identified with plaintiff Sperry Rand. And again, CDC was represented by Allen Kirkpatrick, Sperry by Laurence B. Dodds. Dodds, however, did not choose to cross-examine the witness, so that the testimony was limited to Kirkpatrick's direct examination, with Dodds interceding from time to time (Regenerative Memory Trial Records).

Kirkpatrick began by reading into the record the request for documents as it appeared in the subpoena under which Mauchly was being deposed. This included relevant "writings, photographs, letters, telegrams and records of any form" in Mauchly's possession or control that had "come into existence prior to October 31, 1947," the date of the patent application. He then proceeded to take Mauchly's educational, professional, and business history, concentrating, of course, on the development of the mercury-delay-line memory of the EDVAC and later machines, the patent that covered this and other forms of regenerative memory, and the various contacts with Atanasoff. In the process, Kirkpatrick introduced many documents that he had secured elsewhere, including nearly all of the Mauchly-Atanasoff correspondence, none of which Mauchly had produced (although he evidently had it).

This Mauchly deposition is over 300 pages long. We will cite only those portions relevant to the present inquiry, and will in fact offer little comment, since the reader is already familiar with the story as elicited in the ENIAC trial. We begin with some items in the history prior to late 1940 when Mauchly first met Atanasoff.

The Pre-Atanasoff Period

Mauchly affirmed Kirkpatrick's suggestion that he had become "fairly familiar with vacuum tube circuits" as of the end of 1940, but added that he was "by no means an expert" on such circuits, having pursued them "as another extracurricular study." He cited two semester courses he had taken, and amateur radio and other magazines he had read. He also said that at Ursinus College he had "installed various pieces of experimental equipment and laboratory procedures to acquaint the students with electronic measuring techniques." (Pp. 25–27)

Mauchly testified that in the Ursinus period of 1935 or 1936 he "began experimenting with electronic circuits which might aid the calculating process," as applied to the analyses he was then doing on standard desk calculators (p. 48). He later stated this interest more precisely: first, in terms of increased speed; second, in terms of enhanced storage facilities that would make both data and sequences of operations continually available once they had been entered (pp. 59–61). At the ENIAC trial, of course, Mauchly claimed no such concepts for the mid-1930s.

As to what, specifically, he did about this interest in electronic calculation, Mauchly told of his experimentation with neon lamps, "to see if there were any ingenious ways in which I could throw together at low cost something that I could use myself." He explained that he first thought of these bulbs "as visual output devices for a computer." But later, he said, he saw them as entities "to store information" and "began constructing small experimental devices utilizing both neon bulbs and vacuum tubes to see what I could make with such equipment." Mauchly placed this work "in the years like 1935, '36, '37." (Pp. 62–63)

As we have seen, Mauchly was unable at the ENIAC trial to document the inclusion of vacuum tubes in neon computing designs until 1941, after he had met Atanasoff and discussed his small two-neon circuit with him. And he claimed only to have designed that circuit a year or two earlier.

As for devices actually constructed in the pre-Atanasoff period, Mauchly mentioned only his cryptographic device and his harmonic analyzer (pp. 63–68). He did claim to have "built various kinds of ring counters, [and] pulse generators to activate them," but his only documentation of these was notes he had taken on journal articles (pp. 69–70).

Asked, much later, to state whether, prior to his visit to Atanasoff in Ames, Iowa, he had "ever seen or heard of or thought of a digital computer that would operate in base 2," he mentioned reading about the binary scaling circuits in use in the 1930s. He also noted "that the relay computers designed by George Stibitz are said to have a biquinary system," but did not say at what point he learned this. He then explained that he himself had built "a ring counter which had five elements" which he *could have used* in conjunction with a binary counter "to achieve some economy . . . in the decimal representation of [numbers]"; again, he was unable to produce any documentation. (Pp. 255–57)

Mauchly did not so much as mention, here or elsewhere, the two-neon device that he presented at the ENIAC trial as a binary counter or, alternatively, a relaxation oscillator (toy railroad signal).

The 1940 Meeting

Mauchly testified that he first met Atanasoff at a Philadelphia meeting of the AAAS in December of 1939. This mistake in the year—it had been 1940—was to persist throughout the entire deposition, as Mauchly claimed to have let a year and a half go by before finally visiting Atanasoff, remarked on how little progress Atanasoff had made on his machine in that length of time, and tried to explain his sudden rash of correspondence with Atanasoff in early 1941 as arrangements for his pending visit.

He said of that initial meeting that he had apparently indicated in his talk that he "had been devising electronic and electrical computing devices," and Atanasoff introduced himself afterwards, saying that "he, too, was interested in electronic computation and had been working on a digital computing device." Mauchly said they talked for a short while, but Atanasoff was "somewhat reluctant" to be specific and instead invited Mauchly to come to Ames to see and discuss his machine. (Pp. 84–85)

The Iowa Visit

Mauchly testified of his June, 1941, experience in Iowa that Atanasoff continued "not too free" with his disclosures concerning "the digital computing device which he had told me about in 1939" because of a fear "that certain large corporations might try to take advantage" of him. He was, however, "permitted to see the device" and "told something about [its] principles . . . and how they hoped it would operate when completed." (Pp. 88–89)

In the course of his extended testimony on this visit, Mauchly emphasized the incomplete state of the computer; claimed not to have seen any tests or demonstrations of it, or any evidence that anyone was even working on it; estimated his total time with the machine at no more than an hour and a half; and could not recall reading the 1940 manuscript or taking notes on this or anything else to which he may have been exposed (pp. 100, 140–44, 247–49, et al.).

Nevertheless, he gave two lengthy discourses on its features that together amounted to a truly scholarly overview (pp. 89–94 and 141–44). This included a physical and functional description of the major components and their interaction: input via a decimal card-reader; the main storage as charges on capacitors within a revolving cylinder; communication via metal brushes and wires to the "electronic circuitry"; a rotational speed of "once per second, or possibly some multiple of this"; a "synchronous motor from the 60-cycle supply line"; "electronic circuitry" that both "regenerate[d] the charges on the condensers" and "perform[ed] the necessary arithmetic operations . . . as a kind of binary adder"; and, finally, a "novel" binary output device that used a high voltage spark "to rupture the dielectric paper" and create "a minute pinhole in the card," coupled with an input device that used a "smaller voltage spark" to detect those pinholes.

Mauchly also supplied some rather surprising details, given the limited exposure he was claiming: one terminal of each capacitor was "connected in common with all others to a slip ring insulated from the main frame of the machine"; there were two wire brushes for each ring of a storage cylinder, "so that a charge sensed by the first brush as the cylinder rotated could then be amplified and restored to the same capacitor an instant later, as that capacitor passed the second brush"; and there were special "timing relationships involved in reading the [binary] cards and getting the information on the drum storage," that is, "the card movement had to be synchronized with the rotation of the drum, as I understood the operation."

How much thought Mauchly had given to this presentation and how much he was "carried away" in the process is a matter for speculation. It is the case, however, that he revealed a very clear overall understanding of the computer and a detailed acquaintance with certain of its features, at the same time that he claimed to have been denied access to certain other features. Thus he was open about details of the regenerative memory—and would be throughout this examination—but was guarded about details of the add-subtract mechanisms except as these were involved in the regeneration process. We can only note that in the ENIAC trial he did recall being given demonstrations, and did acknowledge learning enough of the logic and electronics of the add-subtract mechanisms to be satisfied that they worked as intended.

Mauchly also commented, with respect to his Iowa visit, that the name of Clifford Berry had arisen as the man "who had apparently brought [the computer] to the state in which I then viewed it" (p. 94). He even went so far as to volunteer that he really did not know "what part of the Iowa design or plans had been [Atanasoff's] and what part had been Clifford Berry's" (p. 184). He could not recall whether he had met Berry, however, and only realized that he must have when shown his own letters to Atanasoff, one in which he inquired of Berry's future, another in which he sent his regards to "Clif Berry and all the gang" (p. 234).

Mauchly was, of course, questioned at great length on what he learned of Atanasoff's regenerative memory while in Iowa (pp. 96–105, 131–46, 190, and 242–59). As we have said, he acknowledged fully his understanding of it and of its interaction with the add-subtract (and restore-shift) mechanisms. His quarrel was, instead, with whether he had gotten this "mere idea" of regenerative storage from Atanasoff in the first place, whether Atanasoff's was the first regenerative memory, and whether Atanasoff had "exercised diligence" in pursuing his rights! We shall return to these issues when we discuss the post-Iowa period.

Correspondence

Mauchly thought he must have written to Atanasoff to "set up my visit with him," but did not "have the letter at hand"; he could recall no correspondence with him after the visit (pp. 178–79). Subsequently, Kirkpatrick introduced nearly all of their 1941 letters (apparently obtained from Atanasoff as originals or carbons), beginning with Mauchly's of February 24. Of this letter, which closed with his hope "that you have made progress and that I'll get out to see it all," Mauchly said it was "exactly what I was testifying about, in that I met him in 1939 . . . [and] he extended an invitation to me, which I did not accept immediately, nor for some time, and this letter . . . was my correspondence which led to my visit" (p. 222).

As Kirkpatrick introduced letter after letter, Mauchly continued to treat them as preliminaries to his Iowa visit, expressing surprise, for example, at the "extent [to which] the correspondence went in arranging this visit." But he also, increasingly, seized upon items in them for indirect denial; he claimed, for example, that his own request for more information showed "how reluctant the group at Iowa were to make disclosures" (rather than how eager *he* was to learn more). (P. 226)

He continued to stress the "1939" theme, stating of Atanasoff's May 21 letter to him that "it confirms that I had not seen Dr. Atanasoff from the time I first met him until I went to Iowa for the visit in 1941" (p. 228). Kirkpatrick never did reveal his error to him, and, of course, Dodds was unable to—he had only Mauchly's word on the date of the AAAS meeting. Dodds knew nothing, either, of these letters introduced by Kirkpatrick. Mauchly, when asked about his search for them, did not deny that he might have them, or that he understood they were included in the subpoena. He simply said that he had not looked through a folder of "miscellaneous correspondence, where I would only have one letter or two [from an individual]," as against folders tabbed with one name (p. 223). He apparently did produce them for the ENIAC trial.

Mauchly was asked to explain certain specific references or statements in his letters. Concerning his expression " '*Atanasoff Calculator,*' (à la *Bush* analyzer)," in his September 30 letter, he claimed that "again, [Atanasoff] was communicating generalities rather than details with respect to any of these ideas." He went on to suggest, however, that this was perhaps the original concept of "a device known as the Madida, which was a digital integrator" (built on the West Coast in the early 1950s), and to allow that Atanasoff "may well have a claim to [this concept] as embodied in the Madida at this early date." (Pp. 236–37)

It is difficult to see the relevance of a machine built over a decade later to the issue of Mauchly's request to use Atanasoff's ideas. To us, this testimony seems a diversionary tactic, in the guise of a generous gesture, designed to conceal the fact of Mauchly's own serious consideration in 1941 of Atanasoff's basic "logical" approach to electronic calculation.

The Post-Iowa Period

Mauchly several times made clear that he had, as a matter of course, discussed Atanasoff's computer with people at the Moore School, especially with Eckert (pp. 94, 145, 165, 220, et al.). But when asked of any interest in building such a machine there, he cited the ABC's rotating memory as the chief obstacle, claiming a particular concern for its reliability. He said, for example, that "Eckert and I discussed what ideas we had in relationship to possible calculators in the light of what I had seen out there, and the most important thing resulting from those discussions was I believe the fact that we felt that relying upon mechanical commutation . . . in what was otherwise an electronic machine, was a poor way to start out" (p. 94).

As we noted earlier, Mauchly did always acknowledge that Atanasoff's rotating drum memory was a regenerative memory. Asked, for example, whether he had come away from Ames "recognizing that the group there [under Atanasoff] was proposing a regenerative memory for a digital computer," he answered, "Yes" (pp. 145–46).

When pressed to admit that he had gotten the idea for such a memory from this manifestation of it, however, he refused. It was hard to recall, he said, just when "ideas of recirculation" had first come to his notice, or how he himself "might have thought of them without having come to my notice from someone else at all." (Pp. 161–62)

Mauchly now explained that Kirkpatrick had introduced an idea that was new to him, namely, that of separating "the regenerative aspects of a storage system from whether they are all electronic or not." He and Eckert had been "considering practical and economic means for getting electronic regenerative storage," he said, whereas Kirkpatrick was addressing the "mere idea of regeneration of storage," which did not seem to him "the central idea." It was "an interesting question," though, and he "would like to explore it" if he had the time. (Pp. 169–71)

Mauchly's argument here is flawed on several, progressive scores. First, Atanasoff's *was* the original regenerative memory. Second, in all probability it marked Mauchly's first exposure to the concept—he had absolutely no evidence to the contrary. Third, one is not entitled to appropriate the idea of a prior inventor simply because one's application of it is more advanced. But, fourth and most overwhelmingly, Mauchly and Eckert in their patent actually *claimed* the invention not only of their own version of regenerative memory, the mercury-delay-line register, but also of other versions, including Atanasoff's! Indeed, they included such an electrostatic rotating store both in their descriptive exposition, as one "embodiment" (Regenerative Memory patent 2,629,827, cols. 1–2), and in their formal claims (cols. 36–37).

But the matter did not come to rest there. Kirkpatrick next asked whether the BINAC's "acoustic delay line storage device," as Mauchly had termed it, was a form of regenerative memory within his definition, and he replied: "Yes. Well, I have been pointing out that your definition of regenerative memory includes a lot more than mine does, apparently, and that I had originally taken this to mean a regenerative memory which depended on electronic circuits and the acoustic or electrical delay line, but had no mechanical commutation or other such features which we consider to be less reliable" (p. 172).

We believe Mauchly responded in this way because he feared that Kirkpatrick was about to challenge whether the delay-line memories were any more electronic than Atanasoff's: they too relied on mechanical motion, the vibration of the mercury; their "electronics" was, like his, the regenerating circuitry, not the memory device itself. Their advantage was that they could operate at *electronic speed,* and so keep up with that circuitry. Thus was Mauchly induced to change his own definition of "regenerative memory" to exclude Atanasoff's electrostatic memory, after he had repeatedly and forthrightly stated that it *was* regenerative.

A highlight of this whole dispute was a portion of an article written by Eckert in 1953, just after their Regenerative Memory patent was granted, which revealed that he too understood the regenerative character of Atanasoff's rotating memory, as conveyed to him by Mauchly (Eckert 1953). The pertinent paragraph, read into the record by Kirkpatrick, is worth quoting in full (p. 1,394):

> Probably the first example of what might generally be termed regenerative memory was developed earlier than 1942 by Atanasoff in Iowa. He used

a drum with many capacitors mounted on it, and with a commutative method of connecting the capacitors with brushes for reading-in, reading-out and regenerating. The principal feature of the capacitor memory was the use of cheap, reusable elements. The capacitors could be charged, discharged and recharged as often as desired, and they were common and readily available components. The characteristics of small initial expense and almost unlimited reusability have guided the development of memory systems since, and remain two important criteria of the merit of any system at any time. There may have been similar systems prior to Atanasoff's but none was as inexpensive to construct. Unfortunately his development was interrupted by the war and never completed.

Kirkpatrick questioned whether, in the discussions Mauchly had had with Eckert, there was "any suggestion, really, that there had been any earlier regenerative system" (p. 166). Mauchly did not respond directly, but said later, "If we go looking for ideas of regenerative memories in mechanical or other fluid or other systems, and perhaps even part electronic systems, I say I have the feeling we will find some" (p. 170). Kirkpatrick closed this topic with a warning to Mauchly that his own view, and the view of his client CDC, was that Mauchly *had* "first learned of regenerative memory from Dr. Atanasoff" (p. 171). He asked to be informed "if you find any earlier documentation on it"; none was ever brought forth.

We should point out, in passing, that at the ENIAC trial Mauchly had "no direct recall of having told [Eckert]" of Atanasoff's computer, although he conceded that "it's natural that I might have mentioned this" (transcript p. 12,213). It is strange that he should have recalled so well, after a period of thirty-six years, but have completely forgotten after another four—in the course of which he was continually reviewing this very subject matter.

Kirkpatrick's examination of Mauchly also occasioned an unexpected admission of untoward ethical attitudes of the sort we saw in his ENIAC testimony. This was with regard to his contacts with Atanasoff in the Naval Ordnance Laboratory during which he observed that Atanasoff was no longer interested in computing. As he put it, "He himself at that time apparently felt committed to administrative duties, and so far as I could tell had no interest in the design of computers" (pp. 181–82). He learned, too, he said, that Atanasoff's computer had never been completed, and so felt, "rightly or wrongly," that he "was signing an application [for the Regenerative Memory patent] which had nothing to do with any machine on which a patent would ever be applied for" (p. 190).

This was Mauchly's first indication that he *felt safe* in patenting concepts of the ABC, after satisfying himself that Atanasoff had lost interest in pursuing them with the Patent Office. But, of course, inventors seeking patents have a duty to inform the Patent Office of any exposure to prior art, regardless of whether or not the prior inventors have pursued their rights.

Mauchly later took up this line again, this time noting a lack of diligence on Atanasoff's part: ". . . there is something called diligence . . . in patent law, and if it turned out that Atanasoff's machine was never completed and no diligence . . . was exercised in applying for patents on it, this seems to me a matter which the Patent Office has under its jurisdiction." Asked then whether the Patent Office was ever informed of his discussions with Atanasoff, Mauchly said he did not know, but again shifted the burden to Atanasoff: "If he had proceeded, of course, why, they would have had an application concerning" his plans. (Pp. 198–99)

It is the case that Atanasoff and his attorneys ceased at some point to pursue a patent on the ABC, and that the Patent Office thus had no record of its invention with which to compare the Eckert-Mauchly application. But, here again, a lack of diligence by a prior inventor does not void an applicant's duty to report exposure to prior art; nor does it clear the way for him to patent that inventor's ideas.

Mauchly's testimony on this topic ended with the only instance in any of the records we have read in which an attorney actually appeared to lose his temper. Kirkpatrick had been exploring certain circumstances of a meeting of the Association for Computing Machinery in Washington in August of 1967—a meeting in which Mauchly was asked at the last minute to discuss his visit to Iowa in 1941. Kirkpatrick inquired whether an attempt had been made to have Atanasoff himself address this meeting, and Mauchly indicated that he knew of none. He and Kirkpatrick then had the following exchange (pp. 277–78):

> *Mauchly:* But of course, as I testified earlier, I was rather amazed at the fact that once he became embedded in the administrative duties at Naval Ordnance Laboratory he did not actually seem to want to talk about his machine or to have much to do with the technical features of computers, and he has never as to my knowledge ever appeared at any Association for Computing Machinery meetings or any of the related society meetings relating to computers, so I thought that this was a case where he was sort of out of the field, but I recognize, of course, that that does not—should not prejudice exposure of his ideas.
>
> *Kirkpatrick:* In other words, if a scientist describes something and is run over by a truck the next day, that is no reason for the rest of the world to move in and appropriate what they might have learned from him; you would agree with that?
>
> *Mr. Dodds:* I object to that question. I think the witness should not answer it.
>
> *K:* You are not going to answer?
>
> *M:* It is an irrelevant question, as far as I am concerned.
>
> *K:* Well, we will see what the Court thinks about that.

And so it was that Mauchly, once more, volunteered both that he had been observing Atanasoff's interest, or lack thereof, in computers, and that he accordingly felt safe in ignoring Atanasoff's contributions to his (and Eckert's) enterprises.

We should not leave this topic without noting that Atanasoff did not "become embedded" in administrative duties at the NOL. He was, rather, embedded in his work in acoustic mine testing, work that later saw him participating in the atomic tests at Bikini Atoll, from June 1 to July 25, 1946, and monitoring the blowing up of a German depository of ammunition on the island of Helgoland in the spring of 1947 (see chap. 4). Nor should we overlook Atanasoff's own report that he had expressed interest in the ENIAC, but that Mauchly had declined to discuss it (Atanasoff 1984, p. 256). Mauchly had told him of this new machine designed by himself and Eckert, claiming it represented "a new way to compute," but had refused details because it was "classified." Atanasoff learned sometime later, he wrote, that Mauchly could have arranged to tell him about the ENIAC, had he wished to; and he certainly could have.

Odd Items

Half a dozen other minor items in this Mauchly deposition are of interest to us. We present them with brief parenthetical comment.

Item

Mauchly refused to admit that Atanasoff's computer could perform the shifting operation, on the following argument: "Let's distinguish again between his plans for the machine and the machine which I saw, which was inoperative, and according to the article which you have introduced here by Mr. Eckert was never completed; consequently, if the machine was never completed it did not have this ability" (p. 178). (Of course, it was Mauchly who had informed Eckert of the ABC.)

Item

Mauchly recalled that he and Atanasoff had discussed the low cost of condensers, "in other words, the economics of the job," but observed that "this of course had nothing to do with its novel features" (p. 96). (During his ENIAC trial testimony, Mauchly stoutly refused to separate economy from novelty. He would not admit that a machine which could solve large systems of linear algebraic equations electronically "would represent a considerable advance in the computing art," as Halladay had put it, because, he said, "without some consideration of cost, utility, availability, whatever you want to call it, there is no advance really at all" [transcript pp. 12,145–47]. This from a man who had nearly half a million dollars to spend on the ENIAC, as against Atanasoff's seven thousand for the ABC!)

Item

Mauchly commented on his stay in Atanasoff's home in 1941 that his son and Atanasoff's children "took care of each other during that period"; he did recall that there was a Mrs. Atanasoff there (p. 106).

Item

He was unable to recall ever consulting Atanasoff on any aspect of computers while he was at the Moore School. Asked if his answer was that he had "no
present recollection," he replied: "Well, more than that, I feel that this recollection
is strong, and as strong as anyone could have, that I did not ask him for any help"
(pp. 184–85). (At the ENIAC trial, of course, Mauchly did clearly recall how he
and Eckert had consulted Atanasoff on the subject of piezoelectric crystals for their
mercury-delay-line memory.)

Item

At one point in the proceedings, Kirkpatrick introduced a letter written by
Clifford Berry to R. K. Richards on March 22, 1963. It included the following
paragraph (pp. 216–17):

An interesting sidelight is that in 1940 or 1941 we had a visit from Dr.
John Mauchly who spent a week learning all of the details of our computer
and the philosophy of its design. He was the only person outside of the
Research Corporation and the patent counsel who was given this opportunity, and he may still have notes of what he learned from us.

To all of Kirkpatrick's questions on this passage, Mauchly responded that,
because Berry was unsure as to the year, he must be taken to be no more accurate
than Mauchly in his recollection of events in Ames. He further asserted that here
was an example of the way in which Atanasoff and Berry had withheld information
from him: ". . . they never disclosed to me that the Research Corporation that is
named in this paragraph was involved in this" (pp. 219–20). (Aside from the fact
that this is such a poor example of withholding information, we would note that
Atanasoff *had* told Mauchly, in his January 23 letter, that he was seeking help
"from an outside source.")

Item

Lastly, Kirkpatrick became so familiar with Mauchly's style of testifying—
a way of putting the slant *he* wanted on his answers—that he actually anticipated
him in one instance. He had been showing Mauchly photographs of Atanasoff's
computer, asking in each case if it refreshed his memory as to what he had seen
in Iowa—photographs, as it happened, that had been taken for the 1940 descriptive manuscript. Mauchly had been somewhat resistant, particularly regarding
the add-subtract mechanisms, in his replies. There followed this exchange (p.
247):

Kirkpatrick: Is your memory refreshed to the extent that while you were there in Ames with Dr. Atanasoff concerning his computer, that you were given a written description to read?

Mauchly: I don't remember being given such a written description, but on the other hand, I cannot be sure that I was not given one to read. I can say very definitely I was not given one to—

K: To take home?

M: To take home.

The Mauchly-Eckert Link

We have established that Mauchly got certain ideas from Atanasoff, and that he and Eckert patented those ideas jointly. An intriguing question is the degree to which Eckert was aware of their origin in a third party. That is, how much did Mauchly tell Eckert of his experience with Atanasoff? We have only the testimony of the two men themselves to go on, but in combination it is quite enlightening.

As we already saw, Mauchly, in his ENIAC courtroom testimony, had "no direct recall" of having told Eckert of Atanasoff's computer, whereas he had, in his Regenerative Memory deposition, freely admitted discussing the ABC with Eckert and weighing their own ideas against "what I had seen out there." In fact, he said that he discussed Atanasoff's work with others at the Moore School, as well. (It would seem that Mauchly violated Atanasoff's confidence from the very beginning with such talk.)

Eckert, in direct examination at the ENIAC trial, downplayed any such discussions. When shown the paragraph in his 1953 Institute of Radio Engineers *Proceedings* article where he credited Atanasoff with "probably the first example of . . . a regenerative memory," he did acknowledge hearing of that memory from Mauchly; but he claimed he "received less information than this in our earlier [1941–42] discussions," and "went back and talked to him about it just prior to writing this paper" (transcript p. 17,493).

Of course, he was pressed much harder concerning those earlier discussions in cross-examination. There he recalled learning of a capacitor memory with a "rotating switch," and also of Atanasoff's idea for "burning holes in punch cards"; but he seemed not to recall learning of the computer's use of vacuum tubes (transcript p. 17,611). We quote the final exchange on this topic between Eckert and Halladay (pp. 17,612–14):

> *Eckert:* . . . I knew that voltages were stored on these capacitors in an off-and-on or digital fashion for the purpose of storing information, and I knew that this cooperated with some other equipment which I don't think we discussed in any detail, for the purpose of presumably reading these numbers and using them in other ways. And I asked him how Dr.

Atanasoff planned to put information in this machine, and he described a device for burning holes in cards.

Halladay: Well, you did understand, did you not, . . . that Dr. Mauchly had had a machine shown to him which had rotational features and other arrangements from which calculations were made through the use of vacuum tubes; did you not?

E: I do not recall at this point John mentioning that he actually saw a machine or that he actually saw a machine with vacuum tubes or that he actually saw anything rotate. I only remember these ideas from the conversations that I had with him.

H: And then, of course,—

E: I remember no more detail. I cannot deny that he said these things; I cannot confirm it. I only remember these details from that conversation.

H: And so I don't suppose you would deny either that Dr. Mauchly told you that the Atanasoff device had modular units which were pluggable?

E: I have no recollection of his telling me this. I can't deny it.

H: And that Dr. Atanasoff considered that he had come up with a numerical calculating machine in which [certain] mechanical features . . . were combined with electronic devices to produce a speedier and simpler machine, or words to that effect?

E: I remember no such language.

H: You wouldn't deny that things of that character were said to you, would you, by Dr. Mauchly apropos of what he had seen in Ames, Iowa, in June of 1941?

E: It is my impression that we have now discussed this matter longer in this cross-examination than I discussed it with Mauchly at the time. And this is all I can remember.

H: Well, that may be. It takes longer sometimes on cross-examination to get a simple yes answer to a question, too.

E: I am not trying to be rude about it. I am just saying it was a very short conversation in which I learned a very few facts. And I am just trying to get that point across.

Now the recollections of both Eckert and Mauchly at the ENIAC trial are clearly at odds with Mauchly's recollections in deposition for the Regenerative Memory case. But Eckert's testimony is also most peculiar in that he recalled the capacitor aspect of Atanasoff's memory, which he and Mauchly rejected because of its mechanical (rotating) commutator, yet seemed not to recall—indeed, not even to be curious about—its *electronic* regeneration. Mauchly would surely have told him that vacuum tubes in the arithmetic unit refreshed the capacitors of the memory. It strains credulity to think that he did not tell him of their arithmetic functions as well, and that Eckert, an electronics expert, had no interest in this true highlight of the computer: its ability to *compute electronically*. After all, the whole point of their talks, as Eckert himself stressed, was to design an electronic computer. And they did go ahead and use Atanasoff's switching technology in the ENIAC, which they patented as entirely their own. And they did patent "their own" serial binary adder and regenerative memory.

The issue of what Eckert learned of Atanasoff's work from Mauchly is academic so far as patentability is concerned. A jointly held patent can be invalidated if only one of the inventors has used the ideas of a prior inventor. In this case, however, we are inclined to believe that Mauchly talked freely to Eckert and that Eckert knew they were using Atanasoff's ideas. We think they both were too impressed with how far beyond the basic principles of the ABC their own devices went, *and* they felt safe in appropriating those basic concepts because Atanasoff was showing no interest. We think, further, that they were greedy, for fame and fortune, and did not want to acknowledge any prior inventor.

An Interpretation

We have made the point several times that Eckert and Mauchly used Atanasoff's electronic switching technology for the ENIAC—more precisely, they developed a variant of it—but that they were to use his regenerative memory principle (in their delay-line memory) and his logic principle (in their serial adder) only in the EDVAC. The question naturally arises as to why they opted for the principle of counters for the ENIAC's basic arithmetic operations, rather than Atanasoff's principle of logic.

The answer lies in the fact that the new computer was meant from the start to be a high-speed electronic *replacement for the differential analyzer* and that the ENIAC was accordingly patterned on that mechanical device. As against Atanasoff's approach to the ABC through a relatively large though slow separate memory, Eckert and Mauchly thought in terms of high-speed counters—and accumulators based on these counters—that would incorporate sufficient memory along with their arithmetic functions. This is why they made so much, at trial and elsewhere, of the inadequacy of Atanasoff's slow capacitor drum memory for their purpose.

After another year or so, however, Eckert invented a large high-speed separate memory using acoustic delay lines together with Atanasoff's regeneration principle. He and Mauchly then saw, as Atanasoff had seen earlier, that logical arithmetic circuits were superior to counters for use in conjunction with such a memory. At that point they did appropriate these two basic ideas of a regenerative memory and computing by logic, *without* acknowledging their source.

Thus the ENIAC was conceived as an electronic digital "difference" analyzer, with the mechanical device's basic unit, the integrator, replaced by an accumulator made up of counters in the manner of a desk calculator. It was not long then before they realized that this machine would surpass the analyzer in the range of problems it could solve as well as in speed, and they began to call it a general-purpose computer.

Mauchly, of course, had been trying to design electronic counters for half a year before he visited Atanasoff, with the idea of using them in a single desk calculator, so that his mind-set was already toward counters. In fact, he must have been truly amazed to find that Atanasoff did not use counters at all, but had developed a successful add-subtract mechanism based on logic! We know from his letters that he continued to consider counters, as compared to Atanasoff's logical approach, as he prepared to enter the summer defense course at the Moore School. And there he found Eckert, who assured him that electronic counters could be made to compute reliably at 100,000 pulses per second, and who could himself design them to do so.

One question remains, then, and that is the source of Mauchly's predilection to counters. They would quite naturally occur to anyone thinking of an electronic computer, both because the desk calculator was the obvious model and because electronic scaling circuits had already been developed. At the ENIAC trial, however, the work of another man was brought out, work with which it seems inescapable that Mauchly and Eckert were familiar. That man was Irven Travis, of the Moore School.

The Travis Studies

Travis consulted for the General Electric Company (GE) throughout 1939 and 1940, with a view to making recommendations for an automatic computing facility. He spent 1939 investigating company needs and considering alternatives to the mechanical differential analyzer; and, with those alternatives declined, he spent 1940 advising GE on the design and construction of a modified version of the Moore School analyzer.

It was Travis who had been in charge of building the differential analyzers of both the Moore School and the Aberdeen Proving Ground in 1934, modeling them on the original version built at MIT in 1931. (He had earned his master's degree at the school in 1928, and was pursuing his doctorate.) As he explained at the ENIAC trial, where he appeared as a witness for plaintiff Honeywell (transcript p. 6,534):

> . . . I visited Vannevar Bush at MIT when he built the first one and borrowed his chief draftsman, as a matter of fact, and entered into a design incorporating all the changes that he would have done if he was going to do it over again; and on completion of the machine, I carried on a number of computations. As a matter of fact, my Doctor's thesis [1938] had to do with work that was closely related to the differential analyzer.

He went on to say that his thesis work on "the behavior of iron cored reactors in systems involving capacitors" was "of fundamental practical importance to the General Electric Company"; that the work he and Cornelius Weygandt did along with GE engineers on the Moore School analyzer solved a problem they had with "series capacitors . . . in big power systems," and, in his opinion, "saved the

GE capacitor business." This contact led to Travis's consulting for GE in 1939 and 1940. (Transcript pp. 6,534–35)

Travis wrote two reports for GE. The first, dated August 30, 1939, and titled "On the Possibility of Developing an Electronic Differential Analyser," set forth an analog differential analyzer with electronic components. He explained this report (PX 201.3 for trial purposes) as follows (transcript p. 6,538):

> . . . [The device I suggested] used electrical angle to represent quantity, this having the advantage [that] angle could be boundless as opposed to potentials or currents or other electrical quantities which are necessarily bounded and, therefore, [of] limited accuracy.

This report did envision the use of electronic *counters,* with capacitors receiving the counts, for the purpose of counting cycles and dividing frequencies (pp. 5–6); and also the possible use of a "counter-type [electronic] integrator" (pp. 9 and 15). The overall operation, however, paralleled that of the analog differential analyzer.

On the other hand, the second report, dated March 28, 1940, and titled "Automatic Numerical Solution of Differential Equations," set forth an entirely digital substitution for the differential analyzer. This document (PX 363.5 in the ENIAC trial) described a machine comprised of a series of "ganged" or "chained" calculators, as such an arrangement came to be called, to solve differential equations numerically. It began (p. 1):

> A differential equation solver designed to use numerical integration rather than the means commonly employed in differential analyzers would have the advantage of using standard adding machines which can be built at low cost as well as the advantage of great precision.

It then stated:

> There is some doubt as to whether such a device could successfully compete with a mechanical differential analyzer as regards solution time, but with the use of some type of electronic adding machine the solution time might be made very short.

The body of the report described a device in which adding machines replaced the integrators of the standard differential analyzer; included were block diagrams of interconnected components, tables of integration of difference equations, examples of solutions, and comparisons with the analog machine. The conclusion was drawn that it "appears feasible to solve differential equations by an automatic step-by-step method using adding-machine-type calculators" (p. 13). Four "component devices" were then stipulated and "described functionally," namely, "accumulators, multipliers, adders, and an interconnecting system or register transfer device" (pp. 13–15). These functional definitions are worth quoting in full, for their strong parallelism to Mauchly's 1942 memorandum and even to the 1943 ENIAC proposal:

a. An accumulator simply accumulates totals when the increments are supplied to the keyboard. An ordinary adding machine does this.

b. A multiplier must automatically multiply the number set up on the keyboard by the number indicated as the multiplier. This could probably be done by matching the counter register of a machine of the Friden type against a pre-set multiplier register; the multiplying process in each decade being terminated when the registers match in that decade. After multiplying and transferring the product to the next computing device the multiplier must clear.

c. An adder differs from an accumulator in that it must total a certain prescribed number of quantities supplied by other component devices and then clear in preparation for the next operation of summing. The total is transferred to the appropriate registers immediately after each sum is produced.

d. An interconnection system must be available for transferring register readings from one computer to another in accordance with a scheme such as that indicated in Fig. 6. It seems probable that such a system would be complicated and expensive.

Travis explained in court that figure 6, "Diagram for the Solution of the Equations," was a "functional schematic" for the solution of the difference equations corresponding to the "equation of the square wave oscillator," or the "Van der Pol equation" (transcript pp. 6,549–51). It depicted six accumulators, five multipliers, and one adder, together with the formula calculated by each of these components and arrows indicating their interactions.

Thus Travis had anticipated Mauchly's 1942 plan to gang calculators for the numerical solution of differential equations. Moreover, he had anticipated the basic ideas expounded by Eckert and Mauchly in the 1943 ENIAC proposal and had provided a very valuable schematic diagram of such a computer (their fig. 1 is a similar block diagram for the solution of the much simpler difference equation of a sine wave).

Travis's report concluded with two practical considerations regarding a machine made up of *mechanical* calculators, its solution time and its durability (p. 16). He estimated the solution time at perhaps five to fifteen times that of the differential analyzer, but again suggested that "electronic adding machines might well swing the advantage far in the opposite direction." As for durability, he estimated that a computer based on mechanical calculators could be expected to last "not more than four years." This wear-and-tear factor has generally been cited as *the* reason Travis (and GE) decided against the ganged calculator concept. In our opinion, however, a life of four years might not have been prohibitive, given that the individual calculators could be bought for less than five hundred dollars apiece.

It does seem that the time factor alone was sufficient to exclude the ganged calculator concept in the mechanical version, and that GE would have been unwilling to risk the electronic version, with its untried technology and uncertain cost, when the tried-and-true mechanical analyzer was available for a relatively fixed figure. In his report Travis cited only the solution time as a possible disadvantage (see opening paragraph). But at trial he himself recalled the durability issue as a disadvantage, going on to say, however, that he had suggested that "some type of electronic adding machine or counter might solve that life problem" (transcript pp. 6,539).

We should note that Travis testified that he was "generally aware of the possibility of [electronic] counting devices" when he wrote these reports, because GE was developing such instruments for measuring bullet velocities at Aberdeen (transcript pp. 6,540–41).

Curiously enough, Atanasoff out in Iowa had devised a plan to solve systems of linear algebraic equations with ganged desk calculators in the summer of 1937; he had a Washington law firm conduct a search for prior art at that time (transcript pp. 2,383–88). We present this matter further in chapter 4.

The 1940 Moore School Proposal

These two studies for GE led to a proposal by the Moore School, in the fall of 1940, to the National Defense Research Council, for the development of electronic (antiaircraft) fire-control equipment. Travis explained in court that a group at the school (including J. G. Brainerd, Knox McIlwain, S. Reid Warren, C. N. Weygandt, D. B. O'Neill, and himself) met several times to discuss "possibilities based to a fair extent upon the ideas that are in the two GE reports." He said that the thinking had been to develop electronic devices that "would allow mass production with more unskilled manufacturing capabilities and possibly substantially lower weight" than the current devices afforded (transcript pp. 6,544–45).

He stated further (pp. 6,545–46):

> . . . And we felt that if we could develop an integrator, either analog or digital integrator, we were in business in fire control because with an integrator one can couple two together and solve a harmonic equation and get sines and cosines, and by two integrators one can multiply, and if one has sines and cosines in the correct location one has the elements of a fire control system. So our original proposal was to concentrate on the development of an electronic integrator . . . .

The notion of an electronic integrator, adding machine, or accumulator had appeared in Travis's GE reports. His testimony here that the thinking in the Moore School proposal was that two integrators could be interconnected so as to produce sines and cosines should be compared with figure 1 of the 1943 ENIAC proposal (a setup for producing sines and cosines) and also with the actual building in 1944 of a two-accumulator model (a "small ENIAC") to generate sines and cosines (Burks and Burks 1981, p. 343 and fig. 9).

Travis said this proposal to the NDRC "was reviewed by Dr. Warren Weaver and after a considerable period of time it failed of approval and the project was not ever financed" (p. 6,545).

The Travis-ENIAC Link

Now we believe that Mauchly learned of these two reports, and of this proposal, during his visits to the Moore School at the time, and that he appropriated Travis's ideas for his own ends: first, for his initial plan to build a single electronic desk calculator (expressed in late 1940, in letters to two friends, and pursued sporadically at least through September of 1941); second, for his own proposal to build an electronic computer based on interconnected electronic calculators (written in August, 1942); and, third, for the ensuing ENIAC proposal (written in April, 1943).

We believe further that Eckert also had direct acquaintance with this prewar work of Travis, and that he too was influenced by it as he assisted Mauchly with the memorandum and as he and Mauchly jointly wrote the ENIAC proposal. Let us look now at the Travis link to each of them.

Travis-Mauchly

We have already mentioned Travis several times in connection with Mauchly. We presented Mauchly's testimony concerning Travis's 1938 "Bibliography of Literature on Calculating Machines," where questions were raised as to possible influence on his conception of a harmonic analyzer; we noted in this connection that Mauchly had sought to take a course (in computing devices) from Travis in 1939, only to have it canceled (chap. 2). We also presented two letters to Atanasoff in which Mauchly spoke of Travis (chap. 3). Mauchly himself did not say when he first met Travis, but did testify that he had taken a course at the Moore School as early as 1936 (transcript p. 11,804); he said he could not remember whether he had seen the analyzer before June, 1941 (transcript p. 12,179).

Most of these matters came up during direct examination of Travis by the Honeywell side in the ENIAC trial. He testified that he had first met Mauchly in 1937 or 1938 (transcript p. 6,537):

. . . I would guess it was 1938. He was at that time an Assistant Professor of Physics at Ursinus University. He was interested, as I was, in automatic computation and he used to come down to my office at the Moore School from time to time and we had numerous chats on this subject of common interest.

Travis said that these visits by Mauchly lasted until "late spring of '41," when he was called into active duty by the Navy. He could not recall Mauchly's telling of his weather research at Ursinus, although he thought it quite possible they had discussed harmonic analyzers since this was one of his own interests. Nor could he recall whether he had talked with Mauchly about his GE study of the ganged calculator concept for solving differential equations numerically; but, again, he thought he might have.

Of the 1938 bibliography, Travis expressed his amazement "at the attention this thing gathered," with "inquiries from all over the world for copies"; he said that "copies were distributed to anybody who wanted them" (p. 6,543).

Attorney Halladay read into the record the third paragraph of Mauchly's May 27, 1941, letter to Atanasoff. We reproduce this here:

> From your letters I have gathered that your national defense work is unconnected with the computing machine. This puzzles me, for as I understand it, rapid computation devices are involved in N.D. [National Defense]. In a recent talk with Travis, of the E.E. School at U. of Pa. I asked him about this, and the matter seemed the same way to him. But if Caldwell has looked over your plans (I think you said that he was out there) and hasn't seen any N.D. possibilities, I suppose that that means your computer is not considered adaptable to fire control devices, or that they have something even better. Travis (who goes into active duty with Navy this week) pointed out the advantages of lightness and mass-production for electronic computing methods, but said that when he was consulting with General Electric over plans for the G.E. differential integraph they figured it would take about one-half million dollars to do the job electronically, and they would only spend 1/5 of that, so they built the mechanical type with polaroid torque-amplifiers.

Travis was not questioned at this point about this passage, but we would note that it confirms his recollection of conversations with Mauchly right up until his departure for the Navy, at the same time that it reveals the depth and detail of those conversations. It also clarifies an issue that was puzzling when we presented this letter earlier; that is, whereas we wondered then whether Mauchly knew of *Atanasoff's* fire-control project, we can now see that it was mainly *Travis's* connections to fire-control work that were reflected in Mauchly's comments and questions. (Travis also worked on fire control for the Navy throughout the war, first as an inspector of facilities, then as chief of antiaircraft fire control research in the Bureau of Ordnance; he attained the rank of commander.)

Most significantly of all, this Mauchly letter contains information from both Travis's consulting work for GE on the differential analyzer and the Moore School proposal for an electronic integrator to permit lighter and cheaper fire-control devices. Such information conveyed to Mauchly does not show that he actually saw the documents involved, but it indicates that he had some familiarity with their contents.

Travis's estimate of "one-half million dollars," which so startled Atanasoff, also bears out our belief that GE never seriously considered a ganged calculator scheme based on *mechanical* adding machines, but was swayed by the fact that an *electronic* integraph, whether analog or digital, would have been too expensive. For Travis's expression "to do the job electronically" surely encompasses both the analog electronic integraph and the digital integraph based on electronic adding machines that he set forth in his two reports. As it happens, this cost estimate was highly realistic: the partially electronic (analog) analyzer built by MIT, under Bush and Caldwell, cost about $500,000, and the digital ENIAC, based on the ganged (electronic) calculator concept, cost nearly that much. GE opted for a $100,000 mechanical analyzer that was an improvement over the original.

(Travis said that the polaroid torque amplifier was invented by a GE engineer, T. E. Berry, and was also incorporated in the Moore School machine in place of the "wrapping hand type" [pp. 6,556–57].)

The other Mauchly letter, of September 30, 1941, simply had him occupying Travis's office at the Moore School and regretting that he would "have no opportunity of benefiting by his rich experience." We submit that he had already benefited and would continue to benefit profoundly from that rich experience.

There was further Travis testimony, in the ENIAC trial, on his role as director of research at the Moore School after his return from the Navy in February, 1946—just in time for the formal ENIAC dedication. He said that, in the course of his work as a "contracting officer" for the Navy, he had come to admire the manner in which other university research laboratories were conducted, and that he wanted to put the Moore School "in the same good shape." He felt it wrong that under the current policy, "we agreed to supply information on the patentable projects and to assign to the Government certain rights to patents, and yet our employees had no obligation to the University." (Pp. 6,571–72)

Accordingly, with the university's support, Travis "drew up a patent agreement, to be signed by the employees who were working on sponsored projects, which was . . . patterned . . . after copies of agreements" used in those other institutions (p. 6,573). Then, on March 22, 1946, Dean Harold Pender gave Eckert and Mauchly an ultimatum either to sign this agreement *and,* in effect, affirm their first loyalty to the University's interests, or to resign; both did resign by the five o'clock deadline imposed in the dean's letter (pp. 6,576–81).

One further brief item at trial, during the redirect examination of Travis by Halladay, shows that the Honeywell side did strongly suspect a Travis influence on the ENIAC. Halladay asked, "Would you say that the ENIAC ended up being, in effect, the GE differential integraph that you had figured in 1941 would cost half a million dollars to do the job electronically?" Sperry attorney DeLone objected to this question as leading, but Travis responded, "I don't think I could honestly give you a conjecture on that," and Judge Larson let his answer stand. (P. 6,653)

(Travis left the Moore School in 1949 to join the Burroughs Corporation as director of research; he became vice-president for research and engineering in 1956 and held that post until his retirement in 1969.)

Now, Mauchly was not asked in court whether he had learned of Travis's work for GE in 1939 and 1940, or of the fall, 1940, Moore School proposal to devise an electronic integrator for fire-control applications. In his Regenerative Memory deposition, however, he alluded to the former in a long historical recitation of his early Moore School period (p. 76):

> It turned out that the man who would teach the course on design of electronic computers as offered in the catalog was Dr. Irven Travis, but at the very time that I was joining the staff he was leaving for active duty with the Navy.
>
> I was told that he had already investigated the subject of large computers for integration purposes, and although I did not see any reports he had written the impression given by staff members was that little could be done with digital equipment which would surpass what was already being done on the differential analyzer, an analog device.

It would seem that he had hardly met Travis, much less that he had been talking computing with him for over three years—he had even tried to take that same course from him two years earlier. It would seem that he had had to learn from "staff members" of Travis's investigations into integraphs, and that none of them had mentioned possible *electronic* devices. Finally, it would seem that it was only natural for Mauchly to volunteer, in this testimony, that he "did not see any reports [Travis] had written," and then to go on to what *others* at the school had said. The facts of the case suggest, however, that all of this was a maneuver to shift attention from any inkling that Mauchly *had* gotten to know Travis and his work.

Our own opinion regarding a Travis-Mauchly link is that Mauchly *was* familiar with those prewar investigations by Travis, and that he made use of them as circumstances allowed. We offer the following arguments.

First, the timing was right: Mauchly had known Travis for perhaps a year before Travis began the consulting task for GE, so that their rapport was well developed; after Travis had written his two reports, and just after a "group" comprising most of the Moore School faculty had proposed a project grounded in these reports, Mauchly made his own determination to build an electronic desk calculator; after the Moore School had become deeply involved in a wartime project to solve differential equations on its analyzer, Mauchly wrote a proposal for an electronic computer based on Travis's model of ganged calculators; finally, after Goldstine, of the Army, showed an interest, Mauchly and Eckert wrote the ENIAC proposal incorporating ideas described, illustrated, and diagramed in the Travis reports.

Second, the atmosphere, or milieu, of the Moore School was right. This was an institution oriented toward research and development, where the faculty consulted for industry and where discussions were free and open. (Mauchly, by contrast, was teaching at a small college where he was the only member of the physics department, had no record of support for his research, could not secure a research job despite a doctorate from Johns Hopkins, and was notably unforthcoming with ideas of his own.) Moreover, Travis's projects were a very live topic in 1939 and especially in 1940. Finally, among the Moore School faculty, Travis was one of the freest communicators. How else could a stranger from Ursinus College have had all those talks with him?

Third, Mauchly's style fits our interpretation of this link to Travis. As he himself testified, he devoted much of his energy to learning what others were doing in his field of interest, by meeting them and reading their works; in his words, he "wasn't even thinking about inventing computing machines" in June of 1941. Mauchly had a curious mind, a critical mind, an informed mind, but he did not have an inventive mind. At trial, he could not cite one creative effort of his own of any consequence, except his harmonic analyzer, and even there he had to admit to using the ideas of others; indeed, one has to wonder whether it was not *Travis* who had contributed that idea and steered him to the literature via his bibliography.

Fourth and last, Mauchly's manner under examination fits our interpretation. We have seen him, in both the ENIAC and the Regenerative Memory cases, work in his own "messages" in his longer discourses, messages that appear to be prepackaged attempts to divert. We have also seen him burst forth with the truth, albeit very damaging to his own cause, when placed under strain; it was as though he were struggling within himself against what was being proven and had to say, "But of course I did that! There was nothing wrong with that!" We believe, with Honeywell, that he *was* dissembling in court, and that he *had* done things that were wrong.

Travis-Eckert

A Travis-Eckert link, with respect to Travis's prewar work, was made possible by the fact that Eckert was his student; Eckert actually graduated that June of 1941 as Travis was leaving and Mauchly was arriving. He was queried during direct examination in the ENIAC trial as to discussions he and Travis had had about computing, and he replied that these occurred "prior to his departure into the Services." He said he had been "very much impressed" with Travis as a teacher and had "spent a lot of time going into his office when I was still an undergraduate . . . and talking to him." He had even helped Travis "pack up" in early June. (Transcript pp. 17,262–63)

Asked specifically by Sperry attorney Ferrill about their subjects of discussion, Eckert recalled "two ideas that came up," but did not recall "whether he discussed them with me or . . . showed me a piece of paper with them on." These were, in fact, the two ideas on which Travis had reported to GE. The first Eckert described as "an extension of the electric principle to using frequency modulation as the variable in order to achieve greater accuracy"; he commented that he himself had thought about this and had concluded that "the speed of the device would be interfered with badly by the use of such a procedure." (Pp. 17,263–64)

Of the second, he said (p. 17,264):

At some point—and I can't remember now whether he conveyed this idea when I was graduated or whether he conveyed this several years later when I met him again in the Navy—but at some point there was a mention of the possible interconnection of a number of mechanical desk calculators by an electro-mechanical means. And I remember the comment that a study of this idea showed that the machines would fall apart in a fraction of a year with the kind of use they would then receive.

Eckert did not mention the possibility of using electronic desk machines; indeed, he explicitly specified "mechanical," with "electro-mechanical" interconnections. Unlike Mauchly, he did acknowledge learning of Travis's study directly from him, and also possibly seeing "a piece of paper," but he was vague as to the time. In our own opinion, Travis would not have told Eckert, who was already something of an electronics expert, of this ganged calculator study without mentioning his idea for using electronic calculators. (Travis, of course, had estimated a life of four years for the mechanical version, not a falling apart in a fraction of a year.) And it is hard to imagine his explaining this GE study to Eckert during his later visits to the school; at trial he could not even recall seeing the ENIAC as it was being built (p. 6,562).

Eckert then testified that in arriving at the machine proposed by Mauchly in 1942, they had considered and rejected several ways of introducing electronics into the mechanical differential analyzer. He mentioned the possibility of replacing the integrators with electronic devices for counting streams of pulses, and, further, the possibility of controlling and transmitting these pulses electronically (pp. 17, 269–70). As he pointed out, however, a machine based on such counting integrators could not be both as fast and as accurate as the one Mauchly proposed. This is because a device capable of counting only, not of adding and accumulating, would take too long to achieve adequate accuracy. Again, Eckert's mentioning and dismissing this idea, as he did Travis's idea of replacing the analyzer's integrators with mechanical calculators, does not alter the fact that Travis also suggested *electronic* calculators and that what Mauchly proposed *was* just such a replacement—and *was* the basic plan behind the ENIAC.

It is our considered judgment, then, that Eckert and Mauchly were both influenced, in their conception of the ENIAC, by the prior work of Travis, but that they failed to acknowledge his incipient, though basic, contribution, just as they failed to acknowledge Atanasoff's more substantial contributions. There is, in fact, a striking parallel here: like Atanasoff, Travis showed no concern, gave no inkling of suspicion that his early GE work was drawn upon for the ENIAC. And so, in our opinion, Mauchly and Eckert were not about to credit him, either.

The ENIAC Proposal Reference

The reader may recall that earlier in this chapter we noted a reference in the ENIAC proposal of April, 1943, to two previous cases in which electronic difference analyzers had been considered. We expressed our belief then that one of these two cases was Atanasoff's work and ideas, as revealed to Mauchly; in particular, it was his idea for converting his computer into an electronic digital integraph, a project he hoped to return to when he found time. We can now say that we believe the other case was the work of Travis, as revealed to Mauchly, to Eckert, and of course to Brainerd, the "administrative" author of this proposal who had been a member of the Travis "group" at the Moore School; in particular, it was his idea for ganging electronic desk calculators to solve differential equations.

The ENIAC proposal, however, neither identified those two previous cases nor acknowledged any connection between them and the machine it presented; on the contrary, the thrust of mentioning them was to rule out such a connection. There was a curious bit of testimony, however, that bore on one of the cases, both as to what it was and as to its influence. This came from Herman H. Goldstine, the Army man who oversaw the building of the ENIAC from the military's perspective.

Goldstine's Testimony

In his deposition for the ENIAC case, Goldstine was asked about the reference in the 1943 proposal to two previous "electronic difference analyzers." He said that he had discussed this with Mauchly and Brainerd, and that Mauchly had told him of an "electronic scheme for computation" conveyed to him by "a man by the name of Atanasoff" (p. 872). He then explained (p. 873):

> Generally I recall that these were conversations to indicate the general feasibility of doing computation digitally by electronic means since uppermost in my mind at that point was whether or not this made any sense at all.
> There were various discussions to indicate that the proposals weren't without precedent.

> [Mauchly] told me about—my best recollection is that he told me that Atanasoff had designed such a system and had—if I recall correctly, had even built some parts.
> Whether he told me that he saw these parts I don't remember.

Thus it was that Mauchly felt pressed to mention his acquaintance with Atanasoff's work in order to persuade the U.S. Army that his own (and Eckert's) proposal for an electronic computer was feasible. When confronted with Goldstine's statement at trial, Mauchly said that he could not recall telling him about "Dr. Atanasoff's extensions . . . of thought into the realm of differential analyzers," but that he "could have done that," and he had "no reason to contradict Dr. Goldstine's testimony" (transcript p. 12,280).

With regard to this reference to earlier cases in the ENIAC proposal and Goldstine's recollection that one was to Atanasoff's work, we might point out that

Eckert, too, had to know the subjects of the reference: he was one of its authors! And so he, too, *had* to know of Atanasoff's electronic achievements.

We should note that Goldstine did not mention Travis as the other party to whom the ENIAC proposal had reference; instead, he cited the partially electronic analyzer just completed at MIT. Of course, he was mistaken, since that machine was analog. Moreover, it had already been described in the proposal, with a comment that it was not a *difference,* but a *differential* analyzer. There is no way of knowing whether Brainerd did tell Goldstine of Travis's work, but it would seem that he had to inform him of *two* cases. In contrast to Atanasoff, however, Travis had not designed and tested any components, so that his work could not have proven feasibility. On that basis, it would not be strange for Goldstine to have forgotten what was, in our opinion, the second case.

In Conclusion

We would like to close our presentation of the Atanasoff-ENIAC connection with a note on the *gradualism* of Mauchly's progress from his November, 1940, "plan" to build an electronic desk calculator—probably derived from Travis—to the April, 1943, Eckert-Mauchly plan for the general-purpose ENIAC. Mauchly repeatedly claimed that he rejected Atanasoff's computer as a "model" because it was special purpose (as well as partly mechanical). Yet his own movement toward a general-purpose computer did not begin until he had seriously considered Atanasoff's and, in all probability, Travis's ideas for an electronic differential analyzer. Up to then, after all, he had been interested only in expediting his own weather data analysis.

Mauchly did recognize, in August of 1941, *after* seeing Atanasoff's machine and *after* having many discussions with Eckert, that a larger, more general calculator—but still a desktop machine—would probably be a more practical commercial competitor. But his goal, even then, was not to solve *differential equations* electronically; not until October, 1941, did he begin to attack that possibility on paper. And he wrote his first, very spare, memorandum on an electronic differential analyzer, based on the ganged calculator concept, only in August, 1942. Moreover, the formal ENIAC proposal that grew out of that memorandum some eight months later was itself quite sketchy.

We find Mauchly's claim that he was seeking a *general-purpose* electronic computer in 1941, whether before or very soon after he visited Atanasoff, preposterous.

CHAPTER 4

Atanasoff's Day in Court

The ENIAC Case

The lawsuit that has become known as the ENIAC case arose from two actions filed on the same day, May 26, 1967, in different courts. One was a suit by Honeywell, Inc., against Sperry Rand Corporation, filed in the U.S. District Court for Minnesota in Minneapolis. The other was a suit by Illinois Scientific Developments, a wholly owned subsidiary of Sperry Rand, against Honeywell; it was filed in the U.S. District Court for the District of Columbia. On March 5, 1968, Judge John J. Sirica of the Washington court transferred the latter suit to the Minnesota court, where, on May 1, Judge Gunnar H. Nordbye consolidated the two into one. Honeywell was designated plaintiff, Sperry Rand (SR) and Illinois Scientific Developments (ISD), defendants. Judge Earl R. Larson heard this case, with few interruptions, from June 1, 1971, through March 13, 1972, and issued his 420-page Findings of Fact, Conclusions of Law and Order for Judgment on October 19, 1973 (ENIAC Trial Records).

It is a historical quirk that news of the decision on the invention of the electronic computer received very little media attention because it came at the height of the Watergate scandal. October 19, 1973, was just nine days after the resignation of Vice-President Agnew, and just one day before the so-called Saturday night massacre that saw the departures from the Nixon administration of Archibald Cox, Elliot Richardson, and William D. Ruckelshaus. It is a further quirk that the Watergate case was heard by the judge who had elected to send the ISD case against Honeywell to the Minnesota Court. (One might surmise that Judge "Hanging John" Sirica would have reached the same conclusion as did Judge Larson, had he tried the combined case.)

Both suits as originally filed involved only ENIAC patent 3,120,606, which had been applied for on June 26, 1947, and issued on February 4, 1964. Honeywell charged SR with violation of Section 2 of the Sherman Antitrust Act, for monopolizing the electronic data processing (EDP) industry by fraudulently procuring and enforcing an invalid patent (Count 1); and sought a declaratory judgment of invalidity and unenforceability (Count 2) for this alleged antitrust conduct. ISD, in its suit, charged Honeywell with infringement of a valid patent. Both sides sought injunctive relief and damages.

On May 1, 1968, the day the two suits were consolidated, Honeywell filed its

First Amended Complaint. This consisted in an expansion of Count 1 to include a charge of violation of Section 1 of the Sherman Act, for conspiracy with IBM to monopolize the EDP industry; and a new count charging violation of Section 7 of the Clayton Antitrust Act (Count 3), for monopolizing the industry by fraudulently acquiring an invalid patent application. Judge Larson dismissed this last count in the interim between the presentations of plaintiff's and defendants' cases. Since the judge held to that decision in his Findings, we can simplify matters here by also "dismissing" it.

(The judge ruled that Honeywell had failed to show that the acquisition of the Eckert-Mauchly application by SR's predecessor, Remington Rand, had tended to create a monopoly; he also ruled that Section 7 of the Clayton Act had applied only to stock, not to assets, in early 1950 when Remington Rand made that acquisition, and, in fact, applied only to acquisition from corporations, not from individuals.)

Then, on August 29, 1969, Honeywell filed a Second Amended Complaint, introducing the issue of an additional Sperry Rand portfolio of patents and patent applications that SR sought to sell or trade with other corporations, and citing twenty-five of these as especially relevant. This latter collection, nearly all in the names of Eckert and Mauchly alone—or of Eckert alone—became known by the paragraph number, 30A, under which it was inserted into Count 1 of the Honeywell complaint. Honeywell's charge here was that the 30A package was subject to the same infirmities as the ENIAC patent; that is, that SR's procurement and enforcement of these twenty-five patents or applications formed part of a consistent pattern of conduct in violation of the Sherman Act.

The 30A package included the Eckert-Mauchly Regenerative Memory patent (EM-1) and three Serial Binary Adder patents (EM-21, EM-23, and EM-25), as well as patents on the BINAC (EM-22) and UNIVAC (EM-39), both of which incorporated this memory and some form of serial binary adder. We return to these patents in a later section on derivation from Atanasoff.

ISD, in its countercharge, accused Honeywell of infringement of the ENIAC patent on seventeen representative claims (8, 9, 36, 52, 55, 56, 57, 65, 69, 75, 78, 83, 86, 88, 109, 122, and 142). SR and ISD insisted throughout, in fact, that their count of infringement was the basic case being tried, because, as they maintained, the ENIAC patent was valid and they had engaged in no improper activity regarding it or their portfolio of other patents and patent applications.

The Background, in Brief

The two cases arose in the first instance from attempts by Sperry Rand to collect royalties on the ENIAC during the decade or so before its patent was issued. Sperry Rand (later just Sperry and now part of Unisys) was formed in 1955 by the merger of Sperry Gyroscope with Remington Rand, which had in 1950 acquired the Eckert-Mauchly Computer Corporation and, that same year, rights to the ENIAC and other Eckert-Mauchly patent applications. The newly formed Sperry Rand Corporation entered into a cross-licensing agreement with IBM in 1956, whereby the two companies would share all of their current "know-how" in the tabulating and the electronic data processing fields. IBM agreed, as well, to pay SR $10 million on the ground of the greater value of its ENIAC application and its large portfolio of EDP items. Additional royalties were to be paid if and when the ENIAC patent issued; these were later settled at $1.1 million.

Once the ENIAC patent became a reality in early 1964, SR formed ISD, for the sole purpose of assigning that subsidiary the right to the ENIAC so that it could engage in the sale of that right alone. This maneuver, among others, enabled SR and IBM to continue to keep secret (illegally) the depth and scope of their joint cross-licensing agreement of 1956. ISD almost immediately began demanding royalties on the ENIAC from the larger corporations in the EDP industry, namely, General Electric, Burroughs, RCA, National Cash Register, Control Data Corporation, Philco-Ford, and Honeywell. And it claimed that the ENIAC patent, of and by itself, covered *all* the EDP equipment those firms might produce for the seventeen-year life of the patent.

Honeywell then insisted upon including SR, with its attractive EDP patent portfolio, in their negotiating sessions. The royalties SR (and ISD) asked, however, were many times higher than the $11.1 million SR had accepted from IBM. From Honeywell, SR first asked $250 million! Later, it came down to $20 million, but Honeywell found that, too, out of proportion to the IBM figure and unreasonable in view of the comparative values of their respective portfolios. (SR demanded a total of nearly $150 million from the other six companies; none of them ever made any agreement with SR or ISD.)

Thus it was that on May 26, 1967, ISD sued Honeywell for infringement of the ENIAC patent and Honeywell sued SR for violation of antitrust laws—this Honeywell suit later extended to include a charge of similar conduct for SR's entire EDP portfolio. Had SR chosen to ask a royalty fee more in line with that paid by IBM, quite possibly there would have been no court action. The ENIAC patent would then have remained unchallenged, and Atanasoff's contribution to electronic computing undiscovered! For it was only to resist the excessive SR and ISD demands that Honeywell and others delved into the pre-ENIAC history to the extent they did.

We should note, in passing, that SR did enter into a complete cross-license agreement with Western Electric Company, as it had with IBM, *before* the granting of the ENIAC patent. This was signed in 1961.

The Decision

Judge Larson's rulings on the two undismissed counts of Honeywell against Sperry Rand in their consolidated case were as follows. He found against Honeywell on Section 2 of the Sherman Act (part of Count 1), declaring that SR had not successfully created a monopoly of the EDP industry, although coconspirator IBM had.

He found for Honeywell on Section 1 of the Sherman Act (part of Count 1), declaring that SR had conspired with IBM to monopolize the industry, and also that

Honeywell had suffered injury and probable damage; more precisely, he declared that Honeywell had suffered because of the 1956 agreement between the two companies, but *not* because of the existence of the ENIAC patent. On the other hand, he found that Honeywell had failed to act with diligence to protect its interests in the EDP market, and had itself stalled in an effort to improve its own portfolio for licensing purposes. Accordingly, he awarded no damages.

He found for Honeywell on the issue of validity and enforceability of the ENIAC patent (all of Count 2), declaring that the patent was both invalid and unenforceable.

Finally, he found for Honeywell with respect to the 30A package (part of Count 1). He found infirmities in these twenty-five patents and patent applications similar to those of the ENIAC patent. Not having been asked to rule on their validity, however, he found only that they were unenforceable; that is, they could not be enforced in the courts. He also found violation of Section 1 of the Sherman Act, but, again, no injury to Honeywell because of the existence of any one of these patents or patent applications, and so no right to damages.

As for the Sperry Rand count, Judge Larson found infringement of the ENIAC patent by Honeywell, but awarded no damages since that patent was invalid.

There was no appeal by either side. To avert such an action by Honeywell, however, Sperry Rand paid a substantial sum of money and agreed to accept the judge's decision. While details of this settlement were not disclosed, Atanasoff has indicated that the payment was several million dollars—enough to cover Honeywell's costs (Atanasoff 1984, p. 278).

The Question of Validity

As we have seen, Honeywell's count of antitrust violation against Sperry Rand rested not only on a charge of monopoly, which concerned its business practices, but on a charge of fraudulently procuring a patent, which concerned its dealings with the Patent Office. The latter charge, in turn, was twofold, partly antitrust (fraud before the Patent Office) and partly not antitrust (invalidity on a number of grounds other than fraud). To state the matter differently, a patent shown to have been procured fraudulently will be found invalid *and* the procurer in violation of the antitrust laws, while a patent procured without committing fraud may still be found invalid on other grounds.

The issues on which the ENIAC patent was scrutinized for validity and enforceability included, among others: "public use," "on sale," and "publication," all alleged to have occurred prior to the critical date of one year before the actual application date of June 26, 1947; derivation from Atanasoff; coinventors on the Moore School team; and delay before the Patent Office. On all except the last two issues, the patent was declared both unenforceable and invalid. No incorrect exclusion of coinventors was found. Delay *was* found, such as to render the patent unenforceable on that issue, but not *undue* delay, such as to render it invalid. These seven issues were also examined for any commission of fraud on the Patent Office, which is an invalidating offense as well as an antitrust violation. Here, too, "derelictions" were found such as to render the patent unenforceable, but not "willful and intentional fraud" such as to render it invalid.

Larson put the individual items in the 30A package of patents and patent applications through the same rigorous examination given the ENIAC patent, and reached the same conclusions on the issues relevant to them.

Our chief concern in this book, of course, is with the finding of invalidity of the ENIAC patent owing to derivation from Atanasoff, because that effectively establishes him as inventor of the first electronic digital computer. But we are also interested in the finding of derivation from him of later inventions patented by Eckert and Mauchly, because this shows a fundamental influence beyond the ENIAC. We devote the balance of this section to sketching Larson's decision on each of the issues listed above, except derivation from Atanasoff; this last we present in detail later in the chapter.

Public Use, On Sale, and Publication

Larson found the ENIAC patent invalid on three grounds having to do with filing requirements. To secure a valid patent, an inventor must make application no more than one year after certain acts of public disclosure or sale, regardless of who executed those acts. Thus, once an application has been filed on a given day of a given year, its "critical date" is established as that same day one year earlier, and instances of such acts must not have occurred prior to that date. Even though this requirement is strictly a technicality having nothing to do with inventorship, its violation is nonetheless lethal. The patent system exists to promote the public interest in innovation, and diligence in making one's ideas available to others, while protecting one's own rights to them, is essential to that interest.

Since the application date for the ENIAC patent was June 26, 1947, its critical date was June 26, 1946. In Finding 1 of his decision, Larson ruled that the ENIAC was placed in nonexperimental public use as early as December 10, 1945, when scientists from Los Alamos began running a hydrogen bomb problem on it. He noted that this problem utilized 99 percent of the capacity of the ENIAC in the solution of a complex problem in partial differential equations, and that any errors encountered were those of the mathematicians who had designed the problem, not of the machine. He said the Los Alamos problem, completed in February, 1946, delivered a hopeful verdict on the bomb's feasibility.

Larson found many other instances of public use of the ENIAC prior to the critical date. Included were calculations made by Douglas R. Hartree, from April to July of 1946; the generation of sine and cosine waves by Harry D. Huskey in April; and the press demonstration, newsreel, and formal dedication events of February, which, the judge said, were intended to show the whole world a completed and operational development in computing. He ruled that all of these were absolute statutory bars to the valid issuance of the ENIAC patent. He also cited the fact that

Eckert and Mauchly were warned (by Army Ordnance) that display of the ENIAC would foreclose their patent rights if they did not pursue them promptly.

At trial, Eckert and Mauchly maintained that these "uses" had all been purely experimental, for the purpose of completing and perfecting the machine. The judge rejected this contention on the basis of thousands of documents obtained from disinterested sources, as well as testimony by numerous witnesses. Moreover, he pointed out that Eckert and Mauchly themselves had advertised in 1949 that the ENIAC was rendered operational by January, 1946, and had made similar claims to the Patent Office and to another district court.

In a further ruling that reveals the close relationship between "public use" and "on sale" as bars to patent validity, Larson stated that Eckert and Mauchly had taken commercial advantage of the public uses of the ENIAC by others, and had even given demonstrations of their own for solely commercial purposes. They had, for example, demonstrated the ENIAC to representatives of the Census Bureau on April 11, 1946 (after they had left their Moore School employ), and had subsequently been granted a contract to develop the UNIVAC.

Thus Mauchly and Eckert were trapped between, on the one hand, their desire to claim a finished product at an early date, in order to secure contracts and to establish priority over other inventors, and, on the other hand, their failure to file a patent application within a year of that date. It would seem that Mauchly was particularly negligent with regard to the nonfiling, because the Moore School had given him the academic year 1944–45 to work with the Army Ordnance attorneys on that application. (Recall that the design of the ENIAC was essentially set by the summer of 1944.) Given that this was his major "research" assignment, that he had no teaching duties, that he was spending at most one day per week in Atanasoff's facility in Washington, and that the detailed ENIAC reports were at his disposal, one has to wonder why he did not fulfill this obligation.

Larson returned to this issue of public use in Finding 13, where he noted that a ruling of "no public use" had been made in a prior case, *Sperry Rand v. Bell Telephone Laboratories* (BTL), in 1962 before the ENIAC patent was issued; and in Finding 11, where he described that same case in detail. He said in Finding 11 that Judge Archie O. Dawson, of the District Court of Southern New York, had forced the matter into court after a six-year delay by SR, and that Dawson had reached an incorrect decision in favor of SR on the basis of a one-day trial. Larson explained in Finding 13 that by the time the case reached court SR and BTL had executed a complete cross-license agreement, so that BTL had "nothing to gain and patent protection to lose by establishing the invalidity of the ENIAC patent." "BTL," he wrote, "had no real or legal interest in establishing public use" and presented "no live testimony" and only "scanty evidence" of public use.

Eckert-Mauchly proponents sometimes cite Dawson's decision as against Larson's, saying there really should not have been another trial on the ENIAC patent's validity. And Larson himself thought there should not have been one! He noted in Finding 11 that "a one or two month trial in the spring of 1962 would have possibly eliminated a nine month trial in 1971 and 1972." Of course, he meant that a longer trial, pursued vigorously by BTL, would have produced a finding of *invalidity* on this ground of public use.

In Finding 2 of his decision, addressing the issue of on sale, Larson concentrated almost entirely on the fact that the ENIAC, its components, and any experimental models were all built under contract to Army Ordnance, the "customer." The original 4926 ENIAC contract was executed by the Moore School as a fixed-price contract providing delivery of all completed parts to the government f.o.b. the school. A number of successive supplements extended the initial delivery date from December 31, 1943, to December 31, 1945, at which point the machine was deemed completed. Thereafter, it was owned and controlled exclusively by Army Ordnance, and any further contractual provisions applied only to expenses incurred by the Moore School while the machine was being used there.

The judge also found that the two-accumulator system, successfully operated by July, 1944, and dubbed a "small ENIAC" in the official reports, was both on sale and sold to Army Ordnance at that time, under terms of the contract. He said that the Moore School used that model to persuade the Army to contract for the EDVAC, and that this was an act of commercialization of the ENIAC invention. He said further that during the course of prosecution of their ENIAC patent application, Eckert and Mauchly asserted that the two-accumulator system of 1944 constituted a reduction to practice of the invention they were claiming.

The fact that the ENIAC and its small model were completed prior to the critical date of June 26, 1946, and were by contract sold to the customer at the moment of completion, constituted a statutory bar to a valid patent.

In Finding 7, Larson ruled that von Neumann's "First Draft Report on the EDVAC" (Von Neumann 1945), widely distributed in the early summer of 1945, was an enabling disclosure of the EDVAC (which was never patented). This meant that the report was sufficient to enable a person skilled in the art at the time to construct the computer it disclosed. He also found that it contained an enabling disclosure of the ENIAC, and as a publication prior to the critical date was a bar to the ENIAC patent. This report grew out of meetings at the Moore School among Eckert, Mauchly, Goldstine, Burks, Warren (director of the EDVAC project), and consultant von Neumann in early 1945, at which meetings plans for the proposed EDVAC were discussed.

An element of bitterness has invaded this issue because of the duplication and distribution of the First Draft Report by Goldstine, and because of the listing of von Neumann as sole author. It was, of course, written by von Neumann alone, as he worked out the logical design of the EDVAC, and it constituted a stellar achievement in its own right: it detailed the logical structure of the hardware for the first stored-program computer, and it created the first modern program language. Yet it originated from joint discussions centering on Eckert's mercury-delay-line memory. In our opinion, the First Draft Report should have carried the name of Eckert, at least, in addition to that of von Neumann (or acknowledged his contribution in some other way). Had it done so, Eckert would have justly shared the public credit for conceiving the stored-program computer that has largely redounded on von Neumann.

Larson cited nine claims of the ENIAC patent anticipated by this report (8, 9, 52, 55, 56, 57, 65, 75, and 78). He also declared that it anticipated or rendered obvious the inventive subject matter of the "automatic electronic digital computer." Both of these findings require some explanation. First, a patent is made up of expository matter, usually referred to as the description or the specifications, and a list of claims as to what is novel therein. The expository portion may include earlier inventions, but the inclusion of *claims* anticipated by an earlier invention can invalidate a patent.

Second, in this particular case, the Sperry Rand defense chose to assert that the subject matter of the ENIAC patent was *the* automatic electronic digital computer, so that any later machine that could be so described would, ipso facto, be derived from it. (Indeed, a later copy of Atanasoff's original computer would have infringed the ENIAC patent!) But this broad stance meant, as well, that the known existence of an *earlier* automatic electronic digital computer (or the publication of an enabling description of one) would render the ENIAC patent invalid.

There is some question as to whether the report was an enabling disclosure of the EDVAC, since it treated the logical organization and makeup of the computer, not the electronics. Moreover, while it was a major part of the report von Neumann had intended to write, it was incomplete. In any case, the legal issue of the von Neumann First Draft Report is not nearly as clear-cut in terms of factual documentation and testimony as are the issues of public use and on sale.

(Incidentally, Larson noted that Mauchly, at the time this report was first distributed, was given an extra copy to turn over to the Army Ordnance patent department with which he was working on the ENIAC patent application, but that he never carried out this duty.)

Two other publications prior to the critical date, both actually about the ENIAC, were considered but found not to be enabling disclosures.

Anticipation by Phelps

Larson ruled, in Finding 6, that three claims (83, 86, and 88) of the ENIAC patent were anticipated by prior work by Byron E. Phelps, of IBM, and so were barred from patentability. This finding dealt with U.S. Patent no. 2,624,507, issued to Phelps in 1953, for an electronic multiplier. The judge explained that this device consisted of an electronic multiplying unit electronically connected to a mechanical card-feeding, -reading, and -punching unit, and that the arithmetic portion operated at the rate of 8,000 pulses per second, the mechanical portion at 23 pulses per second.

He said that the IBM multiplier had been granted a British patent in 1949, and that the U.S. Patent Office had first rejected claims of the ENIAC application as obviously fully met by that earlier patent. Eckert and Mauchly overcame this obstacle by filing an affidavit to the effect that their invention predated the Phelps application date of September 27, 1945. But Larson was satisfied that the Phelps multiplier had been successfully tested by November of 1942, before the ENIAC project had even begun.

Again, Larson considered other ENIAC patent claims as possibly "readable on" certain prior art in the form of patents and corporate internal reports, but dismissed the Honeywell argument in each case. It is interesting to note that all of the ENIAC claims brought into contention by Honeywell, in this and other instances (with the exception of claims attributable to Atanasoff), were among the seventeen "representative claims" originally raised by ISD in their charge of infringement of the ENIAC patent. We can thus infer that the claims Honeywell contested were selected to counter this infringement charge; that is, Honeywell was arguing, often successfully, that the very claims it was charged with infringing actually read on prior inventions or publications. And so we must take the claims actually addressed by Honeywell in each instance as themselves *representative,* not as the only ones anticipated by prior works.

Coinventors on the Moore School Team

In another finding that has generated considerable criticism, Finding 4, Larson ruled on possible coinvention by members of the Moore School team who worked under Eckert and Mauchly. He found for Sperry Rand and against Honeywell on this issue, declaring that Mauchly and Eckert were "the inventors" and that "they [had] not been shown to have incorrectly excluded as named co-inventors, other members of the ENIAC team."

Four major points are often overlooked as this finding is discussed. The first is that it rested entirely on the evidence presented by Honeywell as to the contributions of those other team members. As Larson explained, Honeywell had the "burden of proving that persons other than Eckert and Mauchly were co-inventors on a claim-by-claim basis," and it failed to meet this burden. He stated that he himself was inclined to the view that "inventive contributions were made by Sharpless, Burks, Shaw, and others," and even listed their respective contributions in terms of the design of particular components of the ENIAC. Nevertheless, he said, these characterizations lacked the required specificity of relevance to individual claims.

The second point generally overlooked is that the issue here was one of *co*inventors, not of *prior* inventors such as Atanasoff. It has been argued that Larson's ruling that Mauchly and Eckert were "the inventors" contradicts (and discredits) his ruling in Finding 3 that they derived the ENIAC from Atanasoff. But he made clear in Findng 4 that he was now speaking of coinventors. And he explicitly repeated that the ENIAC was unpatentable because of its derivation from Atanasoff. Moreover, a striking feature of this entire decision is its contextual and cumulative nature, finding by finding and paragraph by paragraph. The tacit assumption is that at any given juncture the reader should have in mind every issue already presented, as well as the particular issue being addressed. The judge did not forget—and he did not expect the reader to discount—his earlier ruling on Atanasoff.

A third point is that Larson did not find Atanasoff an *inventor of the ENIAC* anyway. Derivation of an invention from a prior inventor means only that that person's original (and crucial) ideas were used for the device as claimed. Larson recognized explicitly, in Finding 13, that the ENIAC was a "monstrous machine" that was very different from the ABC.

The last point is that, as Larson stated, the alleged coinventors should have claimed inventive contribution at least by the time Eckert and Mauchly were being publicized as the inventors, and their failure to do so permitted the inference that claims made later—especially many years later—were unsustainable.

Delay Before the Patent Office

In Finding 11, Larson addressed the issue of delay by Sperry Rand in pursuing the ENIAC patent. He began by observing that this patent had been pending for seventeen years, but also that pioneering achievements such as the ENIAC were often involved in many time-consuming interferences. He went on to state, however, that SR itself had "instigated numerous interferences by copying claims of patents on devices developed by competitors." Indeed, he said that from 1951 to 1957 SR had copied 156 claims from competitors' patents in order to initiate interferences, and that in the resulting proceedings 147 of those claims were held by the Patent Office to be illegitimate. He concluded that these interferences caused delay in the issuance of the ENIAC patent.

Such delay, of course, benefited SR in extending the expiration date of the patent, by postponing the start of its seventeen-year life. But it served another purpose, Larson found, namely, that of buying time for negotiating the cross-licensing agreement with BTL mentioned earlier. SR had filed an interference against BTL in 1952, over the patent application of BTL engineer Samuel B. Williams on an electronic computer. When the Patent Office ruled for Williams in 1955, SR appealed to the District Court for Southern New York, but then stalled for over six years, Larson declared, to suit "the convenience of its overall cross-licensing negotiations."

In the end, though, Larson declared "with reluctance" that there was "no undue delay in the proceedings before the Patent Office or the District Court," explaining that "a party seeking a right under the patent statutes may avail himself of all their provisions and courts may not deny him the benefit of a single one."

Fraud on the Patent Office

Before turning to the issue of fraud, Larson devoted Finding 12 to reviewing all his earlier findings and explicitly declaring the ENIAC patent invalid and unenforceable on the basis of the various "barring" issues. He then declared it unenforceable, but not invalid, on the basis of delay before the Patent Office that fell just short of *undue* delay. He made a similar, advance, ruling on the issue of fraud on the Patent Office to be addressed in Finding 13, declaring that "the various derelictions of Eckert and Mauchly and their counsel" to be recited in that finding rendered the ENIAC patent unenforceable, but that "willful and intentional fraud" had not been proven.

Among the derelictions Larson noted in Finding 13 were: suppressing documents, withholding information, securing misleading affidavits from witnesses, thwarting efforts of competitors to secure needed documents from the government, proceeding with patent applications in the face of informed warnings of infirmities, and reversing legal stances to suit new needs. For each of these, he cited instances and provided details. Nevertheless, in instance after instance, he concluded that the party or parties "may have acted in good faith," for one reason or another. He said, for example, that Eckert and Mauchly may not have understood the meaning of "on sale," or may have truly believed that the demonstrations of the ENIAC did not constitute "public use."

Comment

We have, of course, greatly simplified a very complex case, particularly with regard to antitrust and fraud, matters that actually took up over half of Judge Larson's decision. And we have barely touched on the 30A package, to which he paid detailed attention, both for its barring infirmities and for its role in Sperry Rand's licensing negotiations. Let us just make two generalizations, as we leave this Mauchly-Eckert–Sperry Rand side of the Atanasoff story.

The first is that, while the finding of invalidity of the ENIAC patent was the most dramatic outcome of the trial, the revelation of a long list of highly suspect irregularities on the part of Sperry Rand, both before the Patent Office and within the industry, had a chilling effect. This was probably what caused Sperry Rand not only to forsake an appeal of its own, but also to strike an agreement to forestall an appeal by Honeywell. Honeywell had not won damages, but it had won the case: with Larson's decision, neither the ENIAC patent nor the vital 30A patents and patent applications hung over its head; and, apparently, it had had its costs paid by Sperry Rand.

The second is that Judge Larson's decision was not a borderline case, as its critics contend. It was not a case in which he might have leaned one way or the other, but chose to believe Honeywell as against Sperry Rand. Rather, it was an *extreme* case, in which he refrained from reaching the extreme decision, namely, finding fraud and awarding antitrust damages, as he might have done. Larson appears to have sought to settle the case once and for all, equitably, with a decision that could be seen as reasonable and from which there would be no appeal.

Let us expand a bit on this second point. The ENIAC trial made clear that Eckert, Mauchly, and Sperry Rand, consistently and from beginning to end, "claimed it all"—even when warned of probable legal difficulties, even by attorneys acting in their interest.

This pattern is evident from Mauchly's earliest efforts to ignore the input of Atanasoff into inventions he and Eckert were developing; from his failure to mention Atanasoff's computer in the history *he* wrote for the Moore School dedication of the ENIAC; from his falsely informing Army Ordnance that no others on the Moore School team had claimed coinvention (Sharpless and Shaw had). It is evident from his and Eckert's pressing forward with the ENIAC patent in the face of warnings and court actions on the public use issue; from their filing an irrelevant affidavit in the case of the Phelps multiplier; from their failure to reveal to the Patent Office their knowledge of BTL engineer Williams's work at the inception of the ENIAC project; and from numerous other acts brought out at trial, of which these are just a sampling.

It is evident from Sperry Rand's attempt to monopolize the entire EDP industry, on the basis first of just the ENIAC patent, then of that patent and others of doubtful validity; from its various acts of dissembling before the Patent Office, withholding, for example, the fact of publication of the 1946 Moore School EDVAC Report, at the same time that it relied on that report for proof of priority of invention; from its claiming *the* automatic electronic digital computer as the subject matter of the ENIAC in the trial.

Had they—these three parties—been more modest in their aspirations, the end result might have been different. Suppose Mauchly and Eckert had, in the first place, acknowledged their debt to Atanasoff and claimed, for the ENIAC, only their own very substantial advances over his original work. Indeed, they would not even have had to acknowledge Atanasoff, since he had not filed on his own computer; they could simply have refrained from claiming his ideas. Had they exercised only that much restraint, Mauchly's contacts with Atanasoff would never have become a serious issue in any challenge to the validity of the ENIAC patent. If Atanasoff's basic contribution had been discovered, by him, say, when the patent issued, he would have had no legal recourse. This outcome may not have been entirely ethical, but it would have fallen within the strictures of the patent system.

In our opinion, however, Mauchly and Eckert were not satisfied with such a prospect, and from Mauchly's contacts with Atanasoff, they had discerned that he did not suspect their appropriation of his work. And even here, as we remarked earlier, they might have succeeded had Sperry Rand not made the position of its competitors in the EDP industry intolerable. We do not know what attitude the Eckert-Mauchly Computer Corporation might have taken in this regard, if it had not had to sell out. We do know that after Sperry Rand came into control, those competitors found it the better part of wisdom to search for "infirmities" in the patents they were being offered at such untoward prices.

A similar scenario can be imagined with respect to the Regenerative Memory patent and the Serial Binary Adder patents of the 30A package. Had Eckert and Mauchly claimed only their own advancements over Atanasoff's original version of these—again since he had taken no steps in the Patent Office—Atanasoff's contributions would not have become a serious issue.

The three resulting patents would still have been very valuable. They would, however, have remained subject to challenge by others who thought they had made substantial contributions, and also by competitors who might raise the late-filing technical infirmities. These difficulties, of course, could have been forestalled through diligence in filing, in the latter case, and through some agreement with the other inventors in the former.

The Question of Ethics

Several arguments supporting this pattern of overreaching on the part of Mauchly and Eckert are more ethical than legal. One was that the ENIAC was built for the U.S. government, as a wartime project through which participants were expected to serve their country while others served in the armed forces. Moreover, the entire project was paid for by the government as an extraordinary wartime effort, so that, in the view of many at the time, any patent on its product rightfully belonged to the American public, not to the employees of the University of Pennsylvania with which Army Ordnance had contracted.

Another was that Eckert and Mauchly were privileged to learn of the work of others in the electronics field, and to receive help from them, only because this *was* a war project. They had personal contact with and received reports from Williams of BTL and Rajchman at RCA, among others, and were permitted to modify and install IBM input-output equipment on that basis alone. For them then to take advantage of these opportunities for their own exclusive gain, via the ENIAC patent, seems to us reprehensible.

A third was that they excluded the University of Pennsylvania from any rights to the ENIAC patent through rather dubious means. The university had agreed, in its basic contract, to grant the U.S. government royalty-free rights to the ENIAC, but it had failed to require the designers to assign their rights to the university. Thus the two men who claimed to be the sole inventors were able to deny the university the power to live up to its agreement with the government without their acquiescence! Ultimately, the university saw that it had no choice but to let Eckert and Mauchly have all rights other than those it had contracted to the government.

Recent Events

Finally, let us return very briefly to our remarks at the beginning of chapter 2, to the effect that Mauchly and Eckert basked in the public perception that they had invented the first electronic computer up until the trial we have been reviewing, and continued to enjoy that limelight even after the decision in the trial was handed down. To this day, in fact, the general public has scarcely heard of Atanasoff, and the "computer" public that has heard of him largely discounts the judicial process in this particular case.

Eckert has contributed to the prevailing view with disparaging—and highly inaccurate—comments in the news media about Larson's findings. We cite a few of these from an article "Verdicts: Who Really Invented the Computer?" published in the Autumn, 1984, issue of *ICP Data Processing Management*. On Mauchly's reason for staying in Iowa several days, in 1941, Eckert was quoted as saying that he had to wait for the neighbors to whom he had given a ride (p. 16). He added:

> . . . Atanasoff didn't even have a computer. He had a couple of counting circuits and a description of what he was going to build. And it didn't work. Actually, Mauchly saw what Atanasoff had in an hour's time. He was there the rest of the time because he was too poor to get back any other way.

On why he and Mauchly did not appeal Larson's decision:

> . . . we had wasted a million dollars on the case already and we had to get on with other things. Also, there was a nutty judge who threatened to turn the whole thing into an antitrust case against Sperry and we didn't want to get into that. It had nothing to do with the case.

On the notion that Atanasoff invented the first electronic computer:

> . . . It's such an outlandish exaggeration to consider that he did it—it's a complete joke. He doesn't tell the truth—that's all. He did some little thing which he never finished and which wouldn't have worked if he had finished it. . . .

On historians who are beginning to credit Atanasoff:

> I don't believe history books and history writers anyway.

Similar remarks by Eckert were quoted in the November 3, 1986, issue of *Computerworld.* Most recently, on February 13, 1986, the Computer Museum, in Boston, held a large celebration in honor of the fortieth anniversary of the dedication of the ENIAC. Eckert was the featured speaker. The *New York Times* account of February 17, "30-Ton 'Brain' Evokes Dawn of the Computer Age," did not mention Atanasoff at all, but said that the ENIAC was celebrated as, "for all practical purposes, the world's first computer" (p. 9).

The *Time* magazine account of February 24, "A Birthday Party for ENIAC," characterized the ENIAC as "the first all-electronic digital computer," noting, however, that it might require two days for the "coders" to set the dials and plug the patch cords, and that "a million [IBM] cards were required for the monster's first assignment" (p. 63). As to Atanasoff, it reported that the "final insult" to Eckert and Mauchly, after "business reversals forced them to sell their fledgling computer company to Remington Rand," came in 1973:

> . . . Seeking to invalidate Mauchly and Eckert's patent for "the" electronic computer, Honeywell convinced a federal judge that Mauchly had based his

ideas for ENIAC on the work of a computer pioneer named John Atanasoff. The patent was dismissed, and Mauchly and Eckert lost legal claim to one of the great inventions of the 20th century.

Atanasoff on the Stand

Atanasoff was first approached by Sperry Rand's competitors in the spring of 1967, when in quick succession attorneys representing Control Data Corporation, Honeywell, and General Electric contacted him. He had left his computer in Iowa twenty-five years earlier and, except for his efforts to secure a patent on it and his short-lived computer project at the Naval Ordnance Laboratory, had no longer concerned himself with what was once an obsession.

Allen Kirkpatrick, representing CDC against Sperry Rand's charge of infringement of the Eckert-Mauchly Regenerative Memory patent, had read about Atanasoff in a book by R. K. Richards, an Ames friend of Berry's who had seen the Atanasoff machine in 1941 (Richards 1966). GE knew of him through its attorney Norman Fulmer, another Ames resident who had actually worked on the ABC under Berry.

As it turned out, GE was not sued by Sperry Rand and did not need Atanasoff's services, but both CDC and Honeywell hired him as a consultant. In this capacity, he reviewed the Regenerative Memory and ENIAC patents and discussed their claims with company attorneys; searched his memory and his files for material on the ABC and people connected with it; wrote comments on Mauchly's trial depositions; summarized all his contacts with Mauchly, including a phone call and a visit by him (in the company of Sperry attorney Laurence B. Dodds) in late 1967; supervised drawings and wrote up expositions of the ABC for courtroom presentation; and had models and test equipment built, again for courtroom presentation. Atanasoff has provided many interesting details on this consulting work; particularly a full account, from notes taken at the time, of Mauchly's 1967 phone call and visit (Atanasoff 1984).

He then became a witness in their trials, though it was actually plaintiff Sperry Rand that subpoenaed him in its suit against CDC. He testified very briefly in that case, in 1972; it was effectively abandoned after only a few days, then settled in an undisclosed fashion some nine years later. As we saw in chapter 3, it did produce some revealing depositions, which may have caused Sperry Rand to lose interest—or hope.

Atanasoff's appearance as a Honeywell witness in the ENIAC trial, on the other hand, was extensive, taking nine days and filling over 1,300 pages in the official transcript (in addition to his earlier deposition of nearly the same length). Indeed, Atanasoff was Honeywell's key witness, insofar as the originality of the ENIAC was concerned, just as Mauchly was Sperry Rand's. The burden of proof, however, rested more heavily on Honeywell: the Sperry side did not have to prove that Mauchly and Eckert had proceeded independently of Atanasoff; rather, Honeywell had to prove that they had not. Fortunately, in contrast to Mauchly's performance as a witness, Atanasoff played his part well. Indeed, he surpassed Mauchly by an order of magnitude, as will be evident from the following account.

We have sorted Atanasoff's testimony into three broad categories, or characteristics, that seem to us basic to his credibility, and we present examples illustrating each. These categories are: the quality of his *memory* at the time of the trial, his professional *competence* throughout his life, and his *procedures* as he pursued his goals. They were not delineated as such by the attorneys who did the interrogating, of course, but they were in fact at issue most of the time, with Henry Halladay, of the Honeywell team, attempting to cast them in a positive light and Thomas M. Ferrill, Jr., of the Sperry team, challenging them. Less tangible indicators of credibility, particularly the witness's demeanor on the stand, are also apparent from our quotations.

The examples cited for these broad categories are drawn from arguments not covered elsewhere in this book. Although they are by no means exhaustive, we believe they do accurately reflect Atanasoff's testimony in the ENIAC trial. All references are from the official transcript, in which pages 1,576–2,346 are mainly taken up by direct examination and pages 2,347–2,859 by cross-examination of Atanasoff; pages 2,859–98 by redirect and pages 2,898–2,914 by recross.

His Memory

Atanasoff revealed a sharp memory for details, whether intricacies of a student's dissertation or circumstances of some encounter. Asked what he could remember about being east in the spring of 1941, for example, he recalled particularly "meeting with Dr. W. G. Cady," of "the New England Wesleyan College," and having "a very good discussion about vibrating crystals," a subject on which "some of my students were working" (transcript p. 2,126). He also remembered events from his tenth year, 1913: his study of his father's new Dietzgen slide rule and its accompanying table of logarithms; his reading of a pair of British books on radio and telephony that were a "faint green color" and had their price marked "in sterling on the cover"; and his repair of the one light that refused to work when the "magic of electricity came to our home" (pp. 1,583–90).

The Roadhouse Trip

Atanasoff's recall of the year in which he took a climactic drive to a tavern in Illinois was strongly contested by Sperry Rand, because it signified the jelling of his conception of the electronic computer that he would ultimately build. He had given a vivid description of this trip in direct examination, setting it in the winter of 1937–38 (pp. 1,700–1702):

Well, I remember that the winter of 1937 was a desperate one for me because I had this problem and I had outlined my objectives but nothing was happening, and as winter deepened, my despair grew and I have told you about the kinds of items that were rattling around in my mind and we come to a day in the middle of winter when I went out to the office intending to spend the evening trying to resolve some of these questions and I was in such a mental state that no resolution was possible. I was just unhappy to an extreme, and at that time I did something that I had done on such occasions . . . . I went out to my automobile, got in and started driving over the good highways of Iowa in those years at a high rate of speed.

I remember the pavement was clean and dry, and I was forced to give attention to my driving, and as a consequence of that, I was less nervous, and I drove that way for several hours. Then I sort of became aware of my surroundings, . . . of where I was and I had reached the Mississippi River, starting from Ames and was crossing the Mississippi River into Illinois at a place where there are three cities there, one of which is Rock Island.

I drove into Illinois and turned off the good highway into a little road, and went into a roadhouse there which had bright lights. It was extremely cold and I took off my overcoat. I had a very heavy coat, and hung it up, and sat down and ordered a drink, and as the delivery of the drink was made, I realized that I was no longer nervous and my thoughts turned again to computing machines.

Now, I don't know why my mind worked then when it had not worked previously, but things seemed to be good and cool and quiet. There were not many people in the tavern, and the waitress didn't bother me particularly with repetitious offers of drinks. I would suspect that I drank two drinks perhaps, and then I realized that thoughts were coming good and I had some positive results.

He went on, then, to detail his plan to "jog" the condensers of a memory made up of two parts, through "a non-racheting approach to [their] interaction" in the form of a "black box" whose electronic circuits were yet to be worked out. Thus he had firmed up the idea of regeneration of memory elements from a separate electronic arithmetic unit, and had conceived the idea of computing electronically by logic instead of by counting.

In cross-examination, Ferrill pressed to shift the date forward to 1938–39, questioning Atanasoff's version both in relation to his attempts at mechanical computing devices before the trip and in relation to his progress on the ABC after the trip (pp. 2,389–2,410). That is, 1937–38 was too early, Ferrill argued, because Atanasoff was busy until the fall of 1937 considering the modification of IBM tabulating machines and the ganging of desk calculators, and because he did not commence work on the prototype of his computer until the fall of 1939.

Atanasoff was unshakable. He explained that he had been thinking about building his own computer for years before his drive to Illinois—in fact, from 1935 or 1936—even as he worked on the other devices; moreover, that he had then to work out the logic circuits for the add-subtract mechanisms before he could request funds from the college to build the prototype. He and Ferrill had this exchange (p. 2,402):

> *Ferrill:* Well, how long did you spend working out these logic circuits, as you call them, to use with the condensers?
> *Atanasoff:* I don't know. I suppose my next remark will ruin my reputation utterly, but I expect I spent ten months at it.
> *F:* Working out the logic circuits?
> *A:* Yes, sir, and trying to see—during that period I have some notes somewheres, I believe, that show that I worked out a logic system for the scale of 5 . . . . My memory tells me I did this, and . . . you see, I am still harking back. Maybe I should go back to 10 all this time. I am never certain about anything. I am exploring in all directions at one time.

But, Ferrill pointed out, he could still have taken the roadhouse trip in, say, December, 1938, and had ten months to work out the logic circuits before he and Berry started building the prototype in September of 1939. Atanasoff replied that although Berry did not come to work until school began in the fall, he had been hired the previous spring, *after* the college had granted the requisite funds. There was this exchange (pp. 2,406–8):

> *Ferrill:* Do you know for certain that it was not near the end of 1938 when you made the trip to Illinois to the roadhouse?
> *Atanasoff:* I am very convinced in my own mind, sir. It has to do with my state of mind and the state of mind of other men of that day. It has to do with my feeling of uncertainty about this whole process of development of a machine in which there's a radical change in elements thereof. . . . I knew that when I went—when I requested funds I would be under severe cross-examination and I delayed asking for funds until I had worked out the details and felt quite certain in my own mind about these details. And I remember the pressures of that time and the fact that I several times went back over my calculations and the methods of approach to this problem to insure that I could answer [the questions]. . . . the gentleman who would actually personally consider the funds . . . was a good friend of mine, but he was also known to be a very severe man in regard to the disbursement of funds and I hated to go up and ask him for funds and I wouldn't do it until I was quite certain that I had a logic system worked out and the methods pretty well in detail before I asked him.
>
> Now, furthermore, I distinctly remember when I engaged Mr. Clifford Berry in the spring of 1939 that at that time I was prepared to give him details of the add-subtract mechanism and the rest of the details in regard to construction—not all details, but most of them—so when he commenced working in the fall we would be ready to go.

F: Wasn't it actually your practice to sort of try to work things out as you go, first starting with kind of a framework and then fitting things in?

A: It certainly was my standard practice, yes, sir.

F: To do it that way?

A: Yes. I mixed everything in together but that doesn't mean—that doesn't mean that I didn't feel the pressures of this professor of genetics who had become the disburser of funds there, that I didn't feel very uncertain about my quest for funds from him, and, you know, the quest for, what was it, $450 looked like a mountain to me in those days.

Halladay had the last word on this subject in redirect examination, as he read from Atanasoff's July 10, 1940, request for $5,000 from President Poillon of the Research Corporation (pp. 2,881–82). There Atanasoff said that he had begun to investigate the possibility of mechanizing the solution of sets of equations "about seven years ago" (mid-1933), and that "beginning about three years ago" (mid-1937) he had had "a succession of ideas which have enabled me to design a computing mechanism." This strongly confirmed Atanasoff's recollection that he had begun to consider building a computer of his own by 1935 or 1936 and that his roadhouse breakthrough as to the design of that computer had occurred in the winter of 1937–38.

The Prototype

Atanasoff also revealed a keen memory for details with respect to the prototype, as he termed his original test model of the ABC—it was sometimes called a *breadboard model* in court, and in chapter 1 we called it simply a *model* of the central computing apparatus. He first explained his rationale for building it in direct examination (p. 1,746):

. . . I was continually worried about whether a vacuum tube could be made into a machine that would calculate, and so our first effort was in that direction. Before Mr. Berry came to work, I had developed diagrams for a logic circuit, and immediately upon Mr. Berry's starting work, we commenced throwing together a test prototype to help us evaluate the problems and to convince us that vacuum tubes in a computing machine was not a pipedream.

Then, speaking from a large mounted drawing that he and Honeywell attorney Charles G. Call had helped a draftsman produce, Plaintiff's Exhibit 21,116 (see fig. 5), Atanasoff gave nearly fifty pages of testimony on this model (pp. 1,752–98). Every detail was examined and the operations of addition and subtraction of vectors explicated. We quote just one representative passage, Atanasoff's introduction to the basic elements depicted in the drawing (pp. 1,760–62):

. . . Here is a disk of bakelite. On its two faces were mounted circles of condensers. You can see the circle of condensers on the near side. Notice, they are connected to a common ring in the center so that the inner end of the condensers were all interconnected. The outer ends come out to contact points on the circumference of the disk. You cannot see it, but the other side has exactly the same structure. And this way, without additional material, we have two abaci elements or memories which are going to inter-wrap by the process of addition or subtraction.

Now, you see some brushes in this region here, two brushes on each condenser ring. You see driven here a high speed shaft which is directly driven by the belt from the motor, the high speed shaft and the low speed shaft. Those shafts are interrelated in such a way that as a contact moves from this point to that point this shaft will move forward a unit. Now, whether the unit is one revolution or half a revolution depends upon the structure of the commutators which are thereon. I believe the commutators—memory tells me the commutators had two contacts, so this shaft only moves half a revolution in the time this moves from there to there.

Now, the relation of the brushes on the near and far side of the disk are such that they correspond in time of contact as the big disk rotates. In other words, it is clear that the condensers are offset and the brushes must be offset the same way.

Now, roughly speaking—oh, yes, we better describe one or two more pieces of apparatus. Here is a little device called a one-cycle switch, and the one-cycle switch is designed so that as the thing is continually rotating you can put your finger on the one-cycle switch, and it is that little gadget right there, and push. At first the disk with the notch in it, you see there, will be in the way and it will not snap in. Then when the notch in that disk comes opposite the switch member it does snap in and then if you immediately pull your finger away, by that time this disk will have caught that little member and holds it in and it will stay in until the disk has made one revolution, then it will jump out. Now, this process of switching in and out takes place during the vacant part of the condenser rings. So if you push there will be one complete revolution and then there will be no more of activity.

Now, the abaci on the front and back represent numbers. These numbers are transmitted by the cable indicated here into the add-subtract mechanism. The add-subtract mechanism is powered by this unit. We have two abaci, remember. One of them is regenerated by . . . the add-subtract mechanism and one of them is regenerated by the memory regenerating circuit.

Now, during the process of addition, for instance, if this switch is set in the add position and the one-cycle switch is actuated the entire number on the circumference on the [keyboard abacus] circle will be presented at this box and, of course, the entire number on the [counter abacus] side of this disk, which is the farther side, will also be presented at that box. That logic circuit will do the additions there and return to [the counter abacus] the sum. I can give more details of that operation.

Atanasoff also described in detail how an oscilloscope was used to monitor the activity of the prototype and to demonstrate such activity to trusted people in the physics building; he particularly mentioned Sam Legvold, who had started to work for him on his antiaircraft fire-control project in that same fall of 1939 (p. 1,765).

As to when the model was completed and demonstrated, Atanasoff said (p. 1,752):

> It was late in the year of 1939, in November and December, the prototype assumed this form [as shown in the drawing], without being exact in every detail at any given date, but it was operating by some time in November and performing the functions which it was designed to perform. . . .

Legvold, too, had described the model at trial and testified to seeing it operate in "either October or November of 1939" (pp. 1,327–38).

Both the nature of the prototype and its operating date were of concern to the Sperry side, because it could be said to constitute an "automatic electronic digital computer," the entity Sperry Rand had made the core of the ENIAC case. In his cross-examination, Ferrill challenged the late 1939 date, but Atanasoff felt certain of his recall (pp. 2,411–16). He said his memory was that the prototype progressed rapidly, and that it did so precisely because he had developed the add-subtract mechanism "during the spring of 1939" (p. 2,411). There was then this exchange, in which Ferrill did appear to have proven him off by a month of two (pp. 2,412–14):

> *Ferrill:* Dr. Atanasoff, if I suggest to you the proving out with your tests with this prototype occurred rather in January, 1940, would you say that that by your memory is wrong?
> *Atanasoff:* I wouldn't say there wasn't one chance in a million but I would say there wasn't much more than that that that occurred.
> *F:* Let me ask you to refer, please, to your manuscript [of August, 1940] . . . Page 31. Do you have a copy of it before you?
> *A:* Yes.
> *F:* Would you read the next to the last paragraph, Dr. Atanasoff?
>
> . . . .
>
> *A:* "A test setup of an abacus and add-subtract mechanism and converter was made in January of 1940. The arrangement performed perfectly and allowed actual tests under working conditions to be given to various components."

With that, Atanasoff bowed to his written word at the time as against his present recall, and Ferrill, who had now placed a time chart of the period on an easel, had Atanasoff enter "Testing of prototype" at the January, 1940, slot.

Ultimately, however, it was again Halladay to the rescue, as he brought out in redirect examination that the key word in the manuscript reference had been "converter" (pp. 2,870–72). Only a converter had been added to the prototype in January, so that Atanasoff's memory of successful tests in late 1939 was reaffirmed. Atanasoff described this device as an "additional gadget of Mr. Berry's," for "sticking numbers in the machine while it was rotating," and told how it worked:

> . . . I remember how it looked but I am not sure exactly what the structure is. I guess I could invent it but it consisted of a little commutator that went on the end of the shaft, one free end or the other, and it consisted of a frame with some notches in it, and Mr. Berry inserted a lead with the appropriate voltage in one of these notches and that notch would transfer voltage to the abaci of the rotating elements which corresponded to the notch in which he placed the point. Now, I think that [it] is pretty obvious how that would work.

Actually, there was one piece of written evidence to confirm Atanasoff's vivid memory—and Legvold's—that successful demonstrations of the prototype had been conducted in the last months of 1939. Atanasoff had brought in a December 21, 1939, letter he had written to Aerovox Corporation concerning condensers for the two large drums of the computer that he now fully expected to build; in this letter he stated (p. 1,748):

> . . . At present we have fifty of your paper tubular condensers built into a unit which is operating very satisfactorily. . . .

The Memory Drums

One may wonder at these Sperry efforts to move Atanasoff's progress toward his computer back by so little as a month or two, but, as we have said, the more fundamental issue was the accuracy of Atanasoff's memory. Indeed, our third example seems to have the impugning of his memory as its sole objective.

Before we turn to it, though, let us just note that the Sperry team's failure in these efforts does not necessarily detract from their appropriateness. Ferrill was doing exactly what he should have done: contest Atanasoff's testimony in every way he could. Unfortunately for his side, very little basis for contesting it could be found and nearly all of what was found turned out to be without substance. Honeywell attorney Halladay used the same tactics in challenging Mauchly's account of events, but he had more and better counterevidence from which to argue and so was usually able to prove Mauchly wrong.

In this last illustration of Atanasoff's memory on the stand, however, Sperry

Rand did seem to be clutching at straws. At issue was the simple matter of the material from which the 3,200 contact studs circling the two large storage drums were made (pp. 2,458–62). Atanasoff recalled their being cut from brass rod stock, or spelter, with the ends flattened in a die. In fact, he thought there was "in our records a procurement document from the instrument shop" for the die to do this flattening (p. 2,458). But Ferrill was somehow convinced that they were made of brass tubing, and he set out to wrench an admission to this effect from the witness (p. 2, 459):

> *Ferrill:* Do you know for sure that these contacts were made of rod stock and were not made of a small tubing?
> *Atanasoff:* Yes, I think I do.
> *F:* You are not certain of that?
> *A:* Well, my memory is pretty strong again. I just can't imagine tubing being used in those things. This is the kind of memory I have and I could be wrong. You could pull one of those out and look at it and tell if it's tubing. . . .

Ferrill then produced from among Atanasoff's submitted documents an invoice dated July 15, 1940, that listed "53 feet of hard drawn seamless brass tubing." After reviewing its inner and outer diameters (one-sixteenth inch and three-thirty-seconds inch), he demanded to know what Atanasoff was "getting 53 feet of this hard drawn seamless brass tubing for?" There was this exchange (pp. 2,460–61):

> *Atanasoff:* Let me think—may I think on this a while?
> *Ferrill:* Certainly. I want you to take your time, Dr. Atanasoff.
> *A:* Thank you sir. My memory persists that those [contacts] are made of spelter, and spelter is not hollow, but it's solid. My memory persists strongly that that was true. Now, let's see as to the tubing. . . . You know, the tubing was used for brush holders, I believe. The brush holders—let's reach for, if you please, sir, the exhibit containing the brushes.
>
>   . . . .
>
> *A:* . . . The brass tubing was used for the brush holders. The brushes went through the center of the brass tubing and it furnished an additional support for these brushes and we soldered the outside of the brass tubing and this way made contact with the brushes.

The foregoing had occurred in a morning session. Immediately after the lunch break, Ferrill made a further attempt to prove that the studs had been made from tubing; this time he succeeded only in rousing Atanasoff to one of his rare shows of annoyance on the stand (pp. 2,500–2,505). Noting that the Sperry lawyers had been inspecting the drum, Ferrill asked him to examine the studs and "see if you don't see hollow spaces in the middles of these flattened ends"; he pointed particularly to one on which he had placed a rubber band. Atanasoff replied, calmly enough at this point (pp. 2,501–2):

> Well, the way of a witness is hard. My memory is very clear that the pins were made of rod, . . . of a form of rod which is used for brazing called spelter.
>
> Now, I am faced with a rubber band upon certain pins here and it shows a slight hollow and I myself am unable to ascertain—now, may I say a word about metals? . . . when you take a metal and squeeze it flat certain distortions and disturbances occur in the fibers and the working ratio is pretty high in these fibers. I suspect that the hollows which counsel has so assiduously discovered in the center of these terminals is due to the working thereof. I don't know, and I may be wrong. Goodness knows, I may be wrong. As far as I am concerned, counsel can take a pin out. . . .

He did raise the issue, now, of who actually owned the drum, as a possible obstacle to defacing it, whereupon Ferrill felt prompted to ask if he thought anyone was "about to use it to make a computing apparatus or try to"! When Atanasoff strongly doubted that eventuality, Ferrill turned once more to the invoice for the tubing, seeming now to challenge him about *it* and how it had come into his hands at the time of his deposition a few years earlier. Their dispute over the contact studs ended as follows (pp. 2,504–5):

> *Ferrill:* Well, are you saying that you have been keeping a copy of this and the hundreds of other papers in Plaintiff's Exhibit 266 in your file at home for some years prior to 1967?
>
> *Atanasoff:* Many years prior. I would say thirty years prior.
>
> *F:* And it is still your testimony, is it, Dr. Atanasoff, that you are convinced that the pins in this Exhibit 21408 are rod stock and not flattened tubing?
>
> *A:* It is my best belief, sir, that they are rod stock.
>
> *F:* But you are not certain?
>
> *A:* Well, goodness, I don't have x-ray eyes. My memory tells me it was rod stock. My memory tells me the alloy which was used. I know this alloy to be hard, I know it will work. If it is worked severely, it will fragment inside. I know these things, so it seems rational and proper and also according to the tenets of my memory to state that those are rod stock.

And what of the "procurement document" Atanasoff had alluded to at the outset? Halladay produced it during redirect examination (pp. 2,874–76). Dated February 22, 1940, some five months earlier than the tubing invoice, it showed a charge both for the rod stock and for the dies to form the ends of the contacts. The courtroom examination of this piece of paper ended in an eloquent description by

Atanasoff of the successive steps that had gone into constructing the two main storage drums on which these brass contacts were mounted.

His Competence

The second broad category we have taken to be basic in Atanasoff's court appearance is his professional competence, indeed, his competence in whatever he undertook throughout his life. This was manifest in his testimony, whether the Honeywell side was seeking to display it or the Sperry side to undermine it. We address his education and his teaching and research briefly before turning to three aspects of his computer on which the Sperry side chose to challenge his competence rather severely: its overall operation, the operating speed of its add-subtract mechanisms, and the issue of counters versus logical mechanisms.

His Education

Even as a precocious child, Atanasoff had been unusually industrious and goal oriented. From his study of the slide rule and logarithms at the age of nine, he had gone on to a book of his mother's, Robinson's *Arithmetic,* which treated bases other than ten. He testified in cross-examination that he had "read this book assiduously," and "fooled with" the bases twelve and two in particular (p. 2,393).

Because his father had majored in science at Colgate University and was now studying electrical engineering by correspondence, the young Vincent had access in his home to college texts on physics, chemistry, astronomy, surveying, higher algebra, trigonometry, and analytical geometry as he proceeded through junior high and high school. He testified that he was allowed to "roam in mathematics" in the small-town schools he attended in Florida—the family moved a lot—so that he completed the first and second years of algebra in the sixth and seventh grades, and later passed both trigonometry and solid geometry examinations by studying on his own. He apparently was allowed to roam in other subjects as well, for he spent much of his time in junior high school reading the *International Encylopedia,* and was in high school given a key to the chemistry and physics laboratory. (Pp. 1,585–92)

Atanasoff testified that he was "an indifferent scholar" in high school, with good grades in mathematics and science and poor grades in Latin and English; but that he determined to do better at the University of Florida and with special effort made "reasonably good" grades in English. He earned his way through college, in part by teaching science at a high school in Gainesville, having been certified by examination after three days of intensive study. (Pp. 1,603–6)

Honeywell attorney Halladay pursued the matter of Atanasoff's formal education through his doctorate in physics from Wisconsin (1930), eliciting several down-to-earth expositions of scientific matters for the benefit of the court. His request, for example, that Atanasoff explain "the virial of Clausius," an application of which was the subject of his master's thesis at Iowa State College, brought this account (pp. 1,612–13):

> Clausius was, I believe, a German mathematical physicist who had come up with a very simple and coherent principle of gas—of a gas. Now, we are talking in the field of kinetic theory of gases, or the dynamical theory of gases. You know, a gas is composed of many particles and the pressure of a gas is not—of the thing compressed and then just stretching itself outward. That isn't the reason when you push down on a pump you feel the reverse pressure. You feel the reverse pressure because the molecules of gas are each a little projectile and striking the piston on which you are pushing and, hence, develops the theory of the pressure.
>
> Now, the virial of Clausius, this way in which the pressure and the temperature and the volume varied, is called the equation of space. The principle of Clausius is a general approach to the theory of this equation of space, and it provides a mathematical theory by which one can attempt to derive the equation of space for an unknown gas. . . .

His Teaching and Research

Atanasoff gave similarly concise and lucid discourses on quartz research, on which he had coauthored an article (pp. 1,641–43), and on the relation of mathematical equations to the real world (pp. 1,650–51). We quote one paragraph of his remarks on the latter:

> The theoretical physicist, or the mathematical physicist, or the physicist, in the more elementary sense, is merely an artist, and he's attempting to depict the external world in terms of these formulations of which I speak, and the question is, if the theory is real—take Newton, Newton did this; he was a great painter and he painted the field of mechanics, you know, for us in certain equational form. Now, the interesting thing is, and the powerful thing is that if you manipulate these equations, they behave in an analogy with the physical world so that by examining these equations, you can tell what the physical world is doing—should do.

Atanasoff's teaching ability, of course, was readily apparent from these extemporaneous expositions. Halladay also succeeded in entering into the record a long string of theses by Atanasoff's graduate students at Iowa State. Ferrill objected to these as irrelevant to the issue of Atanasoff's computer, but Halladay maintained that they illustrated the depth of his exposure to problems in physics and mathematics. Judge Larson then questioned Atanasoff on the degree of his involvement (pp. 1,638–40):

Larson: As to each one of these theses, Dr. Atanasoff, did you have personal knowledge of each thesis?

Atanasoff: Yes, sir.

L: Did you review each one?

A: I reviewed each one and initiated each one and I counseled with the student as he was writing each one, and I participated in laboratory experimentation in regard to each of these theses, if it required laboratory experimentation. The subject matter was a matter of my own conception, and this matter motivated me strongly in these years. These theses represent a motivation of my activities in these years.

. . . .

L: The exhibits will all be received.

Halladay further illustrated Atanasoff's expertise by quizzing him at length on his antiaircraft fire-control project at Iowa (pp. 2,111–16) and on his assignments to Bikini Atoll in 1946 and to Helgoland in 1947. Of the Bikini task, in which he covered two atomic bomb blasts some five weeks apart, he said (p. 2,243):

> I had charge of a considerable amount of sonic, and the like, measurements in connection with the explosion, measuring pressures under the water and in the air at various distances, some of them rather close to ground zero of the explosion.

He explained that the purpose of the Helgoland venture, initiated by the British Admiralty, was to destroy this island that had been used by Germany in both world wars to store all manner of explosives. British engineers went in and wired all of them together with shock cord and then set them off in what was the largest nonatomic man-made explosion ever executed. Atanasoff, representing the U.S. Navy, directed the seismic survey of the continent of Europe in the wake of this blast, a survey that laid the foundation for long-range detection of atomic explosions. He gave the following account in court (pp. 2,281–84):

> . . .We were asked to participate in the Helgoland blast after everybody else had been approached and refused to take part [for lack of time to prepare]. As a result, before the date of the Helgoland blast . . . we had perhaps six or eight weeks left. We had no seismographs, or other apparatus to make these tests with, and I stated to the people from the Office of Naval Research that if they would secure a top priority for me, I would build seismographs and instruments for this. Well, before the test took place, it was put off a couple of weeks, and at the end of six weeks or so I shipped to Europe, by way of American Airlines, 22 seismographs . . . [and also] 22 microbarographs . . . ; and during the late summer I revisited the site near Frankfurt in Germany, in the mountains there which I personally manned in taking part in

these tests [in April], and of the 22 measurement sites manned, I believe 21 of them . . . brought back data both of the seismic nature and the barograph nature, and these stations were stretched across Europe from Wesermunde on the North Coast opposite the Island of Helgoland to Goritza in north Italy.

. . . this was, as a matter of fact, the first long distance record of man-made explosives, and it simultaneously served as a first long run record of microbarographic waves traveling through the atmosphere which was excited by man. . . . [This] was the first case by which man by his act excited the waves and then measured them for a distance, which totaled a thousand kilometers, I believe, or more—I guess a little more than a thousand kilometers.

In his recent article, Atanasoff gives a vivid description of his part in this Helgoland operation, for which he received a citation from the chief of the Bureau of Ordnance and also from the Seismological Society of America, to which his group presented a paper (Atanasoff 1984, pp. 259–60).

Finally, Halladay took Atanasoff through the nine categories of problems he had listed in his August, 1940, manuscript, as amenable to solution on the ABC (pp. 2,141–57). We quote from his explanation to the court of just one of these, "Curve fitting" (pp. 2,141–42):

. . . there are various ways you can fit curves, but I am thinking of fitting curves in the terms of the linear ensemble. It means a . . . linear expression in known functions with unknown coefficients, and the unknown coefficients will be so selected that the curve will fit. . . .

Asked to cite a problem from the real world, he chose a problem in the "growth phase of bacteria":

. . . you have formed a culture and innoculated it in a specific way, . . . and then, you know, the bacteria start to grow and you determine their concentration by the process called plating, . . . and then you have a curve of the rate at which the bacteria grow, and you wish to see what the mechanism of bacterial growth is, and so you attempt to analyze this curve in some mathematical expression suitable for some further analytical study.

We turn now to the first of three major aspects of Atanasoff's computer on which Sperry attorney Ferrill strongly challenged him.

Overall Operation of the Computer

Ferrill devoted over half a day, or 100 pages of transcript, to examining Atanasoff on the procedure by which his computer would have solved a set of five simultaneous linear equations in five variables (pp. 2,479–2,578). Atanasoff had not only to give and explain the individual steps in their proper sequence, but to record them on a chart as he went along. His only props, aside from continuous questioning by Ferrill, were the actual memory drum, photographs of the computer and its components, and several schematic diagrams, particularly figure 1 from his 1940 manuscript depicting the central mechanisms (compare our fig. 7).

It is striking that Atanasoff managed to go through this entire performance with only one mistake of any consequence, namely, his initial recollection that the add-subtract mechanisms used complements to subtract; he corrected that after the lunch break, saying that in fact they did straight subtraction, with borrowing. There were several minor items he did not recall, but even for these he could point to different possibilities.

There was just one instance in this lengthy presentation when Atanasoff showed annoyance at the demands being made on him. He had been explaining the part of the elimination process in which the abaci on the keyboard cylinder were shifted and the add-subtract mechanisms were ordered to reverse operations, step 18 on the courtroom chart (pp. 2,534–35):

> *Ferrill:* What is the apparatus that causes these things you have just recited to occur, Dr. Atanasoff?
>
> *Atanasoff:* I don't know—a machine of some kind, I suppose. It's a machine which is built into the thing.
>
> Well, I could build one easily enough. I could build one easier than I could—not easier, because it would take me a week, perhaps.
>
> *F:* Easier than you can describe it? Was that what you were going to say?
>
> *A:* Well, easier than I can describe it if each time I utter three words, I have to stop and write what I have said down, and then I have to reform what I have said into a kind of grammar, and which is peculiar to the written word, and do a hundred and one other things that I'm pursuing at the moment. But I'll do my best.
>
> Now, this is just a temporary complaint—no problem here. (Laughter) I'm having a hard time, and you know it. (Laughter) I don't mind, and I can take this quite a while . . . .
>
> *F:* I want to tell you, quite sincerely, Dr. Atanasoff, that I'm not really intending to make a hard time for you—
>
> *A:* (Interposing) No, and doggone it, I know you could be much worse. That's one of the things that plagues me currently. (Laughter) You're not doing so badly, no.
>
> Pardon me, Your Honor, for giving such remarks.

With that release, he was able to go on and describe the mechanism in question very well, and to continue for the balance of this exercise with an enthusiasm that soon proved contagious. Ferrill actually seemed at times so entranced by Atanasoff's explanations that he slipped from inquisitor to pupil. Or, perhaps, he was just being a gentleman in what was, all in all, a very gentlemanly trial. In either case, when Atanasoff offered to explain the elimination process in the base ten for him, he welcomed the help. And at one point the judge himself was moved to comment on the harmony that had arisen between witness and opposing counsel in his courtroom (p. 2,543):

> *Ferrill:* I think the record ought to show, Dr. Atanasoff, that you did that string of computations on this new sheet which we are about to identify in about seven minutes even with my interruptions to ask you questions about it.
>
> *Atanasoff:* Good.
>
> *F:* I can see you have been in mathematics a long time.
>
> *A:* Thank you, sir.
>
> *The Court:* With all these compliments going around the room, we will take our afternoon recess.

Atanasoff's descriptive expositions *were* remarkable. Take, for example, the passage in which he addressed his difficulty with the binary-card system of the ABC (pp. 2,570–74):

> *Ferrill:* That would be the piece of eight and a half by eleven or tablet sized paper?
>
> *Atanasoff:* Right. Now, there is a further signal which releases the solenoid that keeps the paper back and so during the next cycle of the machine, two fingers that push the paper commence to ride the cam. They have been held back, out of contact with the cam. They commence to ride the cam forward and as they [do], the paper moves forward and an instant later, the paper is seized between the rotating rollers and is fed through in synchronism and during the feed through—now, we have in position right in between those rollers, and it is hard to see in the photographs, we have in position all the pointed terminal tungsten punching points and these punching points are all actuated in unison and at the proper time automatically, and the entire system of coefficients punches out during that operation.
>
> *F:* You are taking them out of the CA [counter abacus] drum or getting, let me say, the stimuli out of the CA drum and out of its various abaci or rings of contact?
>
> *A:* Yes.
>
> *F:* And you are causing them to be tracked out of that card if you could get this base-2 punch to do the job the way you wanted it to do?
>
> *A:* There is no question about the punch. It always did its job. All you had to do was lick it [spark the paper] hard enough and you could make a hole through. You didn't have any trouble licking it hard enough, but the trouble was in the reading mechanism . . . you had to set the level of the reading mechanism so it would read through if there was a hole there and

it would not read through if there was no hole and that was a little critical. It was only the second part of that operation that caused trouble.

F: Did you have to worry about whether that readout system would also do a little hole punching on its own?

A: Well, that is it, you see. If you put too high a voltage to detect the holes, it will punch it on its own, exactly that. You have got to have it low enough so it won't punch on its own and high enough so it will invariably read out and that adjustment was the critical thing that caused the trouble.

.   .   .   .

A: That was the . . . only trouble, really, because the rest of the machine was operative, or so clearly nearly operative, that one could conceive of solutions for anything that came up. . . .

You notice, we had a difficult problem in those days, because we had no interfacing equipment, nothing you could use, and we had to invent, and we did invent, and this part caused us trouble. I suspect if you found the right material, it would still be an operative recording system.

Magnetic systems would be cheaper, wouldn't they? But, remember, they are cheaper in a day when the necessary amplification is available to bring the levels up, and read-heads are available, and we had neither in that day,—neither the read-head, which I can conceive of and had conceived of, but the amplification would take steps of amplification, and amplification was extremely expensive and slow to obtain, and no such thing as transistors, and it was a difficult job.

Thus it was that the Sperry team's challenge to Atanasoff's overall design of the ABC served only to demonstrate his competence as a scientist/inventor. It was fortunate for the Honeywell side, of course, that he was such a good expositor. It was also fortunate that the procedure being described was comprehensible to anyone who had mastered high school algebra; Atanasoff's performance might have been more dazzling if the machine had been meant to solve differential equations (or difference equations), but the trial participants could hardly have been as caught up in the process as they obviously were.

Operating Speed of the Add-Subtract Mechanisms

However disarmed Ferrill may have been during the last portion of this examination on the ABC, he quickly resumed his attack with respect to the time required to solve the set of equations under consideration (pp. 2,578–2,609). He succeeded in showing that Atanasoff was far wide of the mark in estimating that his machine's calculation time would have been "a hundred fold faster" than that of a desk calculator. Atanasoff, for his part, established that a figure of an hour or two to solve five equations on the ABC was not to be taken as typical, because that machine was most efficient with the very large sets for which it was designed (see chap. 1).

Ferrill then shifted to the operating speed of the add-subtract mechanism alone. Atanasoff had given a live demonstration of this component during direct examination and had explained it at length (pp. 2,035–91; see our chap. 3). The cross-examination addressing its speed was carried out in the presence of an enlarge-ment of figure 2 from the 1940 manuscript, a schematic diagram of the mechanism's circuitry (pp. 2,625–48; see our fig. 12). It was meant to show that this electronic feature of the computer was very slow; in particular, that Atanasoff had been wrong in his estimate, made during direct examination, that the add-subtract mechanism consumed only "a few millionths of a second or one-millionth of a second, or a half a millionth of a second" from input to output (p. 2,081).

There was this exchange (pp. 2,625–32):

Ferrill: This machine that you were working on out at Iowa, Dr. Atanasoff, up to the time you left there in September of 1942, was it a high speed, general purpose digital electronic computer?

Atanasoff: I object to answering that question because it depends upon the definition of the age. In its age it was high speed. In its age it was digital, and it was a computer. Now, it's true that a special purpose had been built into the machine, but it could be used for other purposes.

. . . .

F: Did you ever design a digital electronic computer?

. . . .

A: I think I did, yes.

F: When did you do that, Dr. Atanasoff?

A: I think this machine is a digital electronic computer, the machine herein discussed.

F: Dr. Atanasoff—

A: If it is not, why, I did not design a digital electronic computer, but I am certain this is a digital electronic computer.

. . . .

F: Now, that apparatus [shown in figure 2], that doesn't produce at its output terminals an electric signal or result within a microsecond . . . , does it, Dr. Atanasoff?

A: I realized as I spoke that I had not made a circuit analysis. If counselor has and he found out that it takes longer, counselor may be right.

. . . .

F: Do you suppose it might add up to more like a hundred microseconds?

A: Yes, but I am not sure. I am not sure but if counsel has added a hundred microseconds, that might be correct.

F: Please don't assume that I have.

A: Yes.

F: I am just trying to ask you a question and I am going to let you testify.

A: I understand. Counselor is speaking with confidence and I am shaken by his confidence.

(Laughter)

The Court: Shaken by his confidence or competence?

(Laughter)

A: Confidence. However that may be, you realize that this thing was designed for its purpose at hand and its purpose at hand had no such need or requirement. Then I was pressed by counselor for plaintiff how fast will this really by itself work, and I made these guesses and I will have to recalculate if this becomes a matter of an exact requirement here. However, of course, I was speaking also in general about this type of mechanism . . . and I was speaking about what the limitations of this type of mechanism were and then I made my statement.

Ferrill did want him to recalculate, and ultimately the figure of twenty-five microseconds was reached for the actual working time, that is, for the total response or delay time of the add-subtract mechanism. Atanasoff, however, refused to accept Ferrill's suggestion that even this estimate might be much too low (pp. 2,654–56):

Atanasoff: No. Counselor has forced me to calculations I have never made and the reason I didn't make them was very evident. The rest of the . . . machine was relatively slow and I was sure that the speed of this part was high compared to that. . . .

        .   .   .   .

Ferrill: Would you explain to the Court what you meant by that [last]?

A: Well, it happens that . . . I repeatedly attempted to find an immediate and readily applicable method for avoiding using mechanical movements in the machine, and I found none that were practical for my current operation, so the mechanical motion set the speed of calculation. It is true that this machine has thirty parallel channels of calculation. So the speed is enhanced by this factor but otherwise, the machine was restricted in speed by the mechanical switching which is associated with the rotation of the drum in question.

Later, Ferrill returned to this issue, bringing out now that what we have called the pulse time of the add-subtract mechanism was one-sixtieth of a second, or over 16,000 microseconds (pp. 2,669–70):

Ferrill: Expressed in microseconds, which was the term you were using in that answer on direct examination . . . , the time that was mentioned

[there] was a few microseconds . . . . Would this quantity of time you are
now talking about between one little elemental addition and the next little
elemental addition be 16,667 microseconds?

Atanasoff: We have agreed about—I thought we had agreed about the overall speed of the machine. There isn't any use in arguing about that. It adds
two binary numbers in about one second. Now, that is valid and that one
second is equal to a million microseconds, so there isn't any question
about that. We are, on the other hand, at times speaking about the
response time of the add-subtract mechanism, Figure 2, and at that counsel assisted me and I had come up with an answer which was here a
moment ago, but I guess it was 24 microseconds, roughly speaking—on
the order of magnitude of 24 microseconds as it stands for doing the
addition.

F: Would you accept the proposition that the time lag occurring with that
[add-subtract mechanism] chassis may be considerably longer than that?

A: I wouldn't consider it to be an order of magnitude greater, or any such
thing as that.

. . . .

F: So you are saying that at least you are confident it isn't as long as 250
microseconds?

A: Yes. I think it's much less than that, but that's a relative matter and I
hope it is clear to the Court.

It seems probable to us that, whereas Atanasoff could not recall having
made the calculation some thirty years earlier, he did gain an intuitive grasp of the
delay time of the add-subtract mechanism from watching its activity on the oscilloscope. In any case, he made the important distinction for the judge, between that
delay time and the much longer machine time necessitated by mechanical aspects of
the computer.

Let us note, in passing from this second example of Sperry efforts to down-
grade Atanasoff's accomplishment in the ABC, that the speed of its electronics is
irrelevant to the question of what Mauchly learned from Atanasoff that was of use
in the ENIAC and later computers. Mauchly did, in fact, learn how to switch with
vacuum tubes, so far as the ENIAC was concerned, and how to design an electronic
serial adder, so far as the EDVAC was concerned. Again we conclude that Ferrill
was doing the best he could in a situation where he had very little of substance to
contest.

Counters versus Logical Mechanisms

In our third and last example, Ferrill challenged Atanasoff on the difficulty
he experienced in designing a suitable counter for his computer, during the troubled

period before he thought of using logical add-subtract mechanisms in conjunction with his separate store (pp. 2,394–2,401 and 2,656–72). The implication was strong that Atanasoff had failed not only in this regard, but in what would have been the better way to proceed.

Ferrill began by questioning Atanasoff on his acquaintance with the scale-of-two counters used by physicists in the 1930s (pp. 2,394–95):

> *Ferrill:* And I want to be sure my question is understood. First of all, this scale of 2 circuit . . . which you have referred to in your direct examination also by the popular term flip-flop—do you remember that?
>
> *Atanasoff:* Yes.
>
> *F:* All right—this is a circuit you have known from at least 1937 to date, isn't it?
>
> *A:* Yes. . . .

Atanasoff explained that he had never used such circuits to count cosmic rays, but had seen them in the literature; he added, though, that he had built "a few flip-flops" in about 1936 (p. 2,396). Asked then whether he had investigated the use of flip-flop circuits for the purpose of an abacus, he said that he had, adding (p. 2,397):

> My resources were not great, and this verification consisted of building several circuits and trying to make them work. There was some difficulty with instability. You know, there's much modern art in modern flip-flops and we didn't have that art then . . . . They were made to work in cosmic rays in those days, but they were sensitive to the kind of pulses they got and to the form of the pulses, and . . . these few experiments which I did . . . made me unhappy about that direction of approach.

It should be noted here that the term "flip-flop" is being used by both men in the popular sense Ferrill alluded to at the outset: not for *flip-flop* in the technical sense of a simple two-input device but for the more complex *binary counter* (see chap. 2).

Ferrill then pressed Atanasoff as to whether such "flip-flops" would not have fitted in quite naturally with his use of the base two, had he been able to overcome the instability problem. There was this exchange (pp. 2,399–2,400):

> *Atanasoff:* Yes, I would have had some other problems to solve, sir, and I am well aware of that. I would have had the problems of building an adder and other devices. Now you see—I hesitate there, but, you see, there's really two kinds of adders. One of them is like the ratcheting device and the other one presents the data and then answers come out the rear end. These two methods were flopping around side by side in my mind during these years and—well, I finally went one way, didn't I?

Ferrill: I guess you did.
A: Yes, sir.

Ferrill went on to question the form that an "abacus" made of counters might have taken. In the end, he secured Atanasoff's agreement that he would not have had to rotate his abaci mechanically if he had developed a ring counter (and associated circuits); moreover, that in fact he had neither developed such a counter nor planned how to incorporate it in his computer.

The Sperry side dropped this matter here, only to return to it two days later, again stressing the disadvantage of the ABC's dependence on a rotating capacitor memory (pp. 2,656–60). Atanasoff handled this renewed attack well, noting that he had experimented with a variety of ideas, especially the idea of a magnetic drum memory which could have been rotated much faster. He concluded with what we believe is the essence of any radically novel invention (p. 2,658):

> . . . [With rotating magnetic memory elements] I would have to build 30 amplifiers and power them, and this seemed like a heavy additional burden to my machine. I felt that the machine and the course [in] which the Court has seen it in various exhibits here would represent a major improvement in computing speed, and I believe that this is true at the present time . . . .

The idea Ferrill kept pushing, that of using counters, which incorporate their own memory, instead of logical add-subtract mechanisms working with separate memory abaci, is, of course, the basic scheme of the ENIAC. It is a little ironic that so much should have been made of Atanasoff's abandonment of the counter concept, once he had conceived his logical mechanisms; for *his* method was the better one. Mauchly and Eckert themselves adopted it for the EDVAC, abandoning counters in favor of serial binary adders and a separate memory. Moreover, Atanasoff's fundamental approach had been to build a computer with an unprecedented memory capacity, so that arithmetic devices with their own individual memories were a superfluity he could easily discard for purely arithmetic mechanisms.

It is also ironic that in this lengthy attempt to discredit Atanasoff's rotating memory, there was no recognition that it constituted the invention of the *regenerative* memory, a memory continually refreshed by the electronic arithmetic unit.

There is one further flaw in this Sperry argument, namely, that the mechanical rotations of the memory drums were not the major reason that a vector addition took a full second, despite the electronic speed of the add-subtract mechanisms. Ultimately, it was the binary input-output system, a strictly auxiliary system, that determined the overall computation speed (see chap. 1). Had Atanasoff solved his stability problem with counters, he would still have faced this input-output requirement. We submit that, as he maintained in court, he did indeed succeed in designing an extremely well-balanced, highly complex, first electronic digital computer.

In closing our presentation of the Sperry effort to disparage the ABC, let us note that we chose this last example in part because Atanasoff has received so much criticism for his failure to devise a counter for his computer—criticism that is actually based more on a confusion in terminology than on the true circumstance. As we noted earlier, he used the popular term "flip-flop" to mean "binary counter." And, as we explained in chapter 1, his failure was owing mainly to the unsuitability of a binary counter for serial addition. Moreover, in his 1940 manuscript, Atanasoff did employ the technically correct word, "counter," for the device whose stability gave him difficulty when he tried to use it to compute serially.

Atanasoff's critics, however, have glossed over this distinction between popular and technical usage, to give the impression that Atanasoff could not make a simple flip-flop work reliably. John Mauchly was one such critic. He stated in a paper he read at Los Alamos in 1976: "The time I spent in trying to persuade Atanasoff to use flip-flop and scaling circuits instead of his mechanical commutation was to no avail. As I mentioned earlier, he rejected this approach because he had never got a flip-flop to work reliably" (Mauchly 1980, p. 549). Mauchly himself, as it happened, had not been able to get the sort of *counter* Atanasoff had struggled with to work, even though he had been trying hard to do so for at least six months prior to that June, 1941, visit to Iowa. To the best of our knowledge, Mauchly never did design a counter suited to use in any computing context.

His Procedures

The third and last broad category of Atanasoff's testimony that we have taken to be basic to his credibility, after the quality of his memory and his professional competence, is his procedures in pursuing his goals. In this category, too, he was impressive, partly because his oral expositions were so lucid but chiefly because he had the substance on which to draw. A systematic approach to whatever he attempted was consistently revealed. We will present examples of his testimony on three projects that predated the ABC—that, in fact, formed part of his study leading up to it—and on the computer project itself.

The Laplaciometer

In direct testimony, Atanasoff told of a device that he and two others, a graduate student of his and a member of the engineering faculty, had built during his exploration of analog methods of computing (pp. 1,657–59). They called it a Laplaciometer, because it was used to solve LaPlace's equation, a partial differential equation, directly, not by reducing it to a set of linear equations. For a specific application, a two-dimensional representation of the solution was carved from a hundred-pound block of paraffin; that is, a surface was shaved so as to have a LaPlace operator of zero at every point. The value of the dependent variable at any given point, the height of the surface at that point, was then read from a dial.

Atanasoff also described a corollary device he and his two associates built to perform the carving task automatically.

In cross-examination, Ferrill quizzed Atanasoff closely about this analog device, but succeeded only in bringing out the further facts that he had worked on it for the better part of a year, between 1935 and 1937, and that it was then used in the Engineering Experimental Station of Iowa State College; Atanasoff had even been told just two or three years before the trial that it was still there (pp. 2,372–78).

Modification of IBM Machines

Atanasoff spent a large portion of this same 1935–37 period modifying an IBM tabulating machine to solve sets of linear equations by the elimination procedure. A paper he wrote (apparently unpublished), "Solution of Systems of Linear Equations by the Use of Punched Card Equipment," and a drawing for it, "Schematic Sketch of Auxiliary Apparatus," were introduced during his direct examination, along with an April 22, 1937, letter he wrote to IBM concerning this idea (pp. 1,665–75). Atanasoff explained in court that he had had success in modifying an IBM tabulator for the analysis of spectra (in collaboration with statistician A. E. Brandt), and that he was now considering doing it for sets of equations. Ultimately, he said, he abandoned the scheme as impractical, primarily because of the machine's limited storage capacity.

Again, cross-examination brought out only that Atanasoff had worked on the modification of IBM tabulating machines for perhaps a year and a half, and that he turned from this digital approach to another, the ganging of desk calculators, shortly after writing his April, 1937, letter to IBM (pp. 2,378–82).

The Ganged Calculator Concept

Atanasoff's scheme to interconnect a large bank of desk calculators for the solution of simultaneous linear equations was thoroughly aired in direct examination (pp. 1,678–88). His investigation of this idea lasted about five months, culminating in a September 18, 1937, letter to a Washington, D.C., firm requesting a preliminary Patent Office search (with regard both to it and to the IBM tabulator idea). This and the ensuing letters exchanged with that firm were all introduced in court. Halladay read aloud the following paragraph of the original request letter (p. 1,682):

> A sequence of calculating mechanisms (of the Monroe-Marchant type) with the carriages and rotating mechanisms so interconnected that addition or subtraction in all machines is simultaneous and into corresponding columns on each mechanism. The whole arrangement is so controlled as to permit the reduction to zero of the dial readings of one of the mechanisms as in division on a calculating machine. Now if the successive coefficients and

constant term of one equation are set up on the successive sets of dials of the sequence of mechanisms and if a successive coefficient and constant term of another equation are set up on the successive keyboards of the sequence of mechanisms the results of reduction of the dial readings of one mechanism to zero will be that the readings of the dials of the other mechanisms will represent the coefficients and constant term of an equation with one less unknown variable. A repetition of this process enables the solution of a system of equations.

Halladay then brought out that Atanasoff had disassembled a Monroe calculator enough to understand the details of its operation, and had read up on such machines in encyclopedias and other sources in the college library. He also brought out that this idea, too, was abandoned, because, as Atanasoff explained (p. 1,688):

> The array of calculators, although theoretically an approach to the problem—30 calculators of that size put end to end and coupled up some way would require about—well, let me say 60 feet, and it's getting to be a mechanical monster, and by no means inexpensive, either . . . .

Ferrill mounted his strongest attack on Atanasoff's pre-ABC efforts with respect to this last idea, hammering at the point that he had not made any drawings beyond rough sketches, nor any written estimates of the number and kinds of linkages the bank of machines would require (pp. 2,382–88). Atanasoff remained cool and matter-of-fact throughout; there was this exchange (pp. 2,383–84):

> *Ferrill:* Did you go far enough in your planning to work out how you would do the interconnecting of the plural . . . computing machines?
> *Atanasoff:* I have even today an idea how the connections were to be made.
> *F:* Did you draw up some sketches of it?
> *A:* Did I then?
> *F:* Yes, sir.
> *A:* I don't know. I don't remember any, and maybe I did, maybe I didn't, but I had the idea in mind.
> *F:* Well, did you determine how many connections there would have to be— how many different kinds of connections there would have to be running between one of these desk calculators and its twin sitting alongside?
> *A:* Yes. I have a fair idea as I sit here about that.
> *F:* But you never did work it down to the details to see you could connect them, where the connections would go, and that kind of thing, did you?
> *A:* Yes, I had roughed out in my own mind how the connections would be made, but I didn't have the calculators, and it looked—in toto, it looked like a rather expensive method.
> *F:* Well, you did have one of these calculators there, didn't you?

A: I didn't own a one, no, sir.

F: I mean, the school had one? Wasn't one there and available to you?

A: Oh, yes, the school had four or five, but, you see, somebody else had their fingers on them and they weren't accessible to me. I could borrow one for a short space of time if I didn't "booger" it up, but that was the extent of my power in that direction at that moment.

F: They didn't have any objection to you borrowing one and proceeding to make drawings that would lay out the plans of how to—

A: (Interposing) I could have done that.

F: Excuse me. I didn't get my question finished.

A: I'm sorry.

F: But I think you're answering it and perhaps it is clear enough on the record.

   You could have borrowed it and you could have made the drawings as to how this would be planned out?

A: Yes, sir.

F: But you didn't do that?

A: I don't remember any drawings. If there were, they were hand sketches and not mechanical drawings.

It also came out in cross-examination that neither the search Atanasoff hired nor the one he and Berry made in late 1940 turned up the idea of ganging or chaining calculators on the part of any other inventor.

This second digital computing device that Atanasoff considered before deciding to build a machine of his own is of special interest to us, because it is the mechanical device Mauchly used as an analogy to the "electronic computer" he proposed in 1942 (see chap. 3). It is also the device Irven Travis wrote up for General Electric in 1940, except that he also suggested using *electronic* calculators, which could have overcome the solution-time and wear-and-tear difficulties inherent in the mechanical version (see chap. 3). Of course, Travis and Mauchly both intended such a computer to replace the mechanical differential analyzer, that is, to solve differential equations converted to difference equations, whereas Atanasoff meant to solve partial differential equations converted to linear equations.

Let us conclude our presentation of these preliminary efforts, one analog and two digital, with a note on their place in Atanasoff's progress toward the ABC. Although he always had several different computing ideas under consideration at any given time, he seems to have concentrated primarily on the Laplaciometer through most of 1935, on the modified IBM machines from late 1935 until April, 1937, and on the ganged calculator idea from April, 1937, until September, 1937. And while Ferrill did his best to discredit Atanasoff's commitment to the last of these, even though he had gone so far as to secure a Patent Office search, the clear pattern for this period is one of persistence and diligence in the systematic pursuit of solutions to physical problems. Moreover, the two digital efforts were true precursors of the ABC, intended as they were to solve the same sort of equations by the same elimination method.

The Electronic Computer

As for his approach to the ABC, once the 1939 model of the central apparatus had been built and tested out, Atanasoff gave the following direct testimony (pp. 1,819–20):

> One of the first pieces of work that had to be undertaken was redesign of the add-subtract mechanism which was used in the prototype. One of the objects of this redesign was to decrease the size of the . . . add-subtract mechanism. Another objective was to decrease the cost, not only in money but also in filament current and plate current because we knew we were going to add many units together and the total drain which they would take would clearly be substantial.
>
> I have mentioned previously the tube 12SC7. It looked like a natural and then you have had read to you the letter which describes the fact that we had found the tube manual off, the graph in the tube manual off by a factor of ten on the ordinates . . . . Another objective was to remove the pentodes, since the pentodes only came in single envelopes and used quite a little current. The pentodes were placed in the set originally to insure that the impedance of the output of the add-subtract mechanism and the regenerative production mechanism—they each had pentodes in them, remember—would be sufficient to charge the condensers to substantially full level in the few milliseconds that the contacts . . . of the condensers were in contact with the stationary brushes. Recalculation of that situation showed that was no problem at the speed at which we were working so we . . . passed towards dual triodes and after we discovered that the 12SC7 was not satisfactory, we had two dual triodes left for consideration. Those two dual triodes were the 6F8G and the 6C8G.
>
> There was a discussion of the relative merits of these two but the lower filament current, I believe it was approximately half the value, drove us to the 6C8G and we designed an add and subtract mechanism containing seven envelopes, seven dual triodes of the 6C8G variety.

As we saw in chapter 1, this systematic attack prevailed throughout Atanasoff's design and construction of the entire machine. Let us just cite here a few of its more striking aspects. One was, of course, the constant development and use of test equipment (see chap. 3). Such testing, both for his own purposes and for the benefit of special visitors, actually began in the early months of 1940, he testified, since the machine did not have to be very far along to be demonstrated in the way the prototype had been, with the use of temporary controls; indeed, the original prototype demonstrations were first transferred to the computer in its incipient form and then continually expanded to embrace more "fields," with the oscilloscope shifted from one field to another (pp. 1,893–94).

A second striking aspect of Atanasoff's procedure with the ABC was his constant weighing of alternative designs, right through to the last component. He always held open the possibility that the most unpromising choice might prove best in the end—and sometimes, as in the case of the capacitor memory elements, it did! Another example is his extended experimentation with various means of actuating the sparking electrodes that punched the binary cards, before he settled on thyratrons (see chap. 1).

The last aspect of Atanasoff's procedures that we wish to cite was his practice of retaining the records of his research activities. For the ENIAC trial, he produced 1,500 to 2,000 documents, many of them multi-paged; he had even duplicated all of these, numbered and indexed them in his own classification system, and placed the originals in a bank vault. Ferrill expressed incredulity at their sheer number and at the fact that they had all been saved by Atanasoff for more than thirty years—not, say, obtained from Iowa State or elsewhere (p. 2,504). His having kept them would seem to indicate not only that he recognized the necessity of documenting his work at the time, but that he felt a deep and abiding attachment to his computer.

In any case, the existence of such a complete file proved invaluable to the Honeywell side. Atanasoff's achievements could nearly always be established by recourse to a letter, a bill, a memorandum, an article, a drawing, a photograph, rough notes, even payroll vouchers. Perhaps the most valuable document of all was the manuscript he wrote in August of 1940, to which we have referred throughout this book. It constituted the ultimate proof of what Atanasoff had done, and how he had gone about doing it, in the art of electronic computing.

The Decision on Atanasoff

At the beginning of this chapter, we provided an overview of the *Honeywell v. Sperry Rand and Illinois Scientific Developments* case, which was tried before Judge Earl R. Larson from June 1, 1971, through March 13, 1972, and decided on October 19, 1973. In this section we quote and discuss Larson's findings on derivation from Atanasoff, particularly with regard to the ENIAC, but also with regard to the 30A package of Eckert-Mauchly patents. The latter patents, of course, pertained to systems or components of the EDVAC and subsequent stored-program computers. (We explore the *significance* of Atanasoff's contributions to the ENIAC and the stored-program computers in chap. 5, on his place in history.)

The ENIAC

Larson's first ruling involving Atanasoff came in Finding 1, on public use of the ENIAC more than one year prior to the patent application date. There was this series of paragraphs:

> 1.1.1.5 SR and ISD have further characterized the subject matter of the ENIAC patent as "the invention of the Automatic Electronic Digital Computer," and are bound thereby. Conduct with respect to an automatic electronic digital computer is, therefore, conduct with respect to "the invention."

1.1.1.6 Each of the claims of the ENIAC patent reads on the ENIAC machine as it was constructed and placed in operation at the Moore School and described in the Final Reports.

1.1.1.7 The ENIAC machine which was represented by Eckert and Mauchly to be that which "embodies our invention" is identical with the ENIAC invention, however claimed.

1.1.1.8 Counsel for defendants did not object to the Court's statement at trial that there was no dispute about the fact that Eckert and Mauchly claimed to be the two sole joint inventors of the ENIAC, from input all the way through output.

1.1.1.9 For the foregoing reasons, there is no necessity to make specific reference to the individual claims of the ENIAC patent where conduct barring the valid issuance of a patent is conduct involving either the same ENIAC machine (as will be set forth hereinafter with respect to the bars of public use and on sale), or involving a prior automatic electronic digital computer (as will be set forth hereinafter with respect to the bars of derivation from Atanasoff and prior publication by von Neumann).

An important point was made here. Because the defendants were claiming invention of *the* automatic electronic digital computer—and were demanding royalties from all parties producing any such computer—Honeywell did not have to attack particular claims so far as prior inventor Atanasoff was concerned. Honeywell did in fact cite several claims as invalid on account of his work, but it was not required to do so in order to have the entire patent invalidated. Such was not the case, however, for alleged *co*inventors at the Moore School, for whom Honeywell failed to cite particular claims.

This Sperry stance was not simply a mistake, in our opinion; rather, as we argued earlier, it reflected the excesses of Sperry Rand at trial and, in the first instance, the excesses of Mauchly and Eckert. (It is curious that on the very first page of the patent, Eckert and Mauchly termed the ENIAC "the first *general purpose* automatic electronic digital computing machine known to us [italics added]." If they had held to this qualification throughout, they would have had, we believe, a valid patent.)

Larson presented his major ruling on derivation of the ENIAC from Atanasoff in Finding 3 of his decision. We quote this finding in its entirety before commenting on some aspects of it:

3.1 The subject matter of one or more claims of the ENIAC was derived from Atanasoff, and the invention claimed in the ENIAC was derived from Atanasoff.

3.1.1 SR and ISD are bound by their representation in support of the counterclaim herein that the invention claimed in the ENIAC patent is broadly "the invention of the Automatic Electronic Digital Computer."

3.1.2 Eckert and Mauchly did not themselves first invent the automatic electronic digital computer, but instead derived that subject matter from one Dr. John Vincent Atanasoff.

3.1.3 Although not necessary to the finding of derivation of "the invention" of the ENIAC patent, Honeywell has proved that the claimed subject matter of the ENIAC patent relied on in support of the counterclaim herein is not patentable over the subject matter derived by Mauchly from Atanasoff. As a representative example, Honeywell has shown that the subject matter of detailed claims 88 and 89 of the ENIAC patent corresponds to the work of Atanasoff which was known to Mauchly before any effort pertinent to the ENIAC machine or patent began.

3.1.4 Between 1937 and 1942, Atanasoff, then a professor of physics and mathematics at Iowa State College, Ames, Iowa, developed and built an automatic electronic digital computer for solving large systems of simultaneous linear algebraic equations.

3.1.5 In December, 1939, Atanasoff completed and reduced to practice his basic conception in the form of an operating breadboard model of a computing machine.

3.1.6 This breadboard model machine, constructed with the assistance of a graduate student, Clifford Berry, permitted the various components of the machine to be tested under actual operating conditions.

3.1.7 The breadboard model established the soundness of the basic principles of design, and Atanasoff and Berry began the construction of a prototype or pilot model, capable of solving with a high degree of accuracy a system of as many as 29 simultaneous equations having 29 unknowns.

3.1.8 By August, 1940, in connection with efforts at further funding, Atanasoff prepared a comprehensive manuscript which fully described the principles of his machine, including detail design features.

3.1.9 By the time the manuscript was prepared in August, 1940, construction of the machine, destined to be termed in this litigation the Atanasoff-Berry computer or "ABC," was already far advanced.

3.1.10 The description contained in the manuscript was adequate to enable one of ordinary skill in electronics at that time to make and use an ABC computer.

3.1.11 The manuscript was studied by experts in the art of aids to mathematical computation, who recommended its financial support, and these recommendations resulted in a grant of funds by Research Corporation for the ABC's continued construction.

3.1.12 In December, 1940, Atanasoff first met Mauchly while attending a meeting of the American Association for the Advancement of Science in Philadelphia, and generally informed Mauchly about the computing machine which was under construction at Iowa State College. Because of Mauchly's expression of interest in the machine and its principles, Atanasoff invited Mauchly to come to Ames, Iowa, to learn more about the computer.

3.1.13 After correspondence on the subject with Atanasoff, Mauchly went to Ames, Iowa, as a houseguest of Atanasoff for several days, where he discussed the ABC as well as other ideas of Atanasoff's relating to the computing art.

3.1.14 Mauchly was given an opportunity to read, and did read, but was not permitted to take with him, a copy of the comprehensive manuscript which Atanasoff had prepared in August, 1940.

3.1.15 At the time of Mauchly's visit, although the ABC was not entirely complete, its construction was sufficiently well advanced so that the principles of its operation, including detail design features, were explained and demonstrated to Mauchly.

3.1.16 The discussions Mauchly had with both Atanasoff and Berry while at Ames were free and open and no significant information concerning the machine's theory, design, construction, use or operation was withheld.

3.1.17 Prior to his visit to Ames, Iowa, Mauchly had been broadly interested in electrical analog calculating devices, but had not conceived an automatic electronic digital computer.

3.1.18 As a result of this visit, the discussions of Mauchly with Atanasoff and Berry, the demonstrations, and the review of the manuscript, Mauchly derived from the ABC "the invention of the automatic electronic digital computer" claimed in the ENIAC patent.

3.1.19 The Court has heard the testimony at trial of both Atanasoff and Mauchly, and finds the testimony of Atanasoff with respect to the knowledge and information derived by Mauchly to be credible.

We have already argued, over the course of this book, for nearly all of the points made here by Judge Larson. We wish to comment further on just three: his reference to the ABC as itself "a prototype or pilot model" (¶ 3.1.7); his mention of "other ideas of Atanasoff's" discussed during Mauchly's visit to Ames (¶ 3.1.13); and his finding that "the subject matter of detailed claims 88 and 89 of the ENIAC patent corresponds to the work of Atanasoff" known to Mauchly (¶ 3.1.3).

In calling the ABC a prototype or pilot model, Larson was acknowledging that Atanasoff meant to go on to other computers of more elaborate design and function. And indeed he did mean to; as we saw in chapter 3, Atanasoff wrote to Mauchly on January 23, 1941, and again on May 31, 1941, that he hoped to build an electronic digital integraph in the near future. It seems to us entirely natural, not only that he should have seen this possibility while he was still building the ABC— even as Eckert and Mauchly began to envisage the EDVAC while they were still building the ENIAC—but also that, under normal circumstances, he would have proceeded to some such higher level of computing technology. Some writers have seized on Larson's use of the expression "pilot model" to say or imply that Atanasoff built only a *model*, not a real computer; in so doing, they simply overlook the actual character of the ABC *and* the existence of the earlier "breadboard model" cited by Larson in the same paragraph.

The second point to which we want to call attention, the judge's statement that Mauchly discussed "other ideas of Atanasoff's relating to the computing art" while in Ames, is related to the first. For the one "other idea" explicitly confirmed by both men at trial was that of the electronic digital differential analyzer Atanasoff hoped to build when he found the time. In our view, Larson's reference to that idea in this brief historical judgment indicates his acceptance of its influence on Mauchly in the conception and design of the ENIAC as an electronic replacement for the differential analyzer.

The third point we want to address is one to which we have only alluded thus far, namely, Larson's citing of particular ENIAC patent claims as representative of those derived from Atanasoff. In this Finding 3, he cited claims 88 and 89—he was to cite one other in a later finding—even as he re-enunciated his earlier ruling that it was not necessary to show derivation of any particular claims in order to prove the patent invalid. In our opinion, claim 88 was less reflective of actual derivation from Atanasoff than was claim 89. That is, it encompassed a truly novel feature of the ENIAC, but failed to specify that feature in its overly broad language!

To state the matter in modern terminology, claim 88 addressed communication between a computer's input-output devices and its central processor, which for most machines operate at different, unsynchronizable rates of speed. The ENIAC solved this difficulty through a buffer memory that could work at both rates. The framers of the claim, however, did not mention that this memory *was* the means of communicating between the two systems, let alone that it was capable of doing so between *asynchronous* systems. In short, they specified a central processor clocked at one rate, peripheral data handlers clocked at a second rate, and a memory capable of communicating with both, but then claimed only that the processor and the data handlers were thereby "made compatible."

Now Atanasoff's computer had these same three basic elements and means of communicating among them: decimal-card reader transmitting one decimal digit per second, add-subtract mechanisms operating at sixty cycles per second, and both communicating to the counter drum memory. His card reader and internal computing mechanisms, however, were not asynchronous; for he had solved their communication problem by synchronizing them, not by using a memory as intermediary between them.

Claim 89 was a different story. It stipulated the main components of a computer from the point just after initial input through final output, all of which components were to be found in Atanasoff's machine: "memory means for storing data to be processed" (his two main drums), "reading means for reading said data from said memory means and for converting the data into pulse form" (his brushes for reading capacitors of those two drums), "a multiplicity of interconnected pulse responsive processing units for producing output pulses representative of the results of a processing operation, at least one of said units being connected to said reading means to receive the pulses emitted by the reading means" (his add-subtract mechanisms), and "printing means connected to at least one of said units for printing the results of said processing operation" (his binary output punch)—note that both the ENIAC and the ABC "printed" their results by punching numbers on cards.

Larson's citation of a third ENIAC patent claim that was derived from Atanasoff's work, claim 90, occurred in Finding 12, on validity, to which we turn next. Again, we quote relevant portions before we comment on particular aspects of it, including the selection of claim 90 by Honeywell as representative. It is important to have in mind that this finding dealt both with the law on patent validity in general and with the law on the particular issues of public use, on sale, publication, prior inventors, and so on, that had already been explored in previous findings. In other words, having found that the ENIAC patent did or did not meet certain requirements of patent law, Larson had now to state and justify his decision as to validity or invalidity on each of the several scores. He made the following pronouncements with regard to derivation from Atanasoff:

12.2.5 Eckert and Mauchly did not themselves first invent "the automatic electronic digital computer," which SR and ISD contend to be the subject matter of the ENIAC patent, but instead derived that broad subject matter from Dr. John V. Atanasoff, and the ENIAC patent is thereby invalid.

12.2.5.4 The utilization of ideas in a device prior to the time of the alleged invention, whether or not the device was subsequently abandoned, is evidence that when those ideas are incorporated in a later development along the same line, they do not amount to invention.

12.2.7.3 Eckert and Mauchly did not themselves invent the subject matter set forth in at least claims 88 and 90 but instead derived knowledge of that subject matter from John V. Atanasoff.

12.3 I find and conclude that the ENIAC is invalid and unenforceable.

12.3.1 See Findings 1 [Public Use], 2 [On Sale], 3 [Atanasoff], 7 [First Draft Report] and 10 [Pulse] above as establishing invalidity.

We want to comment on two points here: that an alleged invention incorporating ideas from a prior abandoned device did not "amount to invention" (¶ 12.2.5.4); and that "at least claims 88 and 90" of the ENIAC patent were derived from Atanasoff (¶ 12.2.7.3).

It has been argued that Atanasoff, because he did not resolve the difficulty with his binary input-output system, did not "finish" his computer and, in fact, ultimately abandoned it; and that, accordingly, the ENIAC could not have been derived from it. Larson had now addressed this argument, making clear its irrelevancy. Eckert and Mauchly incorporated ideas from the ABC in the ENIAC, and the outcome was therefore not invention by them. As we have noted a number of times, these appropriated ideas were also critical to the electronic computing art; indeed, had they not been, Eckert and Mauchly would not have resorted to them.

Claim 90 of the ENIAC patent stipulated an "apparatus comprising pulse responsive storage means" (Atanasoff's two main drums) having "a plurality of orders for storing received pulses" (orders from an operator pushing buttons, turning switches, and plugging cables), input and output means for that store (his card systems and his writing and reading brushes), and means of transmitting pulses to and fro (his cables between the drums and the add-subtract mechanisms), under "signal responsive control means" (controls of his arithmetic unit).

Honeywell's selection of ENIAC claims 88, 89, and 90 for submission to Larson as "representative" of those derived from Atanasoff does seem to have been a good one. The first concentrated on rendering compatible two differently timed subsystems; the second encompassed the main components of an electronic computer; and the third concentrated on a pulse-responsive memory and its associated input, output, and pulse transmission systems. We have ourselves found 15 other claims, for a total of 18 from the patent's 148, that we believe also read on the work of Atanasoff. We discuss these after presenting the one remaining issue in Larson's decision on derivation of the ENIAC from Atanasoff.

This last issue was that of fraud on the Patent Office, treated in Finding 13.

But because Finding 12 had dealt with the issue of validity in all of the preceding issues, he had closed it with an advance ruling on fraud:

12.3.3 Although not made a basis for a finding of willful and intentional fraud, the various derelictions of Eckert and Mauchly and their counsel before the Patent Office, as set out at Finding 13 below, render the ENIAC patent unenforceable.

This paragraph meant that, whereas the ENIAC patent had been found invalid on other grounds, it was now found only unenforceable, not invalid, on the ground of fraud.

There were then the following paragraphs on Atanasoff in Finding 13:

13.5 M[auchly] and E[ckert] took oaths stating that they were the sole inventors, that there had been no public use of the ENIAC before the critical date, and that the ENIAC was not on sale before the critical date.

13.19 The work of Atanasoff was current and was of great importance to M[auchly].

13.19.1 Detailed findings and conclusions concerning the work of Atanasoff and John Mauchly's knowledge of that work have previously been set forth under Finding 3, above.

13.19.2 Prior to his visit to Ames, Iowa, Mauchly had been broadly interested in electrical analog calculating devices, but had neither conceived nor built any electronic digital calculating device.

13.19.3 In a letter dated June 29, 1941, to H. Helm Clayton, John Mauchly described Atanasoff's work, and its relationship to Mauchly's prior thinking, as follows:

"Immediately after commencement here, I went out to Iowa State University to see the computing device which a friend of mine is constructing there. His machine, now nearing completion, is electronic in operation, and will solve within a very few minutes any system of linear equations involving no more than thirty variables. It can be adapted to do the job of the Bush differential analyzer more rapidly than the Bush machine does, and it costs a lot less. My own computing devices use a different principle, more likely to fit small computing jobs."

13.19.4 After his visit with Atanasoff, Mauchly left his employment at Ursinus College and joined the staff of the Moore School of Electrical Engineering at the University of Pennsylvania.

13.19.5 Mauchly took a short course in electronics at the Moore School and then joined the faculty, during which time he began to consider applying his understanding of the new impulse or digital principles he had been taught by Atanasoff.

13.19.6 On September 30, 1941, Mauchly wrote to Atanasoff from the Moore School:

"A number of different ideas have come to me recently anent computing circuits—some of which are more or less hybrids, combining your methods with other things, and some of which are nothing like your machine. The question in my mind is this: Is there any objection, from your point of view, to my building some sort of computer which incorporates some of the features of your machine? For the time being, of course, I shall be lucky to find time and material to do more than merely make exploratory tests of some of my different ideas, with the hope of getting something very speedy, not too costly, etc.

"Ultimately a second question might come up, of course, and that is, in the event that your present design were to hold the field against all challengers, and I got the Moore School interested in having something of the sort, would the way be open for us to build an 'Atanasoff Calculator' (à la Bush analyzer) here?"

13.19.7 Dr. Atanasoff responded that, while he had no qualms about having disclosed his ideas to Mauchly, he did not wish to have his concepts made public until adequate steps had been taken to obtain patent protection for his ideas.

13.20 The latter had further contact with Atanasoff in 1944 and invited him to the public demonstrations in February 1946.

13.20.1 The April, 1943, proposal for the ENIAC contract referred to the Atanasoff work, but did not identify it.

13.20.2 In August, 1944, Eckert and Mauchly visited with Atanasoff on the same day they began the process of filing patent applications involving subject matters which stemmed from Atanasoff's prior work.

13.20.3 The apparent purpose of this visit to Atanasoff in 1944 was to seek his assistance in the perfection of the recirculating delay line memory for EDVAC.

13.20.4 The purpose of Eckert and Mauchly's visit with Ordnance patent lawyers on the same day was to lay plans for making patent claims to ENIAC and EDVAC inventions, including the recirculating memory.

13.20.5 Neither Eckert nor Mauchly disclosed Atanasoff's work to their attorneys prior to filing the ENIAC patent application.

13.21 If Atanasoff had proceeded in 1942 and 1943 to file a patent application, the information in the application would have been available to the Patent Office.

13.21.1 The ABC was described in a definitive manuscript and in a draft patent application specification which was prepared by Clifford Berry but, because it was never filed, the Patent Office had no means by which it could have become aware of the ABC or of Mauchly's prior knowledge of the ABC.

13.22 Complete candor with and disclosure to the Patent Office is required.

13.23 At the same time, knowledge of the applicant may come from a variety of sources and from many years of education and experience.

13.24 What should be disclosed to the Patent Office as possible sources of invention, prior art or derivation must in some degree be left to the judgment and conscience of the applicant.

13.25 M[auchly] may in good faith have believed that the monstrous machine he helped create had no relationship to the ABC or Atanasoff.

13.25.1 Mauchly may in good faith have believed that he did not derive the subject matter claimed in the ENIAC patent from Atanasoff. In September, 1944, he wrote a summary of the situation as he then saw it: "I though(t) his (Atanasoff's) machine was very ingenious, but since it was in part mechanical (involving rotating commutators for switching) it was not by any means what I had in mind."

13.25.2 Atanasoff saw the ENIAC machine as it existed on October 26, 1945, and in early 1946 extensive publicity was given to the ENIAC project, acknowledging Eckert and Mauchly as the inventors, but Atanasoff did not assert that the ENIAC machine included anything of his until two decades later.

13.25.3 Of the 17 claims of the ENIAC patent at issue in this suit, Honeywell has failed to prove the readability of claims 8, 9, 36, 52, 55, 56, 57, 65, 69, 75, 78, 83, 86, 109, 122 and 142, or any of them, on Atanasoff's machine or any other work of Atanasoff.

13.39 The Court finds and concludes that despite the various derelictions of M[auchly] and E[ckert], defendants and their counsel that the claim of willful and intentional fraud on the Patent Office has not been proved by clear and convincing evidence.

13.39.1 Honeywell has not proven by clear and convincing evidence any willful and intentional fraud on the Patent Office concerning the ENIAC patent.

13.39.9 Honeywell has not proved by clear and convincing evidence that Eckert and Mauchly, their attorneys, successors or assigns, committed willful and intentional fraud on the Patent Office in connection with the work of Atanasoff.

13.39.17 To prove an antitrust violation based on fraud on the Patent Office, the proof of fraud must be by clear, unequivocal and convincing evidence; a mere preponderance of the evidence is not enough.

We will comment on several points here: that the ENIAC proposal referred, implicitly, to Atanasoff's prior work (¶ 13.20.1); that the fact of Atanasoff's not filing a patent application in 1942 or 1943 was pertinent (¶¶ 13.21 and 13.21.1); that Mauchly "may in good faith have believed" the ENIAC was unrelated to Atanasoff or his computer (¶¶ 13.25 and 13.25.1); and that Honeywell had "failed to prove" that sixteen of the "17 claims of the ENIAC patent at issue in this suit" read on the ABC or any other work of Atanasoff (¶ 13.25.3).

As to the first point, we simply want to note that Larson here affirmed our own argument that one of the two "electronic difference analyzers" said in the 1943 ENIAC proposal to have been considered earlier was indeed a reference to Atanasoff (see chap. 3).

On the second point, we would observe that Atanasoff's not actually filing a patent application for the ABC is an intriguing omission, one that he addresses in his article (Atanasoff 1984, pp. 254–57) and that we also plan to explore in a later work, in the context of institutions vis-à-vis individuals.

On Larson's ruling that Mauchly may have truly believed the ENIAC was unconnected to Atanasoff's work, as evidenced by his September, 1944, "diary" notation, we have stated our opinion in chapter 3. This was to the effect that Mauchly was really rewriting the history at a critical juncture in his progress toward two patents (ENIAC and EDVAC) that claimed ideas he had gotten from Atanasoff more than three years earlier. We regard Larson's decision here as one more instance of his bending over backward to be fair to the Sperry side, so as to produce an outcome to the trial that would not be appealed.

The last point we wish to discuss is a puzzling one. Larson found that Honeywell had failed to prove derivation from Atanasoff of all but one of "the 17 claims of the ENIAC patent at issue in this suit." But these seventeen claims were originally cited by Illinois Scientific Developments as representative of claims Honeywell had infringed! We can see no reason why Honeywell attorneys would even have tried to show that this particular set read on *Atanasoff's* work. And, so far as we know, they did not; rather, they showed that a different set of three claims read on his work, among them claim 88, the only one in the ISD list for which Larson (by omission) acknowledged derivation from Atanasoff. In fact, our own analysis, to which we now turn, yielded three others in the list, claims 8, 9, and 36, that we think also read on Atanasoff's work.

Other ENIAC Claims

The eighteen ENIAC claims that we consider derived from Atanasoff are: 8, 9, 10, 11, 36, 39, 81, 82, 88, 89, 90, 100, 115, 131, 132, 137, 138, and 140. Most of them covered electronic digital computers or computer components in fairly broad terms, without regard to number base, type of storage, type of automatic control, type of electronic arithmetic circuit, or particular arithmetic operations. Specifically, they covered means for producing, transmitting, timing, switching, and controlling electrical pulses. All fell within the scope of Atanasoff's conception prior to Mauchly's visit.

We quote in full and discuss six claims from this list that complement the three put forth by Honeywell as representative (88, 89, and 90). Bracketed numbers before components have been added to facilitate reference to them in our analyses.

Claim 8:

In [1] an electronic computing system [2] a multiplicity of pulse responsive units constructed to receive [3] numbers pulses having numerical significance and to receive [4] control pulses having definite control characteristics, [5] means to transmit to said units control pulses and numbers pulses in a correlated order including means to establish a selective relation between certain control pulses and certain predetermined units, said units being constructed to receive numeratively numbers pulses when stimulated by a control pulse, [6] certain of said units being constructed for arithmetic operations and including means responsive to terminal arithmetic operation therein to transmit a control pulse effective on predetermined units for intercommunication of numerical content between one and another of said units.

Counterparts in the ABC were as follows:

[1] The electronic computing system was the ABC.
[2] The pulse responsive units were the add-subtract mechanisms, together with their associated restore-shift mechanisms and carry-borrow drum, and the two memory drums.
[3] The numbers pulses were the binary digits.

[4] The control pulses were the pulses sent to switches to control the transmission of numbers between units; the boost pulses; and the control pulses that indicated a change of sign in the vector elimination process and ordered a reversal of operation, and that also counted the number of shifts in that process and stopped it when it was finished.

[5] The transmitting means were the means for transmitting pulses between the units of [2] and for transmitting control and timing pulses to those units in the order required to cause certain units to work together and perform certain communication functions and arithmetic operations.

[6] Certain of said units were the add-subtract mechanisms, each of which when finished sent a control pulse that caused certain predetermined units to communicate. Specifically, when a variable was eliminated, the controls caused the contents of the counter drum to be punched out.

Notice that the controls we have cited for Atanasoff's computer in regard to claim 8 were electromechanical, whereas the ENIAC's controls were electronic. Because the claim did not specify that all components of the electronic computing system be electronic, however, it did read on the ABC. It is also the case that the 1940 Atanasoff manuscript specified *electronic* controls, which would have been relatively easy to design and build (see chap. 1).

We did reject as not reading on Atanasoff's work many claims where the context suggested electronic controls even though they were not specified. And we included only one that did specify electronic controls, but as a belated shift in terminology. That was claim 39, which we now quote.

Claim 39:

In [1] data processing apparatus; [2] a first network including a plurality of electric circuits passing through a plurality of distinct mutually exclusive states of electrical equilibrium in response to the application of signal pulses thereto, [3] a second electric network having at least two different states of electrical equilibrium and normally in one of said states, and [4] an electric circuit transferring said second network from said one of said states to the other of said states in response to the establishment of predetermined conditions in said first electric network, and [5] data transmission means controlled by said second electronic [*sic*] network.

Translation to Atanasoff's work is as follows:

[1] The data processing apparatus was the ABC.
[2] The first electric network was the add-subtract mechanisms and the carry-borrow drum.
[3] The second electric network was the bi-stable device in the controls that controlled whether the designated add-subtract mechanism added or subtracted. It was set for addition during communication, decimal-

binary conversion, binary-card reading and punching, and idling. (It was used for the vector elimination process and for binary-decimal conversion.)

[4] The electrical circuit was the network that detected sign change.

[5] The data transmission means were the wires from the bi-stable device described in [3] to the add-subtract mechanisms.

Thus claim 39, too, can be said to have read on the ABC itself, if the shift to *electronic* of the second specified electric network is ignored; otherwise, it read on the ABC *and* the 1940 descriptive paper where electronic controls were envisioned.

Claim 36:

In [1] data handling apparatus, [2] a plurality of information receiving and delivering devices, [3] an information interchange network linking said information receiving devices, [4] a signal network delivering a predetermined pattern of control and information impulses to said information interchange network, and [5] apparatus for manually controlling the advance of said signal network through said impulse pattern.

Counterparts in the ABC are:

[1] The data handling apparatus was the ABC.

[2] The information receiving and delivering devices were all four drums, the add-subtract and restore-shift mechanisms, and all input-output devices.

[3] The information interchange network was the communication paths among those devices.

[4] The signal network was the timing commutators, automatic control circuits, and manual controls and plugs, together with the wires that carried their signals throughout the machine.

[5] The manual apparatus was the one-cycle switch.

This, then, was a briefly stated, but extremely broad, claim. It could have covered the entire ABC, or it could have covered a particular system within the machine, as, for example, the decimal-binary conversion system.

Claim 82:

In [1] a data processing system, the combination of [2] reading means to accept data to be processed, [3] intermediate storage means coupled to said reading means to receive data therefrom and temporarily store said data in the form of electrical signals according to data segments wherein each data segment has the same number of order positions, [4] pulse generating means connected to said intermediate storage means to generate one or more pulses in response to data segments stored therein, the pulse or pulses generated in response to each segment being representative of a numerical value of the segment, and [5] data processing means for receiving said pulses and performing a data processing operation thereon.

The correspondence to the ABC is as follows:

[1] The data processing system was the ABC.
[2] The reading means were the decimal- and binary-card reading devices.
[3] The intermediate storage means were the two memory drums.
[4] The pulse generating means were the rotation of the memory drums and the reading brushes associated with those drums.
[5] The data processing means were the add-subtract mechanisms and their associated restore-shift mechanisms and carry-borrow drum.

It will be seen that, like claim 89 cited by Honeywell and accepted by the judge, claim 82 was a good example of a basic ENIAC claim that read very closely on Atanasoff's computer. The next one we quote, claim 100, read very closely on just one aspect of the ABC—so closely that it is unnecessary to spell out the correspondence. Like claim 39 above, however, it included an aspect not incorporated in the ABC but envisioned in the 1940 manuscript.

Claim 100:
   A data processing machine comprising first and second means for manipulating two number quantities for division of one as a numerator and the other as a denominator, each number having a plurality of orders, and program means for controlling the production of a quotient of said two quantities, said programming means including means for subtracting the denominator value from the numerator value until an overdraft occurs, sign indicating means for producing a sign signal indicative of the sign of the remainder, means responsive to said sign signal for shifting the remainder and denominator relative to each other so that the denominator is less than the remainder, means for adding the shifted remainder and denominator until a second change in sign of the result occurs to produce a sign signal responsive to the change in sign of the remainder, means activated by said signal for again shifting the remainder and denominator relatively so that the denominator is again the smaller and subtracting it from the remainder, and means for signalling the number of times each addition and subtraction is performed.

Clearly, claim 100 described Atanasoff's nonrestoring method of elimination exactly, but as a method of *dividing,* with the quotient "counted" out in the process. As we saw in chapter 1, Atanasoff had in his 1940 paper reserved the possibility of dividing by repeated subtraction. Moreover, the ABC incorporated the ability to add and subtract, shift numbers, and sense sign changes—the major design challenges for automatic division; he needed only the final step of keeping track of the quotient. For these reasons, we consider that claim 100 read on the ABC *and* the paper that Mauchly was allowed to peruse. Notice in this regard that Larson often cited Atanasoff's computer *and his other work,* or the manuscript specifically, or simply Atanasoff's work, as Mauchly sources, in just the way we have for claims 39 and 100.

Claim 115:

In combination, [1] a data processor comprising a plurality of component parts, [2] a source of recurrent timing pulses, [3] first means for transmitting timing pulses from said source to said plurality of component parts to advance said data processor through a series of operations, [4] means for disabling said first transmitting means, and [5] manually operable second means for transmitting only a predetermined number of said timing pulses to said plurality of component parts upon operation of said second means.

The translation to the ABC is as follows:

[1] The data processor was the ABC.

[2] The source of recurrent timing pulses was the timing commutators on the fast and slow shafts.

[3] The first means for transmitting timing pulses were the circuits for transmitting the timing signals from the commutators.

[4] The means for disabling the first transmitting means were the controls that stopped the elimination process when it was completed.

[5] The manually operable second means for transmitting timing pulses was the one-cycle switch.

Thus, of these six claims we have quoted, we found that claims 8, 36, 82, and 115 read on Atanasoff's computer, while claims 39 and 100 read on his work more broadly construed. The other twelve we cited all read on the computer itself. Claims 88, 89, and 90 cited by Larson read on the ABC without resort to Atanasoff's 1940 manuscript or ideas otherwise expressed. Claims 11, 131, 132, 137, 138, and 140 were extensions of claim 8. Claim 9 was similar to claim 8 in concept, and claim 10 was an extension of claim 9. Finally, claim 81, which dealt with codes and encoding means, was satisfied by the ABC's decimal-binary conversion system.

We would like to conclude this presentation of ENIAC patent claims by noting that we do not write as patent lawyers but rather from a familiarity with both the ENIAC and the ABC. We do believe that our selection of claims derived from Atanasoff is a conservative one, that many claims not included by us also read on his work. We rejected some, for example, because they were vaguely worded, others because they contained a single stipulation that went beyond what the ABC could do. In short, it is our opinion that the ENIAC patent claims reflected the ideas of Atanasoff in a pervasive and fundamental fashion.

The 30A Patents

In his Finding 14, Judge Larson ruled on the entire 30A package of twenty-five patents and patent applications mentioned at the beginning of this chapter. Like the ENIAC patent, all of these pertained to alleged inventions of Eckert and/or Mauchly, with assignment to Sperry Rand. Because Honeywell did not ask him to rule on their "technical validity or invalidity," Larson found them only unenforceable—after delving into the history of the individual items, however, and citing barring infirmities in each case.

Derivation from Atanasoff was at issue with respect to his regenerative memory and his binary serial adder (add-subtract mechanism), as used in the EDVAC and other computers that succeeded the ENIAC; that is, Honeywell claimed such derivation. We present Larson's responses to these in turn. As we noted at the beginning of this section, we address the significance of Atanasoff's contributions, both to the ENIAC and to the stored-program computers, in chapter 5.

The Regenerative Memory

Regenerative Memory patent 2,629,827, or EM-1 in the 30A list, was applied for in late 1947 and granted in early 1953. It had thus expired by the time of the ENIAC trial decision. As we have seen, it had been the subject of a suit filed some years earlier by Sperry Rand against Control Data Corporation and ultimately settled out of court. We now quote the paragraphs of Larson's finding that bore on derivation from Atanasoff, together with their contextual setting:

14.7 The emphasis in plaintiff's claim of infirmities is on public use and on sale.

14.8 Other claims rest on derivation from Atanasoff, incomplete application at time of execution, and omission of co-inventors.

14.9 As to publication, plaintiff relies on the First Draft Report (already found to be a printed publication before the critical date) [Findings 7.1–7.1.6 above], the EDVAC report of September 30, 1945, the EDVAC report of June 30, 1946, three lectures of the Moore School lecture series, and the report on the UNIVAC.

14.11 I find that the publications referred to above (In Finding 14.9) were printed publications before the critical dates, that the claims as to public use and on sale before the critical dates have been proved, that the claims as to derivation from Atanasoff as to EM-1 have been proved, that the claims as to incomplete execution have been proved, and that the claim as to the omission of a certain inventor has been proved.

14.11.1.22 In addition to the barring of the valid issuance of the ENIAC patent, the publication of the von Neumann First Draft Report also anticipates the claims of the EM-1 patent No. 2,629,827 entitled "Regenerative Memory."

14.11.3 Subject matter claimed in the EM-1 patent was derived from Atanasoff.

14.11.3.1 On October 31, 1947, Eckert and Mauchly filed an application describing various memory systems, designated case EM-1, that resulted in U.S. Patent No. 2,629,827 (the '827 patent).

14.11.3.2 Subject matter claimed in the EM-1 application as the joint invention of Eckert and Mauchly was disclosed to Mauchly by Atanasoff in June of 1941.

14.11.3.3 In one embodiment of the EM-1 application, information is stored in a coded sequence of pulses, the pulses being temporarily recorded on a rotating carrier as electrostatic charges, carried by rotation to another station where they give rise to electrical potential pulses which are handled through an external feedback circuit for replacement or reinforcement of the pulses on the carrier.

14.11.3.4 This subject matter as claimed in the '827 patent was anticipated by the disclosure contained in the Atanasoff manuscript disclosed to Mauchly.

14.11.3.5 Atanasoff's concept of the recirculating or regenerative memory was used in the EDVAC program, with Atanasoff's rotating electrostatic charge carrier being replaced by the recirculation of pulses through an electrical delay line; this delay line version of a recirculating memory was disclosed in the EM-1 application as an embodiment of Eckert and Mauchly's invention.

14.11.3.6 The Atanasoff electrostatic charge version of a recirculating memory was also disclosed in the EM-1 application as yet another embodiment of Eckert and Mauchly's alleged invention.

14.11.3.7 In October 1953, after the '827 patent was granted on the EM-1 application, Eckert stated that prior to 1942, Atanasoff had developed what was probably the first example of what could generally be termed regenerative memory. Eckert's knowledge of Atanasoff's prior work was based on what Mauchly had earlier told him.

14.11.3.8 On April 1, 1964, SR charged Control Data Corporation with infringement of the '827 patent (EM-1).

14.11.3.9 On February 2, 1965, SR charged Potter Instrument Company with infringement of the '827 patent (EM-1).

14.11.3.10 SR also called the '827 patent (EM-1) to other computer manufacturers' attention, including Honeywell, as a part of its basic EDP patent portfolio.

Thus Larson found not only that Eckert and Mauchly had appropriated Atanasoff's basic concept of regenerating a computer memory from a separate electronic source (his arithmetic unit), but that they had claimed *his form* of such a memory, the rotating electrostatic memory (see chap. 3). Their doing so must be attributed, we believe, to a desire to cover every conceivable application of memory regeneration. That Eckert, having been granted this patent, would then go on to call attention in print to Atanasoff's prior regenerative memory seems to us an amazing bit of overconfidence.

It is also noteworthy that Larson, in finding the Regenerative Memory patent anticipated by von Neumann's First Draft Report, has tied Atanasoff's work not just to that patent but to the EDVAC which used it (and so to the BINAC, the UNIVAC, and others that followed).

The Binary Serial Adder

The 30A package actually contained the patents of several serial binary adders and also one computer that made basic use of such an adder. These were: EM-21, Serial Binary Adder; EM-23, Unit Adder; EM-25, Serial Binary Full Adder; EM-26, Parallel Binary Adder; and EM-22, BINAC System. The patent on the BINAC (Binary Automatic Computer), essentially a pair of simplified EDVACs coupled together for checking purposes, was still pending when Larson ruled it unenforceable; the adder patents had all been granted and expired.

Larson did not rule on the issue of derivation from Atanasoff with regard to any of these patents, even though, as we showed in chapter 3, the add-subtract mechanisms of the ABC were as clearly revealed to Mauchly in June, 1941, as was its regenerative memory. We suspect he did not do so because he had more important issues in terms of which he wished to scrutinize these items, issues that took Eckert and Mauchly and their attorneys close to the commission of fraud on the Patent Office. For the regenerative memory patent, on the other hand, he had only prior publication in von Neumann's "First Draft Report on the EDVAC," with no such "disturbing" aspects, as he termed them. Moreover, derivation of these adders from Atanasoff had not been an issue in prior Sperry Rand cases, as derivation of the regenerative memory had.

Larson ruled EM-21, EM-23, EM-25, and EM-26 unenforceable on the ground of prior publication, namely, of the EDVAC Report of June 30, 1946, and of the Moore School lectures of July and August, 1946 (actually published over a period from September, 1947, to July, 1948).

What he found "disturbing" was that a further patent, EM-27, on a binary-coded decimal serial adder, had been rejected by one Patent Office examiner because of prior publication, but that this fact was not called to the attention of a different examiner, in a different division, with regard to the other adder patents. That is, the earlier examiner had, on his own initiative, sought out the EDVAC Report and found it to be a prior anticipatory publication; yet Mauchly and Eckert and their attorneys failed to inform the second examiner of this ruling or even of the existence of that report. The judge cited a similar situation with regard to EM-21, EM-25, and EM-26 vis-à-vis rejected patent EM-28, on yet another binary-coded decimal serial adder (¶¶ 14.13–14.13.14)!

Larson found EM-22, the BINAC System patent, unenforceable on the ground of prior publication, namely, that of the Report on the UNIVAC, and also on the grounds of public use, on sale, and incomplete execution.

CHAPTER 5

Atanasoff's Place in History

A Technological Revolution

The prevalent public view of the origin of modern computer technology is a mistaken one—one that arose with the unveiling of the ENIAC in 1946 and has persisted ever since. That some historians have upheld this view in recent years is a reflection, at best, of their inexcusable ignorance of the work and influence of John Vincent Atanasoff; for they have failed to avail themselves of the huge mass of evidence accumulated and examined in a federal court. The stance of these scholars is the more reprehensible in that they have elected to *oppose* the strong, unappealed decision of that court.

The mistaken view is that the ENIAC, built at the University of Pennsylvania between 1943 and 1946, was the first electronic computer, and that J. Presper Eckert and John W. Mauchly were solely responsible for its inventive substance. The finding of Judge Earl R. Larson in October, 1973, was that the Eckert-Mauchly patent on the ENIAC was invalid, because, among other grounds, Atanasoff had invented a prior "automatic electronic digital computer" and his ideas had been applied by Mauchly and Eckert to the ENIAC and also to the ensuing EDVAC.

In this final chapter we build on the four preceding chapters to elucidate the causal chain of invention from Atanasoff through the ENIAC and the stored-program EDVAC and IAS (Institute for Advanced Study) machine, and on to the computers of today, thus establishing unequivocally that it was he who started the computer revolution. We will show that Mauchly and Eckert began where Atanasoff left off.

Let us make clear that we are addressing the *causal* chain of events, not all electronic computing activity in this early period. We will cite the contributions of Jan Rajchman and Perry Crawford, for example, but will not try to assay the achievements of the British designers of the Colossi or of Helmut Schreyer in Germany. The work of these latter is no less worthy, of course, for not having influenced the causal chain; and it bears out two truisms of the history of invention: that "the time is ripe" may be as significant as "necessity" in stirring creativity, and that causal input is a precarious phenomenon. Indeed, credit for causal input is also precarious; witness the number of years that passed before Atanasoff realized his own influence on the computer revolution, and how near he came to never realizing it at all.

The first section of this chapter is devoted to a brief overview of the history of computing, with emphasis on digital devices; we look at three technologies, the mechanical, the electromechanical, and the electronic. The second section considers Atanasoff's computer from the perspective of his own time; here we consider the quality of its engineering and the question of whether it deserved the court's designation of *automatic electronic* computer. The third section details the causal chain from Atanasoff's work to the modern computer.

Digital Computing Technologies

The modern computer age was brought about by the introduction of electronics into digital computing in the late 1930s. Two automatic computing ages, or technologies, had preceded the electronic, namely, the mechanical, dating from about 1300, and the electromechanical, dating from about 1880. Mechanical computing was based on motion; electromechanical, on motion and electricity. Electronic computing has added electronics to the electrical, and has also retained the use of motion, both for input-output and for slower stores in the form of rotating disks.

We need to review these three digital computing technologies in order to explain Atanasoff's contribution to the last one. We hope, however, that this short survey will also point up how little Atanasoff had to work with when he undertook to design the first electronic computer. Too often his achievement is compared not with what came *before,* but with what came *after* his initial effort. He is—unjustly, we feel—criticized for advanced steps he did not take, rather than credited with the fundamental steps he did take!

Before we begin, let us just mention an "age" that preceded and contributed to these three digital computing technologies, what may be called a pretechnology of manual computing aids. The counting board of ancient times and its successor, the abacus of medieval times, used positional notation and so led to the concept of zero. Napier's bones facilitated decimal multiplication by embodying multiplication tables in easily manipulable form. Finally, tables of trigonometric functions and logarithms facilitated computing at a much more complex level than had previously been possible. The abacus is still used, not only in those cultures that have depended on it for centuries but wherever their people have settled. Mathematical tables prevailed until the 1960s or so, when electronic calculators capable of accommodating them became affordable.

Mechanical Computing Technology

The mechanical computing technology began in the Renaissance with dancing figures driven by pegged drums; these gradually evolved into the music box and various other automata of the late eighteenth century. Although not calculating devices, these clever contrivances for the amusement of the wealthy held the germ of the modern computer program in the preset patterns of pegs that controlled their actions. The wholly utilitarian Jacquard loom followed in the early nineteenth century, basing its fixed read-only program on a sequential loop of punched cards.

Adding machines and calculators began with Wilhelm Schickard, Blaise Pascal, and Gottfried Leibniz, but were first made practical by Charles de Colmar. Digits were represented by the angular positions of toothed wheels, or gears, that went forward for addition, backward for subtraction, and by the linear positions of toothed bars that moved back and forth. Communication was accomplished by other toothed bars moving longitudinally and by rotating shafts. Switching was accomplished by the process of shifting mechanisms into and out of interconnections, in the manner of a gear shift. Lastly, input was accomplished by keys and levers, output by wheels whose positions could be read or printed. Note that the input portion of the mechanical technology remains today, as basic interface elements between the human operator and the internal memory and switching components.

By the 1930s, when Atanasoff was entering the picture, mechanical calculators that added, subtracted, and shifted were common. Some of these could also multiply and divide automatically, and some could print out results in columns. The most advanced machines were electrically powered, but were nevertheless intrinsically mechanical; they did not *compute* electrically.

These automatic devices were cheap enough and reliable enough to compete successfully with hand calculation in both business and scientific applications. The disadvantage of the technology was that it was too clumsy to be extended to more complex systems. That is, the use of rigid elements for the transmission of information and power imposed a practical limitation: communication among many computing parts called for a prohibitive number of long bars moving longitudinally and long shafts rotating. This problem would be resolved only with the substitution of electrical wire for communication between electromechanical computing components.

Charles Babbage was, of course, the preeminent pioneer in complexity of computing mechanics. His attempts—in theory, at least—to extend the digital computing technology beyond that of the adding machine were truly prodigious. He first tried to automate the calculation and printing of various function tables. To this end he conceived and, in the 1820s, worked on a *difference engine* to calculate such tables by accumulating successive differences. A difference engine consisted of a series of accumulators each of which, in turn, transmitted its contents to its successor, which added them to its own contents. He never completed his difference engine, however, and in 1833 left it to design a general-purpose programmable computer, his *analytical engine.*

(A Swedish father-son team, Georg and Edvard Scheutz, and others later constructed modified versions of Babbage's difference engine that were used successfully in scientific and actuarial applications, but they were not economically competitive with the more general desk calculators that were evolving.)

Babbage's analytical engine, which also was never finished as his vision progressed from the grand to the grandiose, had the modern organization of processor or "mill" (control and arithmetic unit) and large memory or "store." The program, which included instructions for branching and indexing, was to be punched on cards that would be chained together in the fashion of the Jacquard loom. Babbage's design was brilliant, an anticipation of our modern central architecture and a hundred years ahead of its time, but it was totally impractical in terms of the mechanical technology available to him. Even a small analytical engine would have pressed the limits of that technology; and it would not have been economically competitive.

It is an interesting sidelight that, in the 1880s, the American philosopher Charles Sanders Peirce saw that Babbage's concept of an analytical engine could be realized electromechanically, with relays (see app. A). This was actually done in the mid-1940s with the Bell Laboratories Model V, and also with the addition of branching to the IBM-Harvard Automatic Sequence Controlled Calculator, the Mark I. Quite ironically, however, the *first* computer to have the full logical powers of Babbage's analytical engine was neither mechanical nor electromechanical; it was the electronic ENIAC, as it went into operation at the end of 1945 (see Burks and Burks 1981, p. 386). (The ENIAC differed from the earlier versions in that it had a decentralized architecture and a much smaller memory; but of course it was almost three orders of magnitude faster than the Model V and the Mark I, which were in turn about two orders faster than the analytical engine would have been.) At this point, stored-program computers were in the design stage, so that almost immediately after a programmable machine of the Babbage type had been realized, the first stored-program computers appeared. Babbage's work had a direct influence on the Mark I, but not on the ENIAC and apparently not on the Model V.

Electromechanical Computing Technology

The active components introduced by the electromechanical technology were electromagnetic switches of two kinds: relays, which were binary, and stepping switches, which had multiple positions and were often decimal. Binary digits were represented by the on or off state of the relay, decimal digits by the position of the stepping switch. Communication was over electrical wires or cables. And input-output was via punched cards or paper tape. The electromechanical technology, then, contributed both faster, more complicated data processing and an improved form of input-output.

As we have seen, it overcame the size limitation of the heavy, rigid shafts of the mechanical technology by substituting lightweight, flexible wires for the transmission of information and power. The electromagnetic switches that processed the information performed the further function of amplifying signals. Whereas mechanical amplification was clumsy, electrical amplification was easily attained: a weak signal could operate a switch, which in turn could control a strong signal. Thus amplification was an important enhancing aspect of this technology.

The electromechanical computing technology developed from telegraphy, which was the first system to use relays and also the first to use punched paper tape for input-output. Telephony, though analog, contributed the concept of plugboards for rerouting electrical signals. The earliest electromechanical computer was invented by Herman Hollerith in the 1880s, with plugboards added by the Austrian Otto Schäffler in 1895. Hollerith's machine was used in compiling and analyzing data in the 1890 censuses of both the United States and Austria. Other machines were developed in the late nineteenth century for accounting purposes.

By the mid-1930s, when Atanasoff was making his initial investigation of current computing instruments, electromechanical computers were well established for various data processing applications, but not for scientific use. These were by now mostly IBM machines that handled stacks of cards automatically. The IBM 601 cross-footing multiplier was the most complicated. It read numbers from punched cards, performed a numerical computation according to a preset formula, and punched an answer on a card; the terms of the formula, which was set by a prewired plugboard, included one multiplication and perhaps three additions or subtractions, with the accumulating total always positive.

(In the late 1930s and early 1940s, IBM developed a partly electronic version of the IBM 601. The Byron Phelps patent, mentioned in chapter 4, was based on this machine, which used vacuum tubes to form the partial products and relays to shift them. The postwar IBM 604 was a further version of this latter. When IBM ultimately entered the modern electronic computing race, however, its main entry, the IBM 701, was patterned after the IAS computer.)

More powerful electromechanical computers also arose in the late 1930s and early 1940s. The time was at last ripe for the development of relay machines for scientific and engineering calculations, which required greater internal complexity but a smaller volume of input data than did data or symbol processing. (With World War II, such calculations had come to include military applications as an urgent consideration.) Development of these machines occurred simultaneously in the United States and Germany. Participants included: Konrad Zuse, with his programmable, floating-point, binary Z-3; George Stibitz, with his complex number computer, and subsequent Bell Laboratories Models II–V; and Howard Aiken of Harvard and C. D. Lake, F. E. Hamilton, and D. M. Durfee of IBM, with their Mark I.

Had these more powerful machines been built twenty years earlier, they would have become a dominant force in the history of computing. But they were too late. The electronic computing technology had already started—indeed, Atanasoff was building his computer even as they were getting underway—and it was growing rapidly, producing machines that not only could do the same work many orders of magnitude faster but could solve problems never before solvable. The historical importance of these larger electromechanical computers, then, lay in certain concepts and techniques adopted by the electronic technology: read-only programming, digit checking, and floating-point calculation.

Electronic Computing Technology

Electronic computing arose as a *discrete* adaptation of the *continuous* electronic technology of high-frequency communication. By the mid-1930s, electronics in this continuous mode was being used for radio (both AM and FM), high-fidelity phonographs, and experimental television; it was also used by physicists for laboratory experimentation, as, for example, by Atanasoff in his study of quartz and other crystals. Thus circuits composed of vacuum tubes, resistors, capacitors, and inductances were highly developed.

The vacuum tubes of the communication technology were used for both transmission and reception. First, because voice and music are of low frequency and only high frequencies can be transmitted efficiently through space, the nonlinearity of the vacuum tube was employed to combine the low-frequency message with a high-frequency carrier into a modulated high-frequency signal for transmission. Then vacuum tubes amplified this signal and powered the transmitting antenna. Finally, because the signal picked up by the receiving antenna was very weak, vacuum tubes amplified the message signal until it was strong enough to be demodulated and to drive a loudspeaker.

But the continuous mode of the communication technology was unsuited to digital arithmetic operations, and the development of a new mode for the vacuum tube was necessary to make these possible. This discrete mode was mostly binary, a tube being operated in either of two states, off or on, except for the transition from one state to the other, which did not affect the computation. Corresponding to the off and on states, the voltage of each wire or signal point of the system had either of two values, relatively high or relatively low.

As we saw in chapter 2, electronic flip-flops, counters, and simple switching circuits had been invented and were being used to count cosmic rays; this digital use of vacuum tubes, however, was relatively rare and was not reliable enough for computing. This earlier adaptation of radio technology to counting did offer the pioneer in electronic computing two things: rudimentary storage, in the form of flip-flops and binary and decimal counters—recall that counters incorporated both counting and storage—and switching, in the form of carryover from one counter stage to the next. (Electronic switches were also capable of sending alternate signals into a cathode-ray-tube screen.)

Atanasoff played a major role in the transition from the continuous to the discrete use of vacuum tubes for computing. Indeed, it was he who took the first step in the causal chain of events, by advancing from the simplest counting and switching devices to logical switching circuits, and so proving that reliable, accurate electronic computation was feasible. We detail his contribution to this transition in the next section. Let us look first at the new technology that was to emerge and evolve in such dramatic fashion.

The electronic computing technology retained the two major advantages of the electromechanical over the mechanical, communication via electric wire and easy amplification. But the moving parts were now electrons, with the further obvious advantage of speed. Vacuum tubes, however, were about as large and expensive as electromagnetic relays; it required the introduction of a number of enhancing features to add economies in size and cost to those of speed in electronic technology.

Thus the ENIAC, although nearly 1,000 times faster than its electromechanical predecessors, was of roughly their size and cost, whereas the first generation of stored-program computers were not only faster than the ENIAC but smaller and cheaper. These latter machines, while still doing arithmetic operations with vacuum tubes, achieved all three economies through their new forms of memory: either mercury-delay-line (EDVAC) or cathode-ray (IAS) memories for the main store and magnetic drum and magnetic tape memories for auxiliary roles.

The second generation, arriving in the 1950s, continued these economies by replacing vacuum tubes with transistors for arithmetic and control, and by introducing the magnetic core for the main memory. Finally, subsequent generations saw integrated circuits gather transistors, capacitors, and wires together on tiny chips for processing and for memory.

And so electronic computers quickly became *faster, smaller,* and *cheaper* than the early machines by many orders of magnitude. And orders of magnitude spell revolution. Our thesis is that Atanasoff, to whose contribution we turn next, was responsible for introducing electronics into the causal chain that produced the computer revolution and ushered in the modern computer age. Others were working on electronic computing simultaneously, primarily at MIT and RCA, and some influenced the ENIAC and the EDVAC, but only after Atanasoff and not in as fundamental a way. Atanasoff was always ahead of the others, and he advanced the art further than they did.

Analog Computing Technologies

Before turning to Atanasoff's electronic digital computer and its contribution to the causal chain, we want to make a few remarks about the analog computing devices of these earlier eras. As we have seen, it was Atanasoff who first distinguished explicitly between analog and digital devices and, in fact, provided the name that came to be associated with the former.

Analog computing devices prior to Atanasoff's time were almost exclusively mechanical. These were: slide rules, planimeters, harmonic analyzers, tide predictors, antiaircraft fire detectors, and differential analyzers. (As with the digital devices, we classify as mechanical those analog devices in which the computing system was essentially mechanical, even though they may have been electrically powered and, in the case of later differential analyzers, electronically enhanced.)

By the mid-1930s, mechanical analog devices were actually in the ascendancy, because they were intrinsically simpler (though also less accurate). The slide rule was the most commonly used instrument for everyday scientific and engineering computations, while the differential analyzer was being introduced for very complex problems. Unlike the digital desk machine, the slide rule could be carried about; and whereas the mechanical technology of a calculator was not extendable to much larger machines, a differential analyzer that filled a room was feasible. It should be noted that both Vannevar Bush and Irven Travis were exploring digital electronic possibilities in the late 1930s and early 1940s, but such a move on their parts was obviated by their calls to government war work.

Atanasoff, however, had begun to explore earlier, firmed his determination, and then moved very fast.

Other noteworthy analog computing instruments in operation as of the mid-1930s were the so-called network analyzers. A network analyzer consisted of a large number of electrical components that could be interconnected to simulate a particular electrical power distribution network.

Electronic analog computers entered the arena only in the wake of electronic digital computers, and lasted only about two decades.

Atanasoff's Computer

In this section we look at Atanasoff's computer in the context of his own time period. We first consider it as an engineering achievement, from the perspective of a visiting expert, then take up two questions that are sometimes raised: was it automatic, and was it electronic.

A Well-Engineered Machine

Imagine that it is early 1942, shortly after the United States has formally entered World War II, and you are an enterprising physicist-cum-engineer doing government-sponsored research. You need to solve a large system of partial differential equations, one that can be adequately approximated by a set of twenty-five simultaneous linear algebraic equations. A differential analyzer is available to you, but it is not really suitable because yours is a multivariable problem. Your only course seems to be to assemble some Monroes or Fridens and hire some computers to do Gaussian eliminations on those machines.

You wonder, though, if the government hasn't access to a faster, better-suited machine. Money, of course, is no object, what with the all-out war effort. Your sponsoring agency tells you of such a machine, a mammoth contraption being built at IBM in association with Harvard's Howard Aiken. It will have about eighty electromagnetic registers, more than enough to handle two of your equations at a time. And you could punch the required sequence of elimination steps on paper tape for entry. Unfortunately, this machine is a long way from completion.

Your contact then mentions, as an afterthought, a small machine being built by an obscure physics professor out at Iowa State College. This one is nearly finished, and it has been designed with precisely your kind of problem in mind. The catch is that its inventor, a J. V. Atanasoff, is trying to use *vacuum tubes* for computing elements, and *condensers* for memory elements, both of which raise serious doubts about its reliability for arithmetic operations. Still, your curiosity is sufficiently piqued that you manage to visit Ames in early May.

Atanasoff and his graduate student, Clifford Berry, show you the machine with marked enthusiasm, pointing out the various "fields," as Atanasoff calls its different systems. You are impressed, all right, as you stand before this compact device, with its synchronous motor whirring and its banks of vacuum tubes lighting up, four memory drums and a timing drum rotating, rows of brushes recording on and reading from their condensers, big cards being fed in, punched, and read with electrical arcs.

Atanasoff waxes eloquent on his choice of the base two, which he claims is not only the most efficient for doing arithmetic, but the best for computing with vacuum tubes—a tube is either on or off—and for punching and reading cards—there's either a hole or no hole. All of the arithmetic is done in these vacuum-tube mechanisms, he explains, with just the two operations of addition and subtraction—and not by counting, but by logic!

You cannot help thinking what a beautifully machined device this is: those two big drums geared into the binary-card reader, so that the "fingers" of the card-feed propel the cards between rollers and the rollers pull them through in sync with the machine's one-second cycles; this gadget for reading five different sections of a decimal card at once, advancing through fifteen sets of columns in succession; the small carry drum geared to rotate fifteen times as fast as the others, because it only has to hold digits for a sixtieth of a second and this design is simpler.

You are elated that this "computer"—another of the inventor's terms—is tailored to your very need, a basic computational need, you remind yourself. But Atanasoff is frowning now, as he shows you a card that holds a newly generated binary equation. In the final shakedown of the machine, Berry has discovered a problem with this punching system, or rather, they hope, with the dielectric. They have not been able to locate the paper card material that worked best in the initial testing, and errors are now appearing often enough to spoil the results. Berry shows you their last scrap of the material they had expected to use.

Well, you say, let's worry about that later. You are more interested in those vacuum tubes and condensers—no problem with their accuracy, apparently. And now you are invited to put the computer through the main steps yourself, to get the feel of it from the operator's standpoint (see fig. 26).

You understand that the most basic operation, the elimination of a designated coefficient from a pair of vectors, has been reduced to addition and subtraction. Atanasoff has done this by modifying Gauss's method to simulate division by repeated subtraction—*without restoring the overdraft,* he emphasizes. What is more, he has applied his variation to both the forward part of the process, which turns out successive sets of equations in fewer and fewer variables, and the backward part, which turns out the final set of single-variable equations. Although it seems an obvious way around multiplying and dividing on the computer, you cannot recall hearing of this particular variation of Gauss's method before.

With a little guidance, you set the controls to designate a coefficient position for the first elimination, enter an already punched—*charred* is more descriptive—binary card and signal the machine to read it onto the "keyboard drum," then enter a second card and signal the machine to read it onto the "counter drum." At this point, you are called to the telephone, but are assured that your coefficients will

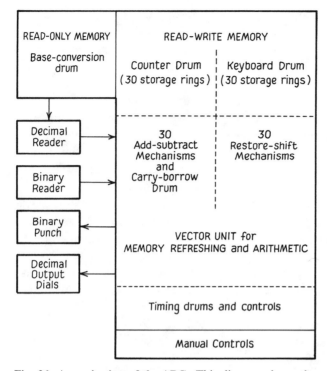

Fig. 26. A user's view of the ABC. This diagram shows the general structure of the ABC and indicates how the input-output devices used by the operator were related to the main parts of the computer. The central operation was to read a binary card onto the keyboard drum and another onto the counter drum, eliminate a designated variable, and punch the result on a binary card.

There were two base-conversion operations: decimal-binary conversion from IBM cards in the decimal reader to a binary card in the binary punch; and binary-decimal conversion from a binary card in the binary reader to the decimal output dials.

remain in place because the condensers inside the drums are recharged on each rotation.

When you return, you push a button to start the elimination of the designated coefficient from the vector on the counter drum; the thirty pairs of numbers flow repeatedly through the separate arithmetic circuits and back to the big drums, with the small carry drum doing its job on the fast axle. In about two minutes, the computer punches out the new equation.

Atanasoff explains that the blank portions of the "abaci" drums allow time for new instructions to be given to the "add-subtract mechanisms" after each operation on the streams of binary digits. And when you ask how those mechanisms work, he shows you one he has on his bench, along with a "holder-shifter circuit"! It seems these parts are all modular! Ingenious, you think, as he hooks in some test equipment and puts them through their paces, explaining the principle of logic behind the adding mechanisms and the principle of "jogging" behind the recharging mechanisms. But how, you ask, does he get around the different voltage levels—doesn't the voltage go way up as these circuits do an operation? Well, he has something he calls a "boost," which jumps the voltage of the condensers in the drums just as the new signals arrive, then drops it again for the next round.

There are lots of nice little touches. Berry explains how they cut down the number of bands on the base-conversion drum from 135 to 61 by taking advantage of pattern repetitions. The precaution has been taken of including two *extra* rings of condensers in the memory drums, in case of failure. And the computer is so synchronized that no matter when you give the start-up signal, action starts at the beginning of a drum cycle. There's also a "one-cycle switch," for the convenience of the operator in testing or demonstrating the machine.

The overall design—the balance among the different fields—is superb, you say: vacuum tubes to compute, condensers to hold numbers, and relays to control the activities, together with rotating drums and card-punching equipment, all working at coordinated speeds. You can hardly believe this has all been accomplished by one professor, one graduate student, a physics shop, and a few hourly workers—for less than $7,000. That machine they are building at IBM, with several engineers and many full-time workers, is going to cost half a million!

When they tell you how long they expect the solution of one system of twenty-five equations to take, you say you could keep their machine busy for several months. You hope they can solve their problem with the binary cards—maybe you can help with that, you have lots of different resources. And you'll see if you can get the project classed as war research, so Berry can be deferred. At any rate, you promise to get back to them after you've talked with your agency.

In the taxi on the way to the train station, you are already making your pitch to those people in Washington: they had better grab this Atanasoff and his ideas about vacuum-tube computing for the war effort.

Was It Automatic?

An automatic operation is one that, once started, carries out a sequence of steps independently of further external command. Any computer has a number of different operations, which may or may not be entirely automatic. (Notice that, as applied to computers, the concept of automatic is to be distinguished from that of programmable; that is, an automatic computer need not be programmable.)

We see as automatic the three basic operations of Atanasoff's computer: elimination of a variable between two equations, which employed the binary-card reader, the arithmetic unit (consisting of add-subtract mechanisms, restore-shift mechanisms, and carry-borrow drum), the counter and keyboard drums, and the binary-card punch; decimal-binary conversion, which employed the decimal-card reader, the base-conversion drum, the arithmetic unit, and the counter drum; and binary-decimal conversion, which employed the arithmetic unit, the base-conversion drum, the counter drum, and the decimal dials.

What made these basic operations automatic, of course, was the ABC's control system. Let us look at the control sequence of the first and most complex of them, the elimination of a variable from a pair of vectors, whose features were similar to those of automatic division (see fig. 27). The operator actuated a control that ordered a sequence of steps to start, be executed, and stop. These included two branching steps, one the repeated lower-level branch decided by the change of sign of the designated coefficient, the other the higher-level branch decided by the completion of the elimination task. (We are here using the term "branch" for *control* branch, as distinguished from *program* branch; the ABC was not programmable.) This operation was further complicated by the requirement of processing thirty pairs of coefficients in parallel.

The binary-decimal operation followed this same pattern, but was simpler overall in that it had to process only one coefficient at a time. The pattern for decimal-binary operation was less complex, because it was performing multiplication, essentially, rather than division; on the other hand, it had to process five coefficients in parallel.

We maintain that these three basic operations were complex enough to qualify the ABC as an *automatic computer*. And our usage is in accord with that of Atanasoff's own time, as can be seen by comparing his machine with others that were regarded as automatic. Although most desk calculators relied on operator-controlled sequencing of steps for multiplication and division, the more advanced ones had what was called "automatic" multiplication and division, operations not nearly as complex as those we have delineated in the ABC.

Likewise, bank accounting machines, the IBM 601 cross-footing multiplier, and the Bell Laboratories complex number computer all executed much simpler operations that those of the ABC. The first two executed formulas of a few straightforward arithmetic steps; the last processed vectors of just two elements, as against Atanasoff's vectors of thirty elements. A major reason why the ABC could do more complicated automatic operations than these others was that it had a much larger storage capacity.

Even when compared with the ENIAC, the ABC stands up well for the automatic character of its operations, so far as it aspired to go. The ENIAC multiplied by referring to a multiplication table, with no lower-level branching step, so that its operation there was similar to the ABC's decimal-binary conversion method. And for division, the ENIAC used the same pattern as the ABC's for the elimination operation. Indeed, all subsequent electronic computers used subroutine control structures similar to the one depicted in the flow chart of figure 27.

The automaticity of the ABC has sometimes been challenged on the basis of the large number of binary-card input-output operations required by its elimination procedure. Ironically, the first problems put on the ENIAC, both by Los Alamos personnel and by Douglas Hartree, involved the solution of large sets of simultaneous equations, and for these it, too, had to rely on an intermediate (IBM) card store. Moreoever, its internal memory was so small—fewer than twenty numbers—that to solve a set of equations in twenty-nine variables would have required several times as many card insertions and removals as were required by the ABC.

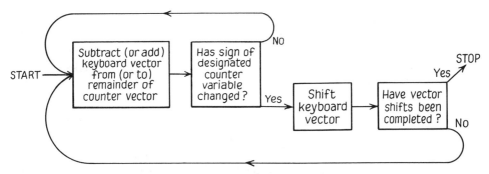

Fig. 27. Flow chart for control of vector variable elimination in the ABC. Depicted here is the step-by-step progression through the double control cycle of the most basic operation of the computer.

Was It Electronic?

A more important question is whether Atanasoff's computer was electronic. For us, it seems sufficient that the ABC made essential, critical, internal use of vacuum tubes for its computations; indeed, that it even effected the transition from radio technology to electronic computing technology. Others, however, have held that this application of vacuum tubes was not sufficient. The Sperry Rand position in the ENIAC trial, for example, was that the ABC was not an electronic computer because it had a mechanically rotated store.

In this subsection we consider these two main internal systems of the ABC, the arithmetic unit and the main memory, and make comparisons with other computers that are routinely classed as electronic. But let us first remove as irrelevant any argument about the computer's initial input and final output systems. Although these systems were electromechanical, so were their counterparts in the ENIAC (and other early electronic computers that used electromagnetic means of reading and punching paper media for input and output). And of course there has always been some purely mechanical human motion in any input-output system—turning switches, writing on a keyboard, moving a mouse, reading print—as the means of interaction between machine and user.

It can be argued in the same vein that the technology of intermediate input-output systems is also irrelevant. Here, though, the ABC deserves credit for the *electronic* recording and reading feature of its binary-card system. This was the first attempt ever to use means other than mechanical for input-output, and, as we have seen, it was adopted to facilitate an overall coordination of the computer's various systems.

The Memory

Because the arithmetic unit of the ABC included a read-write memory, the carry-borrow drum, we will address the computer's main (read-write) memory before the arithmetic unit. This memory had two aspects: first, its elements were capacitors; second, these capacitors were lodged in rotating drums, to be written on and read from electrically. Thus, while the *access to* the memory was part mechanical, part electrical, the memory proper was electronic. A capacitor holds an electrostatic charge, or surplus of electrons, and this serves as an electronic store. (The static charge does leak off, so that periodic refreshing is required.) By comparison, electrical current consists of moving electrons, as from a capacitor through a wire to a vacuum tube; and electrons in a vacuum tube move from cathode to plate.

As we have seen, Atanasoff's great contribution in this memory was regeneration; capacitors are used today as memory elements, with this restoring feature, in the memory chips known as DRAMs. Our interest here, however, is in the fact that he created an *electronic* store to work in conjunction with his electronic arithmetic unit.

The second aspect of the ABC's memory, that it took the dynamic form of rotating drums, does not detract from its essential electronic character. Other computers with such electrically driven rotating stores, both in the early days and today, are considered electronic. The main store of the IBM 650, for example, was a magnetic drum memory—a wise choice for the mid-1950s. The rotating hard disks of today's computers are an application of the same principle at the secondary level.

A variant of the argument that the ABC's memory was not electronic could be brought against the mercury-delay-line memory of the EDVAC. Although the speed of the EDVAC's memory was equal to that achieved by the computer's electronic arithmetic circuits, neither its memory function nor its access was electronic! The memory was mechanical, involving the vibration of a column of mercury. The access was electromechanical, involving, at input, the vibration of a crystal by electrical pulses and, at output, a crystal emitting electrical pulses.

The Arithmetic Unit

Electronics played two essential roles in the ABC's add-subtract and restore-shift mechanisms: processing information, that is, doing the arithmetic with the electronic switching circuits; and providing amplification to compensate for the inevitable degeneration of the electronic signals. Electronics plays these same two roles in computing today.

As we have indicated, Atanasoff's arithmetic unit incorporated an electrostatic drum memory similar in principle to its main memory. But, again, the fact that this carry-borrow drum was mechanically rotated does not detract from its essential electronic character, nor from that of the arithmetic unit itself.

The Matter of Controls

Although the fact that the ABC did not have electronic controls cannot legitimately be cited as an argument that the computer was not electronic, we would like to comment on Atanasoff's achievement in this regard.

He had used vacuum tubes for the arithmetic unit because his goal was to create an electronic computer, a faster computer. And while the overall rate at which his machine could solve large sets of equations did not constitute an earth-shaking improvement over the desk calculator method (see chap. 1), the electro-magnetic relays available to him were too slow for the computing rate required by his binary-card system. The machine's controls were a different matter. As indicated in his 1940 manuscript, he had originally planned to make these electronic, too; but in the end he substituted relays as a cheaper option that would serve his purpose just as well as vacuum tubes. For these had a sixth of a second in which to perform, not just the sixtieth of a second accorded a bit operation.

His choice was consistent with choices made later by others who sought to compute electronically. The Phelps version of the IBM cross-footing multiplier, for example, used relays for shifting its electronically generated partial products. And the ENIAC used relays to interface its input-output with its punched-card machines, only because they were cheaper and were fast enough for the job.

What is perhaps most relevant here is that Atanasoff *did invent* electronic controls. His computer, as built, embodied the requisite electronic switching technology to incorporate vacuum-tube instead of relay controls. And so it is not a question of his merely *intending* to invent such controls. Rather, he already knew how to design them. Had he secured a patent on the ABC, he would have, in the disclosure portion, described his relay controls; and surely, in the claims portion, he would have claimed the invention of their electronic counterpart on the basis of the electronics of his arithmetic mechanisms. Thus, even though the ABC itself did not feature electronic controls, it *and* his manuscript show that he had conceived them in a patentable form.

The Causal Chain

In this section, we detail the conceptual connections from the work of Atanasoff to the computers of today. We begin with an examination of the original features of his computer that had a causal impact on later devices, then turn to his specific influence on the ENIAC and on the first stored-program machines, the EDVAC and the IAS computer. We conclude with a brief statement of the ensuing transition to the modern computer.

Novel Aspects of Atanasoff's Computer

First and foremost in Atanasoff's contributions to the history of computing, of course, was his very invention of the electronic digital computer, for which we have already argued amply. We want now to address particular novel aspects of the ABC that contributed to the causal chain. Let us list these, *in our judged order of importance,* before citing the significance of each one.

1. Electronic digital computation
2. Electronic switching
3. Memory for electronic computer
4. Idea for electronic sequential control of computational processes
5. Clocked digital computer operation
6. Use of binary number system
7. Modular units
8. Idea for magnetic drum
9. Idea for nonrestoring electronic division

Electronic Digital Computation

We observed earlier that analog technology was in the ascendancy for scientific and engineering problems in the mid-1930s, when Atanasoff was casting his lot with digital technology. In fact, MIT in 1935 confirmed its commitment to analog computing by deciding to build its second differential analyzer, using electronics to interconnect its mechanical computing elements; and in 1940, at the conclusion of Travis's studies on its behalf, General Electric chose to build an analyzer of the older MIT type. Thus Atanasoff was moving against a rather formidable tide of accepted technology when he decided to build a computer that was both digital and electronic. Moreover, as we have seen, he later conceived of building an electronic digital computer to solve differential equations.

His decision was perspicacious. The digital method allows for an indefinite extension of accuracy. The attainment of such accuracy does require many steps of computation, but the greater speed of electronic components yields an overall speed greater than that of the pre-electronic technologies, whether analog or digital. Also, the digital method is more general than the analog method, because differential and integral equations can be solved discretely, whereas an analog computer cannot do number theory or data processing. Finally, digital computers can store information internally and use it in the computation, whereas analog cannot; this possibility further expands the scope of problems that can be solved, and it allows serial as well as parallel computation.

As we said before, this move by Atanasoff from analog to digital required adaptation of the vacuum tube, which had been used almost entirely in the continuous mode for communication, to use in the discrete mode for computation. Let us spell out in more detail just what this adaptation involved. It was, in essence, a shift to *binary* logic and mathematics, regardless of the number base of the computation, because a vacuum tube is both nonlinear and variable: the output signal is not proportionate to the input signal; furthermore, the amount of amplification varies from tube to tube and over the life of a tube. Given that digital computing requires reliability, it was best to use the vacuum tube in only two states, either completely on or completely off, not to attempt to use its intermediate states. These same arguments apply to the transistor.

It is of interest that digital methods have taken over not only computing technology, but communications technology, which had been dominated by analog methods from the development of the telephone. Analog computing, however, lasted only into the 1950s, in the form of network analyzers; communications, which began with the telegraph, is just returning to its digital roots.

Electronic Switching

Atanasoff invented a rather complicated electronic switch for addition and subtraction, using the equivalent of truth tables for their analysis. The electronic switching in the ABC was basic to all electronic computers that followed; the technology was derived causally from his computer via the ENIAC, the EDVAC, and the IAS computer, as we show later in this section. A more technical assessment of Atanasoff's contribution to this history is given in appendix A.

Atanasoff's use of binary switching adders was serial, as was their use in the EDVAC. But the IAS computer and most stored-program computers ever since have used random-access stores in which words are transmitted and processed in the bit-parallel mode, with the binary switching adders used in parallel (see app. fig. A.19).

Memory for Electronic Computer

The ABC had an almost centralized read-write block memory (see fig. 18)—*almost*, because its regenerating circuits were in the arithmetic unit. We will distinguish four features of this two-drum memory, every one of which plays a role in contemporary computers.

The first feature of the ABC's memory that we want to single out is its *separation of memory from arithmetic*. Here, too, Atanasoff broke with a strong tradition, establishing a new one that has continued to this day in the separation of memory from processing in general. He had the inventiveness to realize that an adding device need not be combined with its storage device, as it had been in the calculators of the past. As we saw in chapter 1, he made this separation in two steps. Because he wanted an unprecedently large capacity, he attacked the storage problem first, conceiving his read-write capacitor drums. He attacked the computing problem second and, after experimenting with counters, conceived his purely logical switching arithmetic circuit.

We should remark, in passing, that Atanasoff was the first to consider, for the electronic technology, how best to store numbers. He investigated five different materials or devices for his 3,000-element store, including vacuum tubes (used in the ENIAC memory), magnetic material (used in magnetic drum and other magnetic memories), and capacitors (used in DRAMs). All of these, of course, were considered because they were erasable elements, as required for read-write storage.

The second feature of the ABC's memory is this choice of *capacitors for cheap storage*. They were the most economical option because they were passive, simple, and rugged. That is, being passive they did not have to be supplied with much power, being simple they were inexpensive to produce, and being rugged they lasted a long time. The economy was indeed startling, with capacitors costing about a nickel each, as against several dollars for a vacuum tube. Atanasoff achieved a memory of 3,000 bits that would have required not just 3,000 double triodes for storage but many more for switching in and out. Thus the ABC held sixty words, or numbers, three times the ENIAC's twenty.

The third feature of this memory is *regeneration*. With the invention of regeneration of capacitors as storage elements, Atanasoff provided a general technique with implications far beyond the use of this particular element. The engineer looking for possible storage means need not be limited to elements capable of permanent memory, magnetizable materials, for example, but can consider more plentiful materials and devices capable only of transient memory.

Because the capacitor leaks its charge exponentially—in a matter of minutes—Atanasoff did not even seriously consider it until he was near the point of despair. But his mind returned to it as ideal in every other way, and he suddenly realized that a transient device could be made a permanent device if its contents were periodically read out and restored to their original level.

Historically, although this regeneration principle was novel for memory, it, like the concept of electronic digital computing, was actually an adaptation, in this case an adaptation of the relay or repeater idea of communication. In telegraphy, the signal deteriorated over distance, and that problem was solved by inserting (in space rather than in time) a relay or repeater that would respond to the weak signal, before it became *too* weak, and amplify it by switching a fresh battery source in and out. This same idea was carried over to telephone communication lines, including underwater cables.

The fourth and last feature of the ABC's memory that we want to cite is its *rotating drum* form, which was a way of combining mechanical motion with electronic elements for economical switching. A large part of the cost of a memory is in the switching, that is, in the means of selectively writing in or reading out data. Atanasoff invented electronic storage to match his electronic computation and, in order to solve the problem of their interaction, he housed the store in a moving device whose very movement provided the requisite access. The interaction itself was effected by his use of electrical brushes and wires that carried the signals ("1's" and "0's") to the computing mechanisms.

His choice here, then, was something of a tradeoff. The "price" of his cheap storage access was its relatively slow action, as compared to the electronic speed of his computation. And this tradeoff exists today, though at a lower level of the architectural hierarchy, where a certain wait time is justified by the low cost of a moving device that can be used repeatedly to access bits (see app. A).

On the other hand, he was already tied to the sixty-cycle current of his synchronous motor, which had been required for his binary-card input-output system. It should perhaps be said once more for Atanasoff's design that he achieved a remarkable overall balance in the ABC: there was no "bottleneck" in the entire sequence of operations, as there was to be, for example, in the ENIAC, with its slow punched-card input-output contrasting markedly with its internal computing activity—an imbalance overcome in later machines by the introduction of magnetic tapes.

The ABC's rotating drum memory afforded another advantage besides that of cheap access. It facilitated shifting on the keyboard drum with the simple expedient of an extra displaced row of reading brushes working in conjunction with just one triode in each restore-shift mechanism. This same idea of extra reading brushes was used to reduce the number of bands on the base-conversion drum, as noted in the previous section.

Idea for Electronic Sequential Control of Computational Processes

We have already noted, but repeat here for the sake of completeness in our attributions to Atanasoff, that in the aggregate of his computer's electronic arithmetic switches, his electromechanical sequential control switches, and his earlier stated intention to make his controls electronic, he did in fact invent electronic sequential control switching. As with switching specifically for arithmetic operations, the application of electronics to sequential control has been a basic feature of all subsequent electronic computers.

Clocked Digital Computer Operation

In most computers, there are many different systems and subsystems that must work in synchrony. Atanasoff originated the idea of timing all electronic computing activities from a common source, or "clock." His four storage drums and his timing drum were all geared together, and the timing drum provided the appropriate signals for the electronic and relay circuits to work in temporal coordination with these memory devices. Eckert developed this idea into the ENIAC cycling unit, and clocks have been standard in computers ever since.

Historically, the timing system of the ABC grew out of the electromechanical technology, specifically out of the way in which IBM machines geared together the movement of cards, the advancement of stepping switches, the operation of punches, and the generation of timing signals.

Use of Binary Number System

In our presentation of Atanasoff's choice of the digital mode over the analog for electronic computing, we commented that for the sake of reliability it was best to use the vacuum tube (and likewise the transistor) as he did, in only the two extreme states of on or off. Once he had taken this binary option, based on engineering considerations, he saw that it was natural to use the binary number system for the actual computation. Or, more accurately, he was confirmed in his already strong leaning to the base two as the most satisfactory computing base.

His decision to compute in the binary system did entail base conversion at initial input and final output. For this he introduced a highly efficient decimal-binary conversion drum, whose role he coordinated with his binary arithmetic and memory circuits—an elegant solution.

The history of base preference is instructive here. Base ten is the human base, and so long as input-output is dominant over internal computation it is the proper base for a calculator. Both the electromechanical IBM machines and the mechanical desk calculators were decimal for this reason. Moreover, for the mechanical technology the base ten was also the most efficient for internal computation: a gear wheel with ten teeth is not much larger than one with two teeth, two teeth per gear are impractical, and a toothed bar with ten teeth is not much more cumbersome than one with two teeth.

This picture changed somewhat with the introduction of electromagnetic relays. Relays are intrinsically binary, in contrast to the stepping switches of punched-card machines, which are naturally decimal. Nevertheless, machines that used relays as their basic computing elements did not compute in the base two; the Bell Laboratories Model V, for example, computed in the biquinary decimal system.

With the coming of the electronic computing technology, however, the change to binary arithmetic along with binary elements was complete. The ABC was the first to achieve this coalescence. The ENIAC then advanced the utility of the vacuum tube for digital computing, also using it in the binary, two-state, mode; but, to avoid base conversion, the ENIAC performed the actual computations in the decimal system. Finally, with the invention of the stored-program computer, it became possible to program base conversion, so that binary computation was preferable for scientific and engineering applications. The UNIVAC (Universal Automatic Computer), built for data processing, computed in binary-coded decimal. Today, of course, computers generally allow computation in either binary or decimal.

Modular Units

Atanasoff introduced the idea of modular units into electronic computing with his add-subtract mechanisms, some thirty in all. He saw the ease with which they could be installed and removed for purposes of testing, replacement, and repair.

Idea for Magnetic Drum

The list of possible memory elements that Atanasoff provided in his descriptive manuscript included a "small piece of retentive ferromagnetic material, the two states being the directions of magnetization of the material." He rejected this idea because the electrical signal sensed in reading a magnetized area would have been so small as to require amplification by several vacuum tubes in each instance.

Although he did not explicitly mention a drum in connection with these elements, he would have had to move them past the reading heads in some way; since he had drums in mind for his capacitors, it can reasonably be assumed that he had them in mind for these magnetic elements. Thus, we submit, Atanasoff conceived a magnetic drum memory in the course of designing the ABC.

The idea was an important one, conceived independently by Perry Crawford and presented explicitly in his MIT master's thesis of 1942, "Automatic Control by Arithmetic Operations." The magnetic drums used in later computers did require appreciable development. In fact, although the idea of magnetic sound recording had been patented in 1898 by Valdemar Poulsen, it was not until after World War II that wire, tape, and drum recording became useful for either sound or computing.

Idea for Nonrestoring Electronic Division

Finally, Atanasoff is to be credited with conceiving and almost realizing nonrestoring electronic division. We have already argued that the ENIAC patent claim for this division algorithm was not justified in view of Mauchly's familiarity with Atanasoff's use of the concept in his elimination procedure. We have also pointed out both that all the ABC lacked in this regard was a means of accumulating a quotient as the successive subtractions were executed and that he intended to go on to incorporate such means either in the ABC itself or in a later computer based on its principles.

Atanasoff knew of no precedent for the use of nonrestoring division. Nor did he employ the terms "restoring" and "nonrestoring"; our own first experience of them was at the Moore School, in connection with the ENIAC's divider square-rooter. But Atanasoff, although he did not coin these terms that have become standard, did in his manuscript clearly distinguish his algorithm from the usual alternative algorithm, and did call it a form of division. Indeed, he noted that he was using this method in both his elimination procedure and his binary-decimal conversion procedure (Atanasoff 1940/1973, pp. 310 and 318). And he did clearly originate the concept as applied to electronic computing.

Patentability of Atanasoff's Computer

It is often argued—or just assumed—that Atanasoff's computer is to be dismissed because it was never finished and never patented, with the implication that it could not have been patented because of its unfinished state. Although we maintain that it was finished, in that it was entirely built as planned, this is somewhat a matter of semantics. The binary-card system did have a flaw that was never corrected, so that the machine could not do what it was meant to do reliably. On the other hand, the impression given by "never finished" has been that the computer was far from completion. We have even been asked if the single drum that remains and was exhibited at trial was not all that was ever built!

The issue we want to address here is the ABC's patentability, for the notion that an invention cannot be patented if it is not completed is incorrect on several scores. First, it is not only the invention as a whole unit that is claimed in the claims portion of a patent; as we saw in our examination of the ENIAC patent, systems and subsystems within the device, principles, ideas, and other features that are new to the art may also be validly claimed.

Thus, on the basis of a description of the ABC as it stood in May, 1942, Atanasoff could have claimed the following of the novel aspects we have presented: electronic digital computing, separate computer memories, read-write capacitor drum memories, regeneration, shifting by displaced brushes, electronic switching, logical addition and subtraction switches, clocked electronic digital computer operation, use of the binary number system in electronic computing, the initial decimal-binary conversion system, and the use of the one-cycle switch for testing. He could also have claimed vector processing, which we omitted from that list because it had no causal relationship to its use in later machines.

He had valid claim to all of these because they were all new to the art and had all been proven, that is, reduced to practice, in the machine itself. Moreover, as we argued previously, Atanasoff could have claimed electronic controls, on the basis of his relay controls and his idea of how to make them electronic. We would argue similarly for his claim to nonrestoring electronic division and, possibly, a rudimentary form of magnetic drum.

But a second point, and one more basic to the concept of invention, is that the inventor need not have built a complete and working device in order to secure a patent. The applicant need only demonstrate the ideas or principles, either of particular aspects of the device or of the entire device, and explain them in such a way that a person skilled in the art could produce any desired portion. And so, in Atanasoff's case, all of the novel features listed above could have been patented on the basis of his computer as of, say, mid-1941. Indeed, the computer itself, as the embodiment of all those features working together in the fashion described in the 1940 document, could have been validly claimed at that earlier date.

Third and last, it is not necessary even to have the invention proper in *any* state of construction in order to prove principles and secure a patent. As Judge Larson made clear, many of the novel aspects of the ABC could have been patented on the basis of Atanasoff's late 1939 model, which constituted a reduction to practice for them; by the same token, as Larson also made clear, a strong patent on novel aspects of the ENIAC could have been based on the two-accumulator model that worked well in the summer of 1944 (see chap. 4). Of course, the success of the ENIAC in solving its first problem in December, 1945, would have strengthened the patent, just as a partially realized ABC would have strengthened its case for a patent. But nothing was to be gained in either case (and much lost) by delaying the application beyond the time when all aspects had been proven.

In short, the inventor is the creator of the idea, not necessarily the one who makes it practical and ready for sale.

Atanasoff's Influence on the ENIAC

The other issue of prime interest in an invention, after its originality, is that of the inventor's influence on subsequent inventions. We believe we have shown that there would have been no ENIAC if Mauchly had not visited Atanasoff, learned about his digital electronic computing technology, and received the idea of applying this technology to a machine that would do the calculations of the differential analyzer. And had there been no ENIAC, Atanasoff would have had no influence on the history of digital computing and would be unknown today.

Yet the ENIAC looked very different and was very different from the ABC. The ENIAC was large, the ABC small; the ENIAC was fast, the ABC slow; the ENIAC computed with accumulators based on counters, the ABC with logical add-substract mechanisms; the ENIAC's internal computing operations were entirely electronic, the ABC's dependent upon a rotating drum; the ENIAC was programmable, the ABC not; the ENIAC's architecture was distributive, the ABC's centralized. The machines were so different that even Atanasoff failed to recognize the connection between them when he saw the ENIAC at the Moore School in late 1945.

Similarly, the differential analyzer, for which the ENIAC was a replacement and on whose architectural plan it was modeled (though architecture was not yet a conscious design concept), looked very different from and was very different from the ENIAC. The analyzer was analog, mechanical, slow; the ENIAC was digital, electronic, fast. Problems were put on the analyzer by rearranging shafts and gears, and on the ENIAC by plugging cables and setting switches. Both were specialized machines, so that few people knew how to use and maintain them. With the advent of the stored-program computers, the differential analyzer was quickly forgotten—even the new MIT analyzer attracted little attention—and the ENIAC itself was mainly remembered as "the first electronic computer." Only with the trial in the early 1970s was the connection to the analyzer recalled, and almost no one outside of the trial participants noted it even then.

Derivation from Atanasoff's Computer

The electronic technology of the ABC became the basis of the ENIAC's electronics. The most fundamental contribution was twofold: first was the binary use of vacuum tubes, that is, their use in only two states; second was vacuum-tube switching, that is, the combination of these tubes into complex switching circuits for computation and control. Both of these applications, of course, also established that electronic computing was practicable, a factor that must have been paramount in Mauchly's mind during his visit to Iowa, after his own unsuccessful attempts at electronic digital computing the previous six months.

Electronic switches were ubiquitous in the ENIAC, whose circuits were composed of storage devices distributed throughout the machine and interconnected by such switches. The internal storage devices were flip-flops, counters, handset mechanical switches, plugboard connections, and read-only resistor matrices. The basic electronic switches were NOT, NOR, NAN, and OR. The ENIAC's combined switching-memory circuits were used to count and route electronic pulses that originated in the cycling unit and represented either digits or control signals.

(For examples and explanations, see figs. 14–19 and secs. 6, 9, and 10 of our article on the ENIAC [Burks and Burks 1981]).

A third derivation from the ABC was the timing and sequential control of the computer's operations. The ENIAC used a synchronizing clock, its cycling unit, to coordinate all the internal computing activities. As in the ABC, the basic timing unit was an addition time. For sequential control of operations requiring more than one addition time, the ABC had used relay flip-flops and counter, the latter counting the forty-nine shifts of the elimination procedure and also of the binary-decimal conversion (see fig. 27; see also chap. 1). The ENIAC, on the other hand, used electronic flip-flops and counters for these control functions; the counters had electronic gates attached to their stages, so that as they advanced to each stage the gate turned on and controlled the circuits operative during that stage's addition time. The ENIAC also had a one-cycle, or one-addition-time, switch.

A fourth derivation was the use of modular units, adopted wherever practicable in the ENIAC in the improved form of plug-in devices. Lastly, the nonrestoring method that Atanasoff developed for his elimination procedure was used for division in the ENIAC's divider–square rooter.

It should be stressed that the ENIAC designers did develop and advance Atanasoff's basic ideas. Their logical switching circuits, for example, were a great improvement over the ABC's resistor logic. And they applied his concepts to a much larger general-purpose computer. Indeed, they pushed electronic computing to its techological limit, in speed as well as complexity; for the ENIAC was the first machine to do its internal computation completely electronically at full electronic speed.

There were other sources for the ENIAC's digital technology, as shown in figure 28. The ENIAC arithmetic and storage (as well as control) were based on electronic counters, which came from the experimental physicists' use of scale-of-two counters to count physical phenomena (see chap. 2). The ENIAC used resistor-matrix circuits as read-only memories for the multiplication table and the function-table storage units, and these were derived from the earlier inventions by Perry Crawford of MIT and Jan Rajchman of RCA, independently of one another. Electromechanical technology also played an important role. The ENIAC used a plugboard for the numerical communication system, and it used switches and plugboard for entering a program. Peripherally, it used telephone relays and IBM machines for input and output.

Derivation from Atanasoff's Suggestion

The Moore School of Electrical Engineering was already using its mechanical differential analyzer for the analog solution of trajectories, and desk machines for their numerical solution, when Atanasoff suggested to Mauchly that an "electronic integraph" could be designed to solve these same equations, not by analog but by numerical methods; that is, an electronic digital computer could be designed

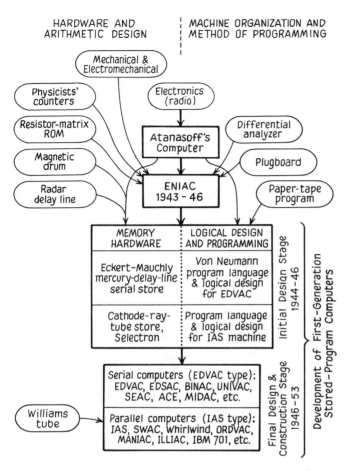

HARDWARE AND
ARITHMETIC DESIGN

MACHINE ORGANIZATION AND
METHOD OF PROGRAMMING

Fig. 28. Origin of the ENIAC and the stored-program computer. This chart shows the technologies, machines, and ideas that led, via Atanasoff's computer and the ENIAC, to the first stored-program computers. Planning for the IAS machine was based on the Selectron cathode-ray-tube store being developed by Jan Rajchman, but the actual machine and most of its successors used standard cathode ray tubes operated by circuits invented by F. C. Willams, of England.

to solve differential equations by solving the corresponding *difference* equations with the four basic arithmetic operations.

Now the ENIAC patent trial has shown, through the testimony of Atanasoff and Mauchly, their correspondence on the subject, and the text of the 1943 ENIAC proposal, that Mauchly's inspiration for the ENIAC did indeed come from this Atanasoff suggestion. And we have just spelled out, more explicitly than was done at trial, the particular features of the ENIAC that were derived from the ABC. We want here to emphasize the degree to which Mauchly and Eckert actually carried out this idea in their structural or architectural design of the ENIAC.

There were various ways in which Atanasoff's suggestion could have been executed. We think that he, had he proceeded to a second machine, would have preserved the essential internal features of the ABC: centralized architecture, serial bit processing of numbers, and logic-based arithmetic. Eckert and Mauchly, however, simply copied the distributive architecture of the analyzer, systematically replacing its analog components by digital counterparts. At the same time, they chose the decimal number system and the accumulator principle of the desk machines, with parallel digit processing, both to exploit the electronic speed potential of counters and to avoid the problems of base conversion. (Eckert and Mauchly were surely swayed in part by Irven Travis's idea of replacing the integrators of the analyzer with ganged desk calculators, even of making these electronic, but Travis had not worked out any electronic circuitry for this purpose. It is also possible that Atanasoff had told Mauchly of his own idea for ganging calculators.)

We detailed this organizational analogy to the differential analyzer in our ENIAC article (Burks and Burks 1981); let us just sketch the main points. Most basic was the replacement of the integrator by the accumulator: multiplication by an increment to achieve integration was accomplished in the ENIAC either by shifting the wires that connected one unit to another (compare the fixed gears of the analyzer) or by using the multiplier. The operator of the analyzer had entered initial or boundary conditions by manually setting the integrators and the starting positions of the input tables; initial values were entered into the ENIAC accumulators from the constant transmitter at the start of each run.

The variable interconnection bay of the differential analyzer was replaced by the pluggable numerical trunk system of the ENIAC.

Plotting tables used in the analyzer for input were replaced in the ENIAC by an IBM card-reading device, the constant transmitter, and the function-table units; plotting tables for output and a printer for recording numerical values were replaced by a "printer" and an IBM card-punching device.

The ENIAC, of course, was more general than the differential analyzer, which was in turn more general than the ABC. The ENIAC was also a much more powerful machine than either the mechanical analyzer or the electronic ABC, by virtue of its programmability and its computing speed of 100,000 pulses per second.

Atanasoff's Influence on the EDVAC

Atanasoff had a crucial influence on the Moore School EDVAC through three distinct developments or stages:

1. An idea Eckert had in 1944 for a magnetic drum memory. Even though such a memory was not used until some years later, this stage was critical, because it was a forerunner of Eckert's conception of the mercury-delay-line memory of the EDVAC. Atanasoff's influence here was both direct and indirect. The idea for a memory in the form of a magnetic drum was probably derived from Perry Crawford's MIT master's thesis, which Eckert read in 1943, but it may also have been derived from Atanasoff's earlier consideration of a magnetic memory, presumably in drum form.

The concept of a separate, relatively large memory surely came directly from the ABC, as explained to Eckert by Mauchly. The indirect Atanasoff influence lay in the fact that the electronic circuitry of the ENIAC, as derived from him, was implicit in Eckert's new plan.

2. The Eckert-Mauchly conception of the EDVAC. This carried forward ideas from the earlier plan, including that of a large separate memory, but replaced the magnetic drum with the mercury delay line. It also used the Atanasoff concepts of regeneration and of logical electronic add-subtract mechanisms as found in the ABC.

3. Von Neumann's variable-address program language for the EDVAC. The Atanasoff influence here was indirect. Von Neumann's contribution was novel, but assumed both the electronics of the ENIAC and the Eckert-Mauchly conception of the EDVAC.

Let us explore each of these three developments in turn, for their derivation from Atanasoff and for their culmination in the stored-program concept.

Eckert's Idea for a Magnetic Drum Memory

Atanasoff and Crawford had basically the same idea for a magnetic memory: use electronic circuits to cause a magnetic writing head to magnetize small spots in either of two states; sense their states with a magnetic reading head; and amplify the sensed signals by means of vacuum tubes. Thus two magnetic heads and associated electronic circuits would be needed for each channel. Because information could be written, read, and then either erased or written over, this magnetic memory was a read-write store. It could operate at electronic speeds, and it could store information far more economically than vacuum-tube counters or flip-flops.

As we have seen, Atanasoff made bare mention in his 1940 paper of magnetic material as a memory element for his computer. It was one of several possibilities that he rejected because they were unsatisfactory with regard to "cost, simplicity, or readiness with which [they] could be interrelated with the rest of the [machine]" (Atanasoff 1940/1973, p. 308). Crawford proposed his version as part of the design of an electronic computer for fire control. Both, then, envisaged a read-write magnetic store for a computer that was to perform a particular, but widely applicable, function. Crawford's idea was actually to use not a drum but metal disks with tapes of magnetizable metal encircling them. Coatings of magnetic material for drums were yet to be developed.

In January of 1944, Eckert wrote a three-page memorandum, "Disclosure of Magnetic Calculating Machine"; it was first published as part of a paper given at Los Alamos in 1976 (Eckert 1980, pp. 537–39). This memorandum brought together two ideas, that of a programmable ENIAC and that of a nonprogrammable Atanasoff or Crawford computer with separate revolving memory.

Eckert's proposed version of separate magnetic memory took the form of drum or disk memories that either had "at least their outer edge made of magnetic alloy," for "temporary" (erasable or read-write) storage, or had their "edges or surfaces engraved," for "permanent" (read-only) storage. Magnetic recording and reading were to be done by magnetic heads and electronic circuits. There could be many disks or drums, either mounted on the same shaft or mounted on different shafts rotating at different rates.

Some possibilities for representing and processing numbers were given in this 1944 disclosure: a basic rotating shaft called the "time shaft" on which at least some drums or disks would be mounted; binary numbers transmitted and processed bit-serially, with multiplication and division done by repeated addition; and a base-conversion table stored on the disks or drums. These are all strongly reminiscent of the ABC, just as the earlier magnetic details are of Crawford's proposed computer. Eckert gave no references in the original document, but did say, in his Los Alamos paper (p. 531), that he had read Crawford's thesis.

The one truly novel idea in this disclosure, expressed in a brief paragraph toward the end, concerned programming. The most important respect in which the ENIAC was superior to Atanasoff's computer was its automatic programming: once a program for any of a wide variety of problems was entered, the ENIAC would run the program automatically—and at the electronic speed of the calculating or numerical part of the computer. This read-only program, however, had to be entered manually into the (distributed) plugboard and switch arrangement; and, when the problem was finished, there was considerable dismantling to be done before a new program could be entered. Moreover, this setup procedure had to be done on-line, so that the computer stood idle for several days between problems. (This is why the ENIAC was most useful for problems of the firing-table sort that required long runs of the same program.)

Notice that the (read-only) programs of the relay machines of that day were punched on paper tape, which provided much easier and faster entry but could not be read at electronic speed. The manual setup of the ENIAC was chosen so that the program could be read at electronic speed.

Eckert wrote (p. 538):

> If multiple shaft systems are used a great increase in the available facilities for allowing automatic programming of the facilities and processes involved may be made, since longer time scales are provided. This greatly extends the usefulness and attractiveness of such a machine. This programming may be of the temporary type set up on alloy discs or of the permanent type on etched discs.

We assume that his reference to "multiple shaft systems" and "longer time scales" meant that he intended to rotate the program disks more slowly than the data storage drums in order to achieve the desired coordination.

The advance being made here over ENIAC programming was that the "temporary" (read-write) magnetic storage would permit the long, tedious on-line setup procedure to be supplanted by a much faster, easier off-line setup. For punched paper tape could be used to enter the (still read-only) program into a magnetic memory, where it could be read and executed at the electronic speed of the computer. After execution, it could be easily and quickly replaced by a new program for the next problem. Such a machine would, of course, have entailed the development of a reliable digital system of magnetic writing and reading. As we explain below, however, Eckert went on to replace the magnetic drum with the mercury-delay-line memory.

Eckert wrote nothing more about the nature of the programming he disclosed in this 1944 memorandum. Nor can present author Arthur Burks recall any discussion of it as an advance over ENIAC programming beyond permitting easier erasure and setup of the read-only program between problem runs. There was certainly no such discussion in the early 1945 meetings with von Neumann, by which time Eckert had conceived the mercury-delay-line memory. It was after these meetings that von Neumann went off and wrote his "First Draft Report on the EDVAC" (von Neumann 1945) encompassing a read-write program that could modify itself during a problem run—meetings attended by Burks, who took notes and rendered the official report on them.

This point is very important for the historical dispute over the respective contributions of Eckert and von Neumann to the stored-program concept, as developed for the EDVAC and utilized in all subsequent computers. We return to it in our section on von Neumann's conception of the *variable-address read-write* program, as an advance over the *constant-address read-only* program of the ENIAC and of Eckert's proposed magnetic drum computer.

The Eckert-Mauchly Conception of the EDVAC

A magnetic drum was superior to Atanasoff's capacitor drum in two respects: speed of operation and, more importantly, information density. (Recall that a magnetic drum was sufficient for the main memory of the IBM 650 ten years later.) Nevertheless, it was clear to Eckert, first, that development of such a memory was a serious problem and, second, that no rotating device could keep pace with the fast arithmetic circuits he was now able to build: serial logical adders operating at a million pulses per second. Fortunately, he had developed a mercury-delay-line timing device capable of that rate during the year before the ENIAC project was launched, and now he brilliantly conceived of turning that into a computer memory. In essence, he applied the cyclic or recurring idea of drum and disk memories to the supersonic delay line.

William B. Shockley, of Bell Laboratories, had invented a noncyclic, or reflective, acoustic delay line; his used a mixture of water and ethylene glycol as the fluid. The Moore School was asked by the MIT Radiation Laboratory to develop a practical form of Shockley's device for timing radar signals. Eckert determined that mercury would be a suitable fluid and, with the crucial help of T. Kite Sharpless, got it to work well. In considering the delay line for a memory, Eckert saw that instead of reflecting a single pulse back from the tube's reflector plate, he could replace the reflector with a second crystal at the end of the tube and so separate the input and output functions of the original crystal. (It was with regard to these crystals that Eckert and Mauchly consulted Atanasoff in Washington in late August, 1944.)

An electrical pulse (or no pulse) entering the transmitting crystal produced an acoustic pulse (or no pulse) that traveled down the tube and stimulated the receiving crystal, which then transduced it into an output pulse (or no pulse). Because the output pulse was much weaker than the input pulse, Eckert planned to use Atanasoff's principle of regeneration. (He had the further problems of loss of pulse shape and timing, which he solved with a pulse-standardizing circuit and synchronous clock as he had in the ENIAC.)

Thus, in addition to the ideas Eckert and Mauchly derived from Atanasoff for the ENIAC and carried forward to the EDVAC, they also adopted for the EDVAC an arithmetic unit using a logical binary serial adder, and a separate memory using a regeneration by vacuum-tube circuits. Recall that Eckert and Mauchly went so far as to take out their own patents on both the adder and the regenerative memory, in various forms of each, but that these were declared unenforceable by Judge Larson. Moreover, they switched from the decimal number system of the ENIAC to the binary number system, and from the distributive architecture of the ENIAC to a centralized architecture.

Von Neumann's Variable-Address Program Language

In early 1944, when Eckert wrote his memorandum, "Disclosure of Magnetic Calculating Machine," the only known programs were the read-only programs of the ENIAC (set manually with switches and cables) and those of the relay machines (punched on paper tape). No programs were stored in read-write memories, because their size and cost were prohibitive: the ENIAC had 18,000 vacuum tubes as it was, and to add a read-write memory that could hold its program would have required several times as many tubes. As explained above, Eckert expected his proposed computer to allow the same "automatic programming" that the ENIAC allowed, but also to provide a "temporary" (erasable/read-write) memory into which the read-only programs could be both easily entered and easily replaced with new read-only programs.

Again, as we said earlier, Eckert did not in that memorandum—or even in the three meetings with von Neumann in March and April of 1945—suggest any further advantage to such programming. He *did not* suggest its being capable of altering the program in the course of a problem solution. In more technical terms, Eckert had seen the possibility of making the read-only, constant-address type of program used in the ENIAC and its electromechanical contemporaries easily enterable and replaceable. It remained for von Neumann to see that a read-write memory would also allow a read-write variable-address program, that is, a program that could change its own address references as the problem was being run, and to work out a variable-address program language for the EDVAC in his June, 1945, report.

This whole topic has been heatedly debated in connection with the question of who invented the stored-program computer. The Eckert-Mauchly side claims that Eckert's magnetic drum memorandum was the first disclosure of a stored-program computer, and that his mercury-delay-line memory constituted an improved memory for such a machine. The von Neumann side acknowledges Eckert's invention of the acoustic memory, but claims that von Neumann's 1945 report disclosed the first stored-program computer.

Neither side, unfortunately, has made the critical distinction between constant-address and variable-address programs; nor has either side discussed explicitly the relative roles of memory hardware and computer architecture. Finally, they both ignore the fact that many steps were taken in succession, with many people besides the two principals involved in those steps.

Burks presented his views on these matters at the 1976 Los Alamos Conference (Burks 1980), and we intend to address them more fully in a future work. Our topic here is the contribution of Atanasoff. For this, we must conclude that his work made possible the transition from the ENIAC to the EDVAC (and the IAS computer), because of his concept of separate arithmetic and memory, his concept of regeneration of memory, and his concept of a cyclical memory coordinated with electronic arithmetic circuits.

The Institute for Advanced Study Computer

The EDVAC was to have a centralized architecture with memory, arithmetic unit, control, and a single switch interconnecting all three. The memory consisted of mercury delay lines with restoring circuits and access switches, an address switch, and timing circuits. Von Neumann outlined this architecture in his EDVAC report, and he also suggested a new form of computer storage based on cathode ray tubes.

This electronic memory would be superior to Eckert's cyclic acoustic memory, because the individual storage points on the inside front surface of a tube could be accessed directly. Thus it could be used as a random-access memory, whereas the EDVAC memory was necessarily word-serial. Moreover, because handling the bits of a word in parallel rather than serially eliminated many timing circuits, the architecture was further simplified.

In 1946, von Neumann worked out the general design of the IAS computer in a paper, "Preliminary Discussion of the Logical Design of an Electronic Computing Instrument," with Herman Goldstine and Arthur Burks (Burks, Goldstine, and von Neumann 1946). This paper gave the paradigmatic form of what is now known as the von Neumann architecture—an architecture, however, to which significant contributions were made by both Atanasoff and Eckert (see fig. 28).

Transition to Today

Let us close this section on Atanasoff's contribution to the causal chain of invention—and the book—with a few remarks on the further transition to the computers of today.

Processing Technology

The components of the first electronic computers were vacuum tubes, resistors, capacitors, sometimes electromagnetic coils, and the wires to interconnect all of these. The active components were the vacuum tubes, which were used for both amplification and logical switching. They required an auxiliary source of power for their heaters.

In the next electronic computers, solid-state diodes were used for switching, with vacuum tubes reserved mainly for amplification. (Put simply, the solid-state diode is to the transistor as the vacuum-tube diode is to the vacuum-tube triode.) These vacuum tubes were physically smaller than the earlier ones, and the diodes were much smaller still. Moreover, the solid-state diodes did not have heaters to take extra power.

Next, transistors were used for both switching and amplification. These, again, were much smaller than their predecessors, and, like the diodes, had no heaters.

There now ensued a revolutionary development: the incorporation of a large number of transistors, capacitors, resistors, and wires into a single circuit. Fabrication of such *integrated circuits,* or "chips," was made possible by a radical extension of the printing art. Over time, their size has become smaller, even as more and more components have been included in them.

Throughout all of these technological advances in processing, Atanasoff's principle of using the active element in only two states has prevailed, as have his general concepts of electronic switching and of performing arithmetic operations logically.

Memory Technology

Atanasoff's principle of memory regeneration had over a decade of disuse after magnetic core memories replaced the acoustic and cathode-ray-tube memories, which had been expensive, clumsy, and unreliable. Magnetic cores were developed in the early 1950s and soon became the basis for the the main random-access stores of computers. They were much smaller per bit, much cheaper, and much more reliable. In addition, because they held their states indefinitely, they did not have to be refreshed; in fact, they held their states even when the computer was turned off.

As integrated circuits were developed, however, to the point that thousands of transistors and capacitors could be included on a single chip, capacitor memories with regeneration reappeared in the form of dynamic random access memory chips

(DRAMs). As their name implies, these are organized as random-access rather than serial memories. Their reappearance makes clear that it was not the basic memory element of Atanasoff's computer—with regeneration—that had been unacceptable for high-speed computation; rather, it was the mechanically rotating drum in which the capacitors were housed that had been unacceptable.

But even his principle of mechanically rotating storage with easy access has proved valuable for modern computing. The magnetic drums that had served chiefly as auxiliary stores were replaced by magnetic disks, first in the rigid "hard" form, then in the nonrigid "floppy" form; these latter have come to constitute a significant economy for the so-called personal computer.

Cost Considerations

Cost factors have profoundly affected the course of electronic computer history, in a rather curious way. In Atanasoff's time, vacuum tubes were relatively expensive, capacitors and resistors inexpensive, and wires cheap. The fact that he could not afford a large number of vacuum tubes led him to invent the drum store, the capacitor store, and the principle of regeneration. This concept of a separate store led him to invent the electronic switching add-subtract circuit. And the concept of interacting store and arithmetic unit led him to devise a compact efficient computer with a centralized architecture.

Today an integrated circuit is manufactured as a single unit, the transistors, resistors, capacitors, and wires being constructed by a sequence of operations applied to the whole chip. The cost of each component therefore depends mainly on the area it occupies, so that in fact they are all of about the same cost. Thus the electronic computer revolution has at last trivialized the economic variations that stimulated its beginning in Atanasoff's design of his computer!

It is ironic that at about the time Atanasoff was grappling with severely limited resources to build his electronic computer, two other groups had almost unlimited resources to build machines that did their computing in the traditional technologies. Although the new MIT differential analyzer introduced electromechanics and electronics for communication between computing units and for problem setup, it computed mechanically in the fashion of the original analyzers. And the large IBM-Harvard Mark I was an entirely electromechanical computer.

The financing of these two machines came from foundations and corporations; in each case, the cost was at least seventy times the cost of Atanasoff's computer. Yet because of the ABC and the rapidity with which it influenced subsequent computers, these massive and expensive machines became obsolete shortly after they were completed! For, as we noted earlier, electronic components were orders of magnitude faster than electromagnetic ones, but cost about the same. And so, once electronic computing was invented by Atanasoff and proved both very fast and sufficiently reliable by the ENIAC, electromechanical computers were doomed.

As we saw in chapter 3, the existence of the ABC was used to persuade the military of the feasibility of the ENIAC. Even with that assurance, however, no foundation or corporation would have risked half a million dollars on a new computer technology. Indeed, neither would the government have done so except for the circumstance of World War II, which provided the real impetus for support of the ENIAC. It was, then, a wartime gamble that allowed the monstrous machine built at the University of Pennsylvania to serve as the critical link from Atanasoff's machine to the first stored-program machines, which in turn opened the way for the economical manufacture of electronic computers.

Appendixes

Logic of Electronic Switching

Logic and Electronics

The electronic computer has added a new dimension to the subject of logic: the logical structure of computer organizations. Excluding the input-output system as conceptually distinct, we can divide an electronic computer at its highest organizational level into *memory, processor,* and *communication system.* These three blocks are arrangements of subsystems, which in turn are composed of lesser subsystems, and so forth, all the way down to the level of hardware components. The components of the memory and the processor, then, or small circuits of them, perform the basic functions of storage and switching, respectively, and are interconnected by means of communication components.

The new dimension of logic has arisen as a tool to the understanding and design of these ever more complex structures. But while all computer architecture is logical in nature, the earliest application of logic to electronic computers was to switching, and Atanasoff was the first to make that application. That is, he was the first in our *causal chain* of historical development; as we noted in chapter 5, Helmut Schreyer was designing electronic switches at about the same time. Appendix A presents Atanasoff's achievement in the context of the basic principles of the logic of electronic switching that has developed since his time. It is actually a technical extension of chapter 1, and in particular its section on the electronic design of the add-subtract mechanism for which he invented electronic switching.

As we have already indicated, Atanasoff's major contributions were the invention of both electronic switching and the regenerative memory. The latter, however, being much more readily grasped, needs no elaboration on our earlier presentation, whereas the former warrants considerable elaboration for the more technical reader. Such further treatment is, in fact, essential for a full understanding of Atanasoff's contribution to electronic computing. Moreover, it adds strength to the outcome of the court case, which was necessarily limited in its technical perspective. Thus appendix A is meant to be read or skipped over, according to the requirements of the individual reader.

We saw in chapter 1 that Atanasoff used capacitors for his electronic storage elements, as is done in some modern dynamic memories, that he used vacuum tubes for switching at the standard alternating current rate of sixty cycles per second, and that he used electrical wires for communicating binary signals between tubes. We also saw that he designed other means than electronic for switching: the rotation of drums with respect to rows of brushes for reading into and out of a serial store, electromagnetic relays for switching that did not require the electronic speed, and plugboards and manual switches for operator controls. Finally, we saw that his design of the add-subtract mechanism involved an explicit application of logical notions to electronic switching, in the form of his (p,q) notation where p was the number of driving triodes in a given circuit, q the minimum number of those triodes that had to have low plate voltages to turn off the driven triode and produce a high voltage at the output.

It should be clear from the foregoing that there are two distinct, yet necessarily intertwined, aspects to any discussion of the logic of electronic switching: the logic that is the tool toward a goal, and the electronics that realizes that goal. This distinction is critical insofar as the engineering of circuits is an art in itself, with its own "language," techniques, and physical entities. Yet it is also one that has gradually broken down, as the isomorphism of the engineering design and the logical design has reached a point where the language of logic readily translates into the language of electronics. Thus it is that we begin our presentation with an explication of the logical forms of switching primitives and networks, translate these into their electronic counterparts, but then find that for the actual overall design of Atanasoff's add-subtract mechanism we need give only the logic.

We should emphasize, however, that the "translation" from logic to electronics is no simple matter; it is more easily said than done. We can illustrate the process here by referring to the simplest electronic switch, the NOT of figure 10. Atanasoff represented binary digits on wires and capacitors by two voltage levels, the higher level signifying a binary "0" and the lower level a binary "1." Logically speaking, the circuit of this figure is a NOT because the polarity of the output signal is opposite to the polarity of the input signal. That is, a low level at the grid causes a high level at the plate, and a high level at the grid causes a low level at the plate.

Electrically speaking, the situation is much more complicated. The two voltage levels at the grid input are not the same as those at the plate output, and the *difference* between the two levels on the input side is not the same as the *difference* between those on the output side. Thus the simple logical notation, a tilde, translates into the conglomerate of figure 10, with representation of the triode's plate, grid, and cathode, positive and negative power sources, a ground, an input from a driving triode, an output to a driven triode, and connecting wires and properly chosen resistors.

It is, in fact, because the electronics is so much more complicated than the logic that we reverse the natural historical order of development and present the logic of switching primitives and networks before the electronics toward which it was aimed. The logic, too, is best ordered from the simple to the complex, that is, from the primitives to the networks into which they are combined. We shall for the most part, however, shift to an equivalent terminology more descriptive of the process, namely, that of "atomic switches" and "compound switches." Note here not only that "atomic switch" and "compound switch" are interchangeable with "primitive switch" and "network," respectively, but also that all four expressions apply to both logical and electronic entities.

We begin, then, with, a section describing some atomic switches and explaining what they do. We follow this section with one showing how some relatively simple compound switches are formed from these atomic switches and with another analyzing a number of adding and subtracting switches.

Later sections present the electronics of switching, with emphasis on Atanasoff's circuitry. We describe how he achieved his primitive switching circuits by combining vacuum tubes and resistor networks. We explain how information was represented by voltage levels on wires and in capacitors, resistor networks, and vacuum tubes, and what specifically happened to the voltages when a circuit switched. And we analyze the logical structure of Atanasoff's add-subtract mechanism and evaluate it.

The sections on the logic of switching, although they refer to Atanasoff's applications on occasion, are more or less independent of the rest of the book. The sections on the electronics of switching are amplifications of our earlier presentation of Atanasoff's add-subtract mechanism in chapter 1, and in its figures 10, 11, 13, and 14.

Last, we devote a section to reviewing the history of switching circuits and discussing Atanasoff's role in it.

Atomic Switches

In the previous section, we used Atanasoff's NOT circuit to illustrate the complexity of electronic switches as compared to the logical formulas they are built to realize. We start our explication of atomic switches by noting that the exact correspondence of logical switch to hardware circuit varies from form to form within the electronic technology. In Atanasoff's case, a typical atomic switch, e.g., a two-input NOR or a two-input NAN, consisted of a single vacuum-tube triode and an associated resistor network. In the ENIAC, a two-input NOR had *two* triodes and some resistors, and a two-input NAN was equally complex. In modern transistor-transistor logic, a two-input NOR has approximately the transistor equivalent of *four* triodes and some resistors.

Figure A.1 presents symbols for the most commonly used switching primitives. These are: negation (NOT); inclusive disjunction (OR) and its opposite NOT-OR (NOR); conjunction (AND) and its opposite NOT-AND (NAN); inequivalence or exclusive-or (XOR) and its opposite, equivalence (EQV); and also two threshold switches, "at least two are true" and its opposite "at least two are false."

Figure A.1 actually gives five alternative representations for these atomic switches. From left to right, these symbolisms are: some abbreviations of ordinary language, the corresponding symbols of mathematical logic, a circle-based notation for electronic switches, engineering symbols, and Atanasoff's (p,q) notation.

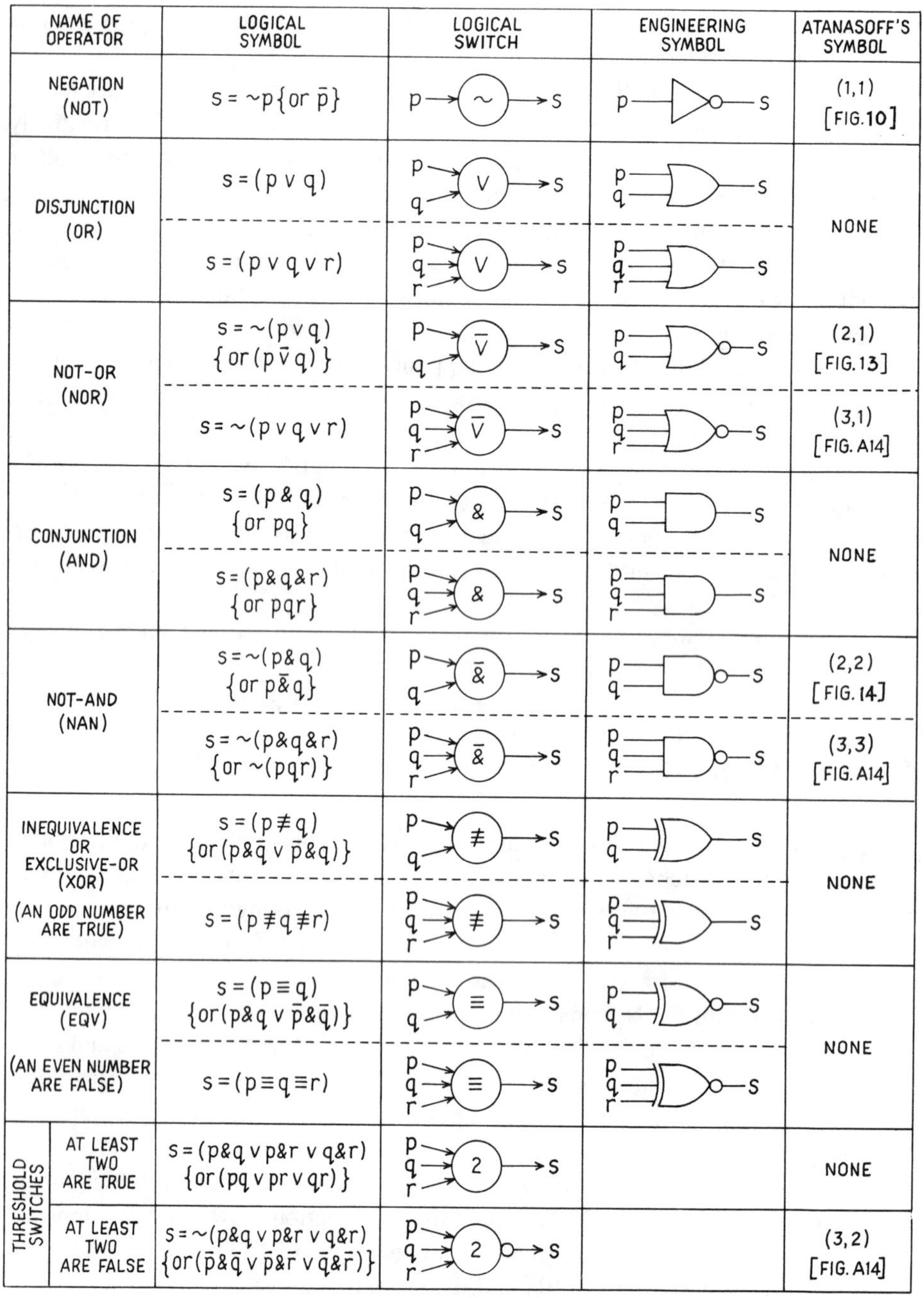

NAME OF OPERATOR	LOGICAL SYMBOL	LOGICAL SWITCH	ENGINEERING SYMBOL	ATANASOFF'S SYMBOL
NEGATION (NOT)	$s = \sim p \{ or\ \bar{p} \}$	$p \to \sim \to s$	$p \to s$	(1,1) [FIG. 10]
DISJUNCTION (OR)	$s = (p \vee q)$	$p,q \to V \to s$	$p,q \to s$	NONE
	$s = (p \vee q \vee r)$	$p,q,r \to V \to s$	$p,q,r \to s$	
NOT-OR (NOR)	$s = \sim(p \vee q)$ $\{ or\ (p\,\bar{\vee}\,q) \}$	$p,q \to \bar{V} \to s$	$p,q \to s$	(2,1) [FIG. 13]
	$s = \sim(p \vee q \vee r)$	$p,q,r \to \bar{V} \to s$	$p,q,r \to s$	(3,1) [FIG. A14]
CONJUNCTION (AND)	$s = (p \& q)$ $\{ or\ pq \}$	$p,q \to \& \to s$	$p,q \to s$	NONE
	$s = (p \& q \& r)$ $\{ or\ pqr \}$	$p,q,r \to \& \to s$	$p,q,r \to s$	
NOT-AND (NAN)	$s = \sim(p \& q)$ $\{ or\ p\,\bar{\&}\,q \}$	$p,q \to \bar{\&} \to s$	$p,q \to s$	(2,2) [FIG. 14]
	$s = \sim(p \& q \& r)$ $\{ or\ \sim(pqr) \}$	$p,q,r \to \bar{\&} \to s$	$p,q,r \to s$	(3,3) [FIG. A14]
INEQUIVALENCE OR EXCLUSIVE-OR (XOR) (AN ODD NUMBER ARE TRUE)	$s = (p \neq q)$ $\{ or\ (p\&\bar{q} \vee \bar{p}\&q) \}$	$p,q \to \neq \to s$	$p,q \to s$	NONE
	$s = (p \neq q \neq r)$	$p,q,r \to \neq \to s$	$p,q,r \to s$	
EQUIVALENCE (EQV) (AN EVEN NUMBER ARE FALSE)	$s = (p \equiv q)$ $\{ or\ (p\&q \vee \bar{p}\&\bar{q}) \}$	$p,q \to \equiv \to s$	$p,q \to s$	NONE
	$s = (p \equiv q \equiv r)$	$p,q,r \to \equiv \to s$	$p,q,r \to s$	
THRESHOLD SWITCHES — AT LEAST TWO ARE TRUE	$s = (p\&q \vee p\&r \vee q\&r)$ $\{ or\ (pq \vee pr \vee qr) \}$	$p,q,r \to 2 \to s$		NONE
THRESHOLD SWITCHES — AT LEAST TWO ARE FALSE	$s = \sim(p\&q \vee p\&r \vee q\&r)$ $\{ or\ (\bar{p}\&\bar{q} \vee \bar{p}\&\bar{r} \vee \bar{q}\&\bar{r}) \}$	$p,q,r \to 2 \to s$		(3,2) [FIG. A14]

Fig. A.1. Logical symbolisms for atomic switches

Corresponding symbols are given in rows. Atanasoff's notation was explained in full in chapter 1 and will be reviewed as required in this appendix.

Each switch has inputs and outputs, and in a switching network the output of one switch drives one or more inputs of other switches. For convenience we call these input, output, and interconnecting junctions *switching points*. At a given moment, each switching point is in one of two states. Atanasoff chose two voltage levels for the two states, and he assigned the binary digits, or bits, "0" and "1," to the high and low voltages, respectively. He could have made the opposite assignment, of course. Today his assignment is called negative logic, and the dual (reverse) assignment positive logic. Although positive logic seems more natural to us, for the sake of consistency we use Atanasoff's conventions throughout this book.

The bits "0" and "1" can also be interpreted as truth values, with "0" usually taken to mean "false," and "1" to mean "true." On Atanasoff's convention, this assignment amounts to associating the statement "the voltage at this point is low" with each switching point. Such an association with "true" and "false" connects electronic switching theory to a well-developed branch of mathematical logic, known variously as Boolean algebra (after its founder, George Boole), the statement or propositional calculus (since statements or propositions are true or false), and truth-function theory (since the truth-value of a compound statement is a function of the truth-values of its atomic statements and only of those).

The concept of negation (NOT) is fundamental, and it is symbolized in many different ways. Its most common representation is a tilde ($\sim$) prefixed to a variable (e.g., $\sim p$) or a formula [e.g., $\sim(p \lor q)$]. But it is sometimes symbolized with a bar over a variable (e.g., $\bar{p}$) or an operator ($\overline{\lor}$, $\overline{\&}$), or by a slash through an operator ($\neq$). In switching theory, it is often symbolized by a small circle attached to a larger symbol. In figure A.1, the small circle is used on the output side only (the engineering symbols for NOT, NOR, NAN, and EQV). But the small circle may also be used on the input side, as in figure A.4a. The purpose of these alternatives is simplicity of representation.

The meaning of each symbol of figure A.1 can be specified by a *finite* table in which the binary output is given for each binary combination of inputs; such a table is appropriately called a *defining truth table*. See tables A.1 and A.2 for examples. Notice that the number of inputs n of a switch is a crucial parameter because the number of rows of the associated truth table is 2^n.

The terms "input state" and "output state" are used in these tables to emphasize that for each physical representation of "0" or "1" that is applied to an input of a switch, the physical action of the switch will cause the output to take on the state shown in the table. For example, in Atanasoff's NOR switch ($\overline{\lor}$), when both inputs are set high (input state 0,0) the vacuum-tube circuit will cause the output to become low (output state 1). Thus it is desirable to treat the states of all input wires as a single binary word. These concepts of input and output state also apply to compound switches, where there may be many output wires as well as many input wires.

Logical states have typically been associated with physical states in three

**TABLE A.1.   Truth Table for the
One-Input Switch, NOT**

Input State (p)	Output State (~p)
0	1
1	0

TABLE A.2.   Truth Table for Six Primitive Two-Input Switches (fig. A.1)

Input State		Output States					
p	q	$p \overline{\vee} q$	$p \vee q$	$p \overline{\&} q$	$p \, \& \, q$	$p \not\equiv q$	$p \equiv q$
0	0	0	1	0	1	0	1
0	1	1	0	0	1	1	0
1	0	1	0	0	1	1	0
1	1	1	0	1	0	0	1

ways, corresponding to the three computer technologies discussed in chapter 5 (although computer designers did not systematically design switching circuits until they worked with electricity). For mechanical calculators, two different physical positions of a shaft or lever could be used to represent a bit. For electromagnetic relays, the difference between an open and a closed circuit was used. And for electronic computers, two different voltages are used.

We have already noted that the lower of two voltages can be associated with either "1" (the convention of negative logic) or "0" (the convention of positive logic). It is not even required that the two voltage levels be the same at every switching point. For example, in Atanasoff's machine the voltage transferred from a drum capacitor to an input tube of the add-subtract mechanism was +30 ("0") or −40 ("1"), whereas the corresponding plate voltages of an output tube were +120 ("0") and +50 ("1"). (See fig. 11.) But at each switching point the high voltage represented a binary "0" and the low voltage represented a binary "1." Similar statements hold for ENIAC switches.

The defining truth table for those switches of figure A.1 that have three inputs is an obvious eight-row extension of the defining truth tables just given. More generally, all of the switches of figure A.1 can be generalized in a natural way to make n-input switches, for any finite n. It will be instructive to show how this is done. When the binary truth-function OR is iterated, the truth value of the result is independent of the order and grouping of the statements disjoined; thus "$[(p \vee q) \vee r] \, s$" and "$(p \vee s) \vee (r \vee q)$" are each true when one or more of the four atomic statements p, q, r, s are true. In mathematical terms, disjunction is associative and commutative. Hence an atomic disjunctive switch with any number n of inputs is well defined: its output is "1" if and only if at least one of its inputs is "1."

Similarly, binary NOR, binary AND, and binary NAN are associative and commutative, and so atomic NOR, AND, and NAN switches with any number of inputs are well defined. Binary inequivalence ($\neq$) and binary equivalence ($\equiv$) are also each associative and commutative; this is not obvious, but it is easily proved by mathematical induction. Consequently, atomic inequivalence and equivalence switches with any number of inputs are well defined.

When a logical operator is both associative and commutative, switches based on it are *symmetrical* in the sense that the output of a switch depends only on the *number of inputs* that are true, and not on which particular inputs are true. All of the switches of figure A.1 are symmetrical. The symmetry of the atomic switches used in computer design reflects the symmetry of the basic hardware circuits used to realize them. Of course, not all truth functions are symmetrical (e.g., "if p then q" is not), and most compound switches used in computers are not symmetrical.

We come finally to the last two switches of figure A.1. These are called threshold switches because in each case the operation can be stated in terms of whether or not the sum of the input signals reaches a stated threshold. Thus for the last switch of figure A.1, the output s is true if and only if *at least two* of p, q, r are false; equivalently, s is false if and only if *at least two* of p, q, r are true.

The threshold notation is interesting here for two reasons. First, NOT, OR, NOR, AND, and NAN switches can all be expressed in threshold notation. Thus an n-input AND has a true output if and only if *at least n* inputs are true, and any OR has a true output if and only if *at least one* input is true. Second, Atanasoff's switching notation (p,q) was a threshold notation, because q gave the number of inputs that had to be true (have a low voltage) to turn the driven triode off and thus make the output false (have a high voltage). Moreover, Atanasoff physically achieved his logical switches by a threshold method (adding voltages and setting a threshold level for switching), as we shall see.

Compound Switches

Atomic switches can be combined in evident ways to make a compound switch or switching network. Figure A.2a illustrates the basic rules for building compound switches. The output of an atomic switch may drive any number of inputs; in the figure, there are none at switching points s and c_{out}; one at switching points d, f, and h; and two at switching points e, g, and i. The inputs to atomic switches can be identified; points a, b, e, g, c_{in}, and i are examples. Finally, there is no cycle from any switching point through successive atomic switches and back to the starting point.

Just as an atomic switch has inputs and outputs, so does a compound switch. Switching points not driven by other atomic switches are the inputs to the compound switch—in this case a, b, and c_{in}. Certain switching points may be designated as outputs; we chose s and c_{out} to be the outputs of figure A.2a. The remaining switching points are called "internal." Note that whereas an atomic switch has a set of input states and a set of (two) output states, a compound switch has three sets of states: input, internal, and output.

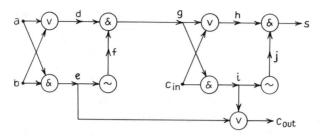

(a) A compound switch with inputs a,b,c_{in} and outputs s,c_{out}

	Input state			Internal state							Output state	
	a	b	c_{in}	d	e	f	g	h	i	j	c_{out}	s
	0	0	0	0	0	1	0	0	0	1	0	0
	0	0	1	0	0	1	0	1	0	1	0	1
	0	1	0	1	0	1	1	1	0	1	0	1
	0	1	1	1	0	1	1	1	1	0	1	0
	1	0	0	1	0	1	1	1	0	1	0	1
	1	0	1	1	0	1	1	1	1	0	1	0
	1	1	0	1	1	0	0	0	0	1	1	0
	1	1	1	1	1	0	0	1	0	1	1	1
Height	0			1		2	3	4		5		6

(b) Calculating truth table for the switch, showing the height of each switching point

Fig. A.2. A compound switch and its truth table

Using the defining truth tables for atomic switches, one can easily calculate the truth table for a compound switch, as in figure A.2b. The calculation proceeds step by step according to the heights of the various switching points. The height of a switching point is the maximum number of switches on any path from the inputs to it. In theory a switch responds instantly to a change of input state, but in actuality a switch takes time to respond, and the height of a switching point is a rough measure of the time delay from the inputs to that point.

Comparing the inputs and outputs of figure A.2 with table 1 for binary addition in chapter 1, we see that the circuit of figure A.2 is actually an adder: a and b are the binary digits in, c_{in} is the carry in; s is the sum out, and c_{out} is the carry out. This is standard adder design, the circuit being composed of two half-adders (each of the square blocks being a *half-adder*) and a two-input OR. We develop some other designs in the next section, for purposes of comparison with Atanasoff's add-subtract mechanism.

There are two basic grammatical rules for compounding switches from atoms: a switch output may be connected to one or more switch inputs, provided this does not produce any cycles; and free switch inputs can be identified. These rules guarantee that every switching point has a unique height, so that a *calculating truth table* can be constructed, and hence that each input state will determine a unique internal state and a unique output state. Correspondingly, when a compound switch is realized by an electrical circuit, that switch will react to a given input state (a pattern of voltages imposed from the outside) to produce an output state (the pattern of voltages specified in its truth table).

A cycle in a compound switch is a path from switching junction to switching junction that always goes forward through switches and eventually returns to the switching junction from which it started. If a switch contains a cycle, the heights of the junctions in that cycle are not uniquely determined, so that a calculating truth table cannot be constructed. Moreover, if a switch contains a cycle an input state may not determine a unique output state. Indeed, the switch may not even have any inputs. For example, if the output of a NOT switch is fed back into its input, the result is a contradiction ("$p \equiv \sim p$" is false whether p is true or false). Similarly, if the output of an OR switch is connected to all its inputs, the result is indeterminate ["$p \equiv (p \vee p)$" is true whether p is true or false].

The nature of switch cycles is illustrated by the story of the engineer who designed an electronic switch with input from a microphone at the end of a wire and output to a bell at the end of another wire. The equipment was designed to ring the bell whenever the microphone sensed no sound, and not to ring the bell when the microphone did sense sound. In this way, it could be used to signal that it was quiet in a given room. But what happened when the microphone and the bell were placed in the same room? In terms of switching theory the situation is undefined, whereas in actuality it takes time for a switch to respond to a change of input state. (As we mentioned earlier, the height of a compound switch gives a rough measure of the time for the switch to react to a change of input state.) Thus, given an adequate delay, the engineer's equipment might alternate between ringing the bell and not ringing the bell!

There is a paradox here. The verb "to switch" means *to change,* and for a physical system to change takes time; yet the formal theory of switching does not encompass time. Moreover, we will need to take account of time when we give a logical analysis of Atanasoff's add-subtract mechanism, for it was serial, combining three bit streams to produce the sum or difference bit stream and remembering each carry or borrow bit for one pulse time.

To represent memory in our logical diagrams, we use a discrete time frame t = 0,1,2,3, . . . , and add another atomic element. With this time frame every switching point acquires a history, for at each moment t it is in either state "0" or state "1." In this context, it is simplest to think of "0" and "1" as the absence and presence of a pulse, respectively. The new primitive is a unit delay, shown in figure A.3a. Its initial output is "0" (no pulse); its output for any later moment t is its input at time $t - 1$. Thus the unit delay delays a pulse stream by one pulse time.

There is a whole theory of automata based on logical nets that are compounded from switches and delays. Here we note only that in this broader theory cycles are permitted and, indeed, are necessary for indefinite memory. Figure A.3b shows a cycle consisting of a NOT switch and a unit delay. The behavior of this

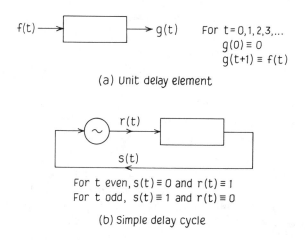

(a) Unit delay element

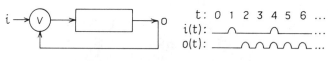

(c) A pulse trap and initial history

Fig. A.3. Adding a delay element to switching theory

cycle is not paradoxical but is well defined, the output of the delay oscillating between "0" and "1." Figure A.3c shows a simple cycle that will remember indefinitely if a pulse ("1") has ever appeared at its input and will so indicate on its output. That is, $o(t) \equiv$ [for any pulse time τ less than t, $i(\tau)$].

After this brief excursion into automata theory, let us return to switching theory and establish some results that will be useful in subsequent sections of this appendix.

Calculating truth tables can prove useful logical equivalences or identities. Table A.3 proves two forms of Augustus De Morgan's theorem. The last two formulas are called *tautologies* because they are true for all values of their variables. Since they are equivalences, the left- and right-hand sides of each are exchangeable in formulas or switches without changing truth values or switch behavior. These two forms of De Morgan's theorem are shown in figure A.4a.

De Morgan's theorem illustrates a general principle underlying truth-function theory, namely, that the roles of "0" and "1" are symmetrical. We saw earlier that there are two conventions for associating logical states with physical states: negative logic, in which a high voltage represents "0" and a low voltage represents "1"; and positive logic, in which a high voltage represents "1" and a low voltage represents "0." Notice that the correspondence of logical switches to physical switches depends on the convention chosen in assigning the truth values "0" and "1" to physical states. And so if we interchange the "0's" and "1's" in table A.3, the column for OR becomes the column for AND, and vice versa.

TABLE A.3.   Calculation of De Morgan's Theorem

Input State		Subformulas of Height 1				Subformulas of Height 2		Two Forms of De Morgan's Theorem	
								$(p \vee q) \equiv$ $(\bar{p} \& \bar{q})$	$(p \& q) \equiv$ $(\bar{p} \vee \bar{q})$
p	q	$\bar{p}$	$\bar{q}$	$p \vee q$	$p \& q$	$\overline{p \& q}$	$\overline{p \vee q}$		
0	0	1	1	0	0	0	0	1	1
0	1	1	0	1	0	1	0	1	1
1	0	0	1	1	0	1	0	1	1
1	1	0	0	1	1	1	1	1	1

Thus OR and AND are duals. Similarly, NOR and NAN are duals, and so are inequivalence (XOR) and equivalence (EQV).

De Morgan's theorem generalizes easily to any number of inputs. There are also forms of it involving threshold switches; examples are given in figure A.4b. Moreover, by combining the law of double negation with De Morgan's theorems, one can obtain a myriad of additional forms; some examples are given in figure A.4c.

The address decoding switch of figure A.5 is a commonly used switch that can be used to illustrate several points in switching theory. There are three input variables a_1, a_2, a_3 and hence 2^3 or eight input states $\bar{a}_1\bar{a}_2\bar{a}_3$, $\bar{a}_1\bar{a}_2a_3$, $\bar{a}_1a_2\bar{a}_3$, . . . , $a_1a_2\bar{a}_3$, $a_1a_2a_3$. (For convenience the conjunction signs have been dropped, as is often done.) There are eight output wires, one for each input state, so that at each moment exactly one output wire is on ("1") and all others are off ("0"). Hence for each three-bit address this switch chooses a memory cell that stores a byte or word. This structure is easily extended to binary addresses of any length, but for longer addresses there are more efficient switching structures.

Atanasoff did not need an addressing switch, because his computer was a vector machine that normally processed simultaneously the thirty numbers on the counter drum and the thirty numbers on the keyboard drum. (For base conversion, only a few numbers were involved, but this selection was made mechanically by the operator.) The first address switch in an electronic computer was that of the ENIAC function table, where two decimal numbers were used to choose one of a hundred function values. The EDVAC used a binary address of two parts, the first part choosing one of thirty-two storage lines by means of an address switch, the second part choosing one of thirty-two words from within a storage line by means of a clock. The electrostatic stores of the IAS computer used address switches for each of the two spatial dimensions of the memory.

Each of the input variables a_1, a_2, a_3 of figure A.5 is needed in negative form as well as positive form. When the variables are stored in flip-flops (two-state memory devices with set and reset inputs), a variable is available on one side of the flip-flop and its negation on the other side. NOT switches, then, are not needed on the input lines of the address switch.

The eight conjunctions produced by figure A.5 are called the *complete basic conjunctions* of the variables a_1, a_2, a_3. Since these correspond to the rows of a truth

(a) De Morgan's theorem

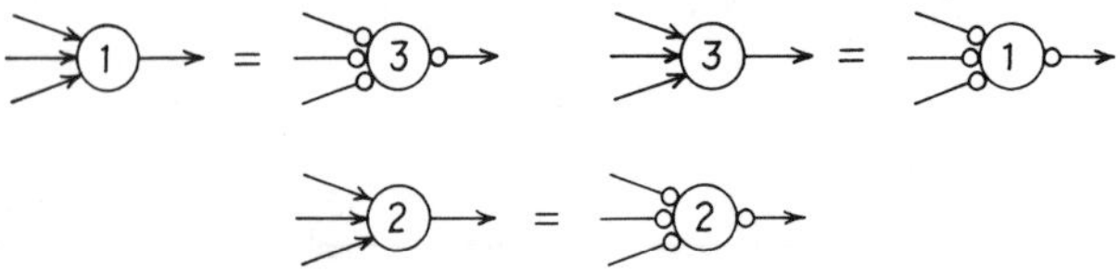

(b) Threshold forms of De Morgan's theorem

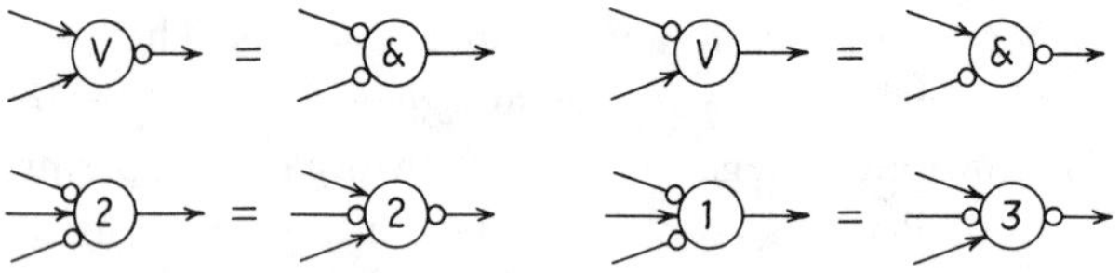

(c) Other forms of De Morgan's theorem

Fig. A.4. De Morgan's theorem

table, any truth function other than a contradiction can be expressed as a disjunction of the complete basic conjunctions of its variables. (A contradiction can easily be expressed by conjoining any variable with its negation.) Some examples are given in table A.4 (p. 311). A basic conjunction has a single "1" in its truth column, so that to express a function in *disjunctive normal form* one merely disjoins the basic conjunctions for which the function is true.

Because any truth function can be expressed in disjunctive normal form, the sum and carry bits in binary addition can also. Figure A.6 shows how to use this method to design an addition switch.

Various sets of atomic switches are used for computer switching. The designers of the ENIAC used mainly the set {NOT,NAN,NOR,AND}, whereas Atanasoff used the set of primitives {NOT,NAN,NOR, and the threshold switch (3,2)}. It is obviously very important that the set be complete in the sense of being sufficient to construct absolutely any switch. Clearly, the set {AND,OR} is not complete: when all the inputs of any switch composed of these atoms are "0," the output will be "0"; thus a simple negation switch cannot be built from this set.

On the other hand, the fact that every truth function can be expressed by a formula or a switch in disjunctive normal form shows that the set {NOT,AND,OR} is complete. De Morgan's theorem (fig. A.4a) then shows that the sets {NOT,AND} and {NOT,OR} are complete. Figure A.7a shows how to construct NOT and OR from NOR; hence NOR is sufficient by itself for the construction of all switching functions. The similar construction of figure A.7b shows that any switch can be constructed from NAN, as well. It follows, then, that Atanasoff's set of switching primitives was complete.

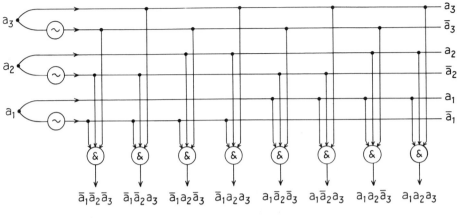

Lines to memory cells

Fig. A.5. Address decoding switch

Logical Structure of Adding and Subtracting Circuits

In designing a switch, one begins with a specification of what the switch is to do. As we saw in chapter 1, Atanasoff began with a truth-table specification of addition and subtraction, then proceeded intuitively to design a circuit that would satisfy that specification. The ENIAC's switching circuits were designed even more intuitively.

Somewhat later, logical switching diagrams were developed to represent the logical essence of a circuit, and these gradually came to be used as an intermediary step in the design process. For this procedure, one began with a logical characterization of the switching functions needed (a defining truth table, a formula, or a verbal statement), next developed a switching network that performed those switching functions, and finally translated that network into a circuit diagram. Although Atanasoff did not design his add-subtract circuit in this manner, a good way to study and evaluate its logical structure is to abstract from it a switch diagram and compare that diagram with others developed by a more formal procedure.

Serial addition and subtraction presuppose that two bit streams are available at the rate of one bit per pulse time, and that a means exists to store the carry or borrow bit for one pulse time. For addition, the two input streams are the addend and the augend, and the output stream is the sum. For subtraction, the subtrahend stream is subtracted from the minuend stream to produce the difference.

Because Atanasoff used fifty-bit numbers, we start with a formal analysis of fifty-bit addition. Atanasoff's machine was fixed point, as were the ENIAC and the first stored-program computers. Without any loss of generality we can restrict the number range to numbers whose absolute values are less than unity. This means that the leftmost bit is the sign bit, with the binary point positioned to its right (and indicated as a single dot in the sequence).

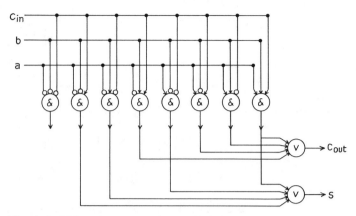

Fig. A.6. Disjunctive normal form adding switch

Let the fifty-bit addend and augend be $a = a_0.a_1a_2 \ldots a_{48}a_{49}$ and $b = b_0.b_1b_2 \ldots b_{48}b_{49}$, respectively. The subscripts progress from left to right so that each one indicates the value of a binary position. Thus a "1" in the position a_i has the value 2^{-i}. For example, a "1" in position 2 has the value 2^{-2}, and a "1" in position 5 has the value 2^{-5}. It will be clear from the next paragraph that this rule holds for the sign bit, as well; that is, a "1" in position 0 has the value 2^0, or 1.

Following Atanasoff, we represent negative numbers as two's complements. Thus, $-x$ is represented by

$$C(x) = 2 - x = \bar{x} + 2^{-49},$$

where $\bar{x}$ is the bitwise complement of x; that is,

$$\bar{x} = \bar{x}_0.\bar{x}_1\bar{x}_2 \ldots \bar{x}_{48}\bar{x}_{49}.$$

For example, $+\frac{5}{16} = 0.010100 \ldots$, the bitwise complement of 0.010100 is $1.101011 \ldots$, and $-\frac{5}{16} = 1.101100 \ldots$.

We now define the sum, s, of a and b, where $s = s_0.s_1s_2 \ldots s_{48}s_{49}$. The simplest way to do this is to define it step-by-step with the carry number $c = c_0.c_1c_2 \ldots c_{48}c_{49}$, where c_i ($i = 0$ to 48) is the carry-bit *into* the ith position from the preceding position and $c_{49} = 0$. Addition is then defined by

Carry: $c_{i-1} \equiv (a_ib_i \vee a_ic_i \vee b_ic_i)$ for $i = 49$ to 1,
Sum: $s_i \equiv (a_i \neq b_i \neq c_i)$ for $i = 49$ to 0.

These rules give the correct sum for both positive and negative numbers, provided that the absolute values of a, b, and s are all less than unity.

Consider next a serial binary adder that is fed two bit streams and processes one bit at a time to produce an output bit stream. Since the least significant bits of our numbers are on the right, and since carrying takes place from right to left, the

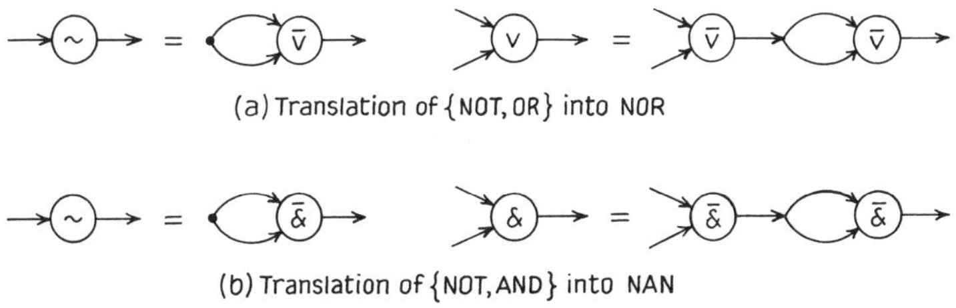

(a) Translation of {NOT, OR} into NOR

(b) Translation of {NOT, AND} into NAN

Fig. A.7. Proof that all switches can be constructed from a single primitive

addend a and augend b must be fed into it in the order of decreasing subscripts $a_{49}, a_{48}, \ldots, a_2, a_1, a_0$ and $b_{49}, b_{48}, \ldots, b_2, b_1, b_0$, and the sum will appear in the same order $s_{49}, s_{48}, \ldots, s_2, s_1, s_0$. In other words, i now becomes a time variable proceeding backward, the first step of addition or subtraction taking place at time $i = 49$ and the last step at time $i = 0$. This convention holds for figures A.8 through A.10.

A *serial binary adder* contains a switching circuit and a unit delay. The switching circuit has three inputs a_i, b_i, c_i and produces two outputs, a sum output $s_i \equiv (a_i \neq b_i \neq c_i)$ and a carry output $c_{i-1} \equiv (a_i b_i \vee a_i c_i \vee b_i c_i)$. The carry output c_{i-1} is fed back to the carry input c_i through a unit delay (a delay of one pulse time or one bit position). Figure A.8 shows a serial binary adder based on the two switching atoms, "exclusive-or" (for the sum bit) and "at least two" (for the carry bit). This simple representation of an adder, with just one primitive for each output, will serve our purpose here. The simplicity does not translate into hardware, however, because XORs and threshold switches generally require more circuitry than the other primitives of figure A.1.

In the case of subtraction, it is essential to distinguish the minus sign that symbolizes a negative number from the minus sign that symbolizes the operation itself. Although the binary adders of figures A.6 and A.8 handle negative numbers when these are in complement form, they cannot form the complement of a number or subtract one number from another. In the context of addition, complementation and subtraction are equivalent—with an adder one can either subtract by complementing or complement by subtracting. We can therefore modify the adder of figure A.8 so as to develop an adder-subtracter for each method (figs. A.9 and A.10).

To subtract by complementing, we prefix a complementer to an adder, as in figure A.9. This circuit is the same as that of figure A.8 except for the two-input XOR at lower left and the disjunctive element on the right. When the control lines p_i and q_i are held at "0" throughout (that is, for $i = 49$ to 0), figure A.9 reduces logically to figure A.8.

We have changed the symbolism of a, b, c, and s used in figure A.8 to that of α, β, γ, and σ so as to have a notation that covers subtraction as well as addition. Thus α, β, γ, σ are the addend, augend, carry-word, and sum, respectively, for addition, but are the minuend, subtrahend, borrow-word, and difference, respectively, for subtraction. Hence in figure A.9, when lines p_i and q_i are held at "0" throughout, $\sigma = \alpha + \beta$.

Consider next subtraction in figure A.9. Here the control line p_i is held at

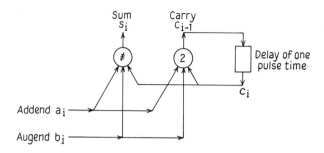

The numbers $a = a_0.a_1... a_i...a_{49}$ and $b = b_0.b_1...b_i...b_{49}$ are supplied serially, the least significant digits (a_{49} and b_{49}) first.

Fig. A.8. A serial binary adder

"1" throughout, so that $\beta'_1 \equiv \bar{\beta}_i$ for all i and the bitwise complement of β is accordingly fed into the adder. The control line q_i is set to "1" for the first time-step of the subtraction only, causing 2^{-49} to enter the adder. As a consequence of these control actions, the complement $C(\beta)$ of the subtrahend β is added to minuend α, so that $\sigma = \alpha + C(\beta)$, or $\sigma = \alpha + (2 - \beta)$. This is the correct answer, because if $\alpha \geq \beta$, the extra 2 is lost through a carry bit γ_{-1}, which is ignored, and $\sigma = \alpha - \beta$. If $\alpha < \beta$, then $\sigma = 2 - (\beta - \alpha)$ and $\sigma = C(\beta - \alpha)$, which is the correct difference when $\alpha - \beta$ is negative.

Now while figure A.9 shows subtraction executed by first complementing the subtrahend and then adding that complement to the minuend, it is also possible to subtract the subtrahend directly from the minuend. This is the method Atanasoff used. As we have seen, he used voltage tables for addition and subtraction that are equivalent to truth tables. The first table is in accord with the ordinary carry rule, the second with the ordinary borrow rule (see chap. 1). Figure A.10 represents another adder-subtracter we have designed along these same intuitive lines.

It is apparent by inspection of the truth tables for addition and subtraction that the sum bit is the XOR of the addend, augend, and carry bit, *and also that* the difference bit is the XOR of the minuend, subtrahend, and borrow bits. Hence both sum and difference can be calculated by the formula

$$\sigma^*_i \equiv (\alpha_i \neq \beta_i \neq \gamma^*_i),$$

where we have added asterisks to the outputs σ^*_i (sum-difference) and γ^*_i (carry-borrow) to associate them with figure A.10.

One can also see by inspection of the truth tables that if the minuend bit is reversed (complemented), the rule for the borrow bit is the same as the rule for the carry bit. Using the symbol ② for the threshold function "at least two are true," we have

For addition: $\gamma^*_{i-1} \equiv ② (\alpha_i, \beta_i, \gamma^*_i),$
For subtraction: $\gamma^*_{i-1} \equiv ② (\bar{\alpha}_i, \beta_i, \gamma^*_i).$

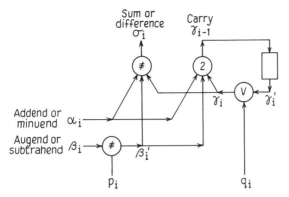

For addition,   α is the addend and β the augend. Control signals must hold p_i and q_i at "0" throughout (i.e., for $i = 49$ to 0).

For subtraction, α is the minuend and β the subtrahend. Control signals must hold p_i at "1" throughout and q_i at "1" for $i = 49$ and at "0" otherwise.

Fig. A.9. Adder-subtracter made from adder and complementer

These formulas are achieved in figure A.10, because the two-input XOR is a complementer; that is, when r_i is held at "0" for addition, $\alpha^*_i \equiv \alpha_i$, whereas when r_i is held at "1" for subtraction, $\alpha^*_i \equiv \bar{\alpha}_i$.

Compare now the structure of the adder-complementer (fig. A.10) with the structure of the adder-subtracter (fig. A.9). Each has an adder on the top level. Each has an XOR on the bottom level, but the adder-complementer also has a disjunction that needs to be pulsed at the beginning of serial subtraction. The adder-subtracter is thus a simpler circuit, and Atanasoff was wise to choose it over the adder-complementer.

The adder-complementer and adder-subtracter also differ as to where the controlled XOR is placed, one having it in the augend-subtrahend input line, the

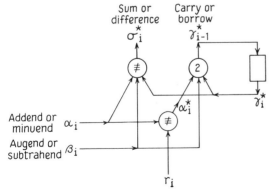

For addition,   hold r_i at "0".
For subtraction, hold r_i at "1".

Fig. A.10. Adder-subtracter with carry-borrow circuit

other in the addend-minuend line. In view of these differences, one might ask how we know that these two circuits are equivalent. Both are derived by seemingly valid processes from an intuitive knowledge of binary arithmetic, but their logical equivalence can also be proved directly. The situation is obvious for addition, where both diagrams reduce to the adder diagram of figure A.8. We need only prove equivalence for subtraction, then, that is, that $\sigma^* = \sigma$, or $\sigma^*_i \equiv \sigma_i$ for $i = 0$ to 49. This proof is offered for the interested reader in the next two paragraphs.

We first prove by induction that, during subtraction, the borrow output of figure A.10 is always the opposite of (inequivalent to) the carry output of figure A.9. The first bits to be processed are the least significant bits ($i = 49$). From the figures it is easy to see that the borrow output of figure A.10 is $②(\bar{\alpha}_{49},\beta_{49},0)$, while the carry output of figure A.9 is $②(\alpha_{49},\bar{\beta}_{49},1)$. These two formulas are inequivalent by a form of De Morgan's theorem (cf. fig. A.4). For the inductive step, assume that the borrow output γ^*_i is inequivalent to the carry output γ_i, that is,

$$\gamma^*_i \not\equiv \gamma_i \qquad \text{(inductive step assumption)}.$$

From the figures we get

$$\gamma^*_{i-1} \equiv ② (\bar{\alpha}_i,\beta_i,\gamma^*_i) \qquad \text{(fig. A.10)},$$
$$\gamma_{i-1} \equiv ② (\alpha_i,\bar{\beta}_i,\gamma_i) \qquad \text{(fig. A.9)}.$$

Substituting $\bar{\gamma}_i$ for γ^*_i in the first formula, we get

$$\gamma^*_{i-1} \equiv ② (\bar{\alpha}_i,\beta_i,\bar{\gamma}_i) \qquad \text{(fig. A.10)},$$
$$\gamma_{i-1} \equiv ② (\alpha_i,\bar{\beta}_i,\gamma_i) \qquad \text{(fig. A.9)}.$$

Again, by a form of De Morgan's theorem,

$$\gamma^*_{i-1} \not\equiv \gamma_{i-1} \qquad \text{(inductive step conclusion)}.$$

Hence, by mathematical induction,

$$\gamma^*_i \not\equiv \gamma_i \qquad \text{for all } i = 49 \text{ to } 0.$$

Our goal is to show that for subtraction the binary ouput σ^* of figure A.10 is the same as the output σ of figure A.9. From the figures we get

$$\sigma^*_i \equiv (\alpha_i \not\equiv \beta_i \not\equiv \gamma^*_i) \qquad \text{for all } i = 49 \text{ to } 0,$$
$$\sigma_i \equiv (\alpha_i \not\equiv \bar{\beta}_i \not\equiv \gamma_i) \qquad \text{for all } i = 49 \text{ to } 0.$$

But $\gamma^*_i \not\equiv \gamma_i$ for all i, and hence

$$\sigma^*_i \equiv (\alpha_i \not\equiv \beta_i \not\equiv \bar{\gamma}_i),$$
$$\sigma_i \equiv (\alpha_i \not\equiv \bar{\beta}_i \equiv \gamma_i).$$

TABLE A.4.   Some Disjunctive Normal Forms

Input State		Corresponding Basic Conjunctions	Disjunctive Normal Forms of		
			$p \vee q$	$p \not\equiv q$	$p \equiv q$
p	q		$(\bar{p}q \vee p\bar{q} \vee pq)$	$(\bar{p}q \vee p\bar{q})$	$(\bar{p}\bar{q} \vee pq)$
0	0	$\bar{p}\bar{q}$	0	0	1
0	1	$\bar{p}q$	1	1	0
1	0	$p\bar{q}$	1	1	0
1	1	pq	1	0	1

Since inequality is a symmetric truth-function, these last two formulas are equivalent, so that $\sigma_i^* = \sigma_i$ for all i, and hence $\sigma^* = \sigma$. This completes the proof that figures A.9 and A.10 produce the same result for subtraction.

We have now completed our analysis of binary addition and subtraction. In the previous section, we designed two adding switches (figs. A.2 and A.6). Either of these could be made into serial adders by feeding the carry output back into the carry input with a unit delay. In the present section, we designed an adding switch from an XOR and a threshold-two element, and incorporated a feedback delay to make a serial binary adder (fig. A.8). We then designed two serial adder-subtracters, one based on complementation (fig. A.9), the other on subtraction with borrowing (fig. A.10). We turn next to the use Atanasoff made of his atomic switching concepts, that is, how he realized them electronically.

Let us note in passing that we have concentrated on serial addition and subtraction in our presentation of switching theory because that is what Atanasoff used. Our analysis of binary arithmetic as well as the five different switching circuits we developed can also be used for parallel addition and subtraction. We return to this fact when we discuss the structure of Atanasoff's add-subtract mechanism in a later section (see fig. A.19).

Atanasoff's Atomic Switches

The importance of Atanasoff's contribution to electronic switching lies in his having defined a set of primitives and shown how to realize them with vacuum tubes, how to interconnect them, and how to compound them for arithmetic operations. His set, included in figure A.1 and defined by his (p,q) notation, consisted of NOT (1,1), two-input NOR (2,1), three-input NOR (3,1), two-input NAN (2,2), three-input NAN (3,3), and the threshold switch (3,2). The present section explains these in terms of the electronics. The following section then assumes the electronics and explains Atanasoff's particular compounding of atomic switches for addition and subtraction in terms of the logic.

It will be helpful to begin with a brief description of a vacuum triode and its mode of operation. A typical triode consisted of a cylindrical *cathode* with a heating element inside, a cylindrical *grid* of wires surrounding the cathode, and a solid cylindrical *plate* surrounding both. The cathode was heated by a wire loop, the loop itself being heated by an alternating current driven by 6.3 volts, obtained from a transformer. This heating element made the cathode so hot that electrons (negative charges) boiled off from it. If the grid was sufficiently negative with respect to the cathode, the electrons stayed near the cathode. But if the grid was at about the voltage of the cathode or positive with respect to it, the electrons traveled from the cathode, passed through the grid (since it was a wire mesh), and went to the plate. Because electrons are by definition negative charges, a flow of electrons from the cathode to the plate was the same as the flow of a current from the plate to the cathode.

Now a binary application of the triode requires two limiting cases: one in which the grid voltage was 0 or positive (in which case there was a substantial current from plate to cathode) and one in which the grid was substantially negative (in which case no current flowed from plate to cathode).

Atanasoff achieved these two conditions for the 6C8G double triodes he used for all his electronic switching by the following arrangements. He connected each plate through a "load" resistor to a direct-current power supply of +120 volts, and he connected the cathode to 0 volts (i.e., to a grounded terminal in the power supply). See figure A.11 for an example. Under these circumstances, the triode operated as follows. When the grid was at 0 volts or more (i.e., at the same voltage as the cathode or higher), the triode conducted current from the plate to the cathode and acted more or less as a fixed resistance of about 22.5K (22,500) ohms. When the grid was at −5 volts or lower (i.e., at 5 or more volts below the cathode), essentially no current flowed through the tube and it acted as an open circuit. For reliability reasons the circuits were designed with a safety factor. The grid was driven 10 volts or more negative to turn the tube off and raised at least to 0 volts to turn the tube on.

Thus the grid of the triode was used to switch it on and off. When the grid was high (representing "0"), current flowed from the +120-volt power supply terminal through the plate resistor and on through the tube, returning to the power supply through its 0-volt terminal and making the voltage at the plate of the triode low (representing "1"). When the grid was low ("1"), no current flowed through the tube and the voltage at its plate was high ("0"). We shall see later that the exact values of these two plate voltage levels varied from case to case, depending on the switching circuit driven by the triode.

It is interesting to note that for the continuous (or analog) mode of radio, audio amplification, and television, the operation of the triode in the range between the limits of off and on is significant. The amount of plate current in this range is a nonlinear function of the grid voltage and the plate voltage (both being measured relative to the cathode voltage). Basically, a vacuum-tube switch is a nonlinear device: as the voltage on the grid moves from positive to extremely negative, the tube starts to cut off, cuts off rapidly, and then remains off over a large range of negative voltages.

Now in both discrete and continuous applications, the vacuum tube was an amplifier and so a control element. A small amount of power expended at the grid controlled a relatively large amount of power at the cathode. In Atanasoff's use of vacuum tubes for switching, the voltage swing at the plate was much larger than the voltage swing at the grid, and this amplification served to counter losses such as the leakage of electrons from capacitors.

In most usages, whether discrete or continuous, the role of the triode has been taken over by the transistor, a solid-state descendant of the vacuum tube. It is characterized as controlling the flow of positive charges (holes, or absences of electrons) as well as the flow of negative charges (electrons).

Let us turn now to the details of Atanasoff's use of triodes to execute logical functions. We do not have access to precise resistor values, such as would be given on detailed circuit diagrams, but we supply some plausible figures in order to explain the general principles. As it happened, resistors and tubes varied considerably from one to another, and Clifford Berry chose the actual resistances for the add-subtract mechanisms by laboratory measurement.

We begin our study of Atanasoff's novel resistor-network method of logical switching with the basic NOT shown in figure 10 (see chap. 1) and now in more analytical detail in figure A.11. Logically speaking, the NOT circuit is a negation switch because the polarity of the output signal is opposite to that of the input signal. Electrically speaking, however, it does two distinguishable things, the first of which is desired, the second an undesirable by-product. The circuit amplifies the input signal, converting a grid difference of 5 volts (0 to −5 volts) or more into a plate difference of 65 volts. But the circuit also shifts the voltage levels upward, from near 0 volts to an entirely positive range (+35 to +100 volts). Since atomic switches must be iterated to make compound switches, and since computing requires the cyclic transfer of information from memory devices to switches and back again, these output voltage levels have to be reconciled with the input voltage levels.

This reconciliation can be accomplished in a variety of ways. We discuss Atanasoff's approach here, then contrast it with others later in this appendix. As we saw in chapter 1, Atanasoff used a different method for each of two junctures in his add-subtract circuit: between one switching triode and the next, and between an output tube and a drum capacitor. In the former case the output levels were shifted down at the cost of loss of signal, whereas in the latter the adjustment was made through a "boost" circuit with no loss of signal.

Let us review these two methods briefly, starting with the boost circuit (see fig. 11). Before a writing brush contacted a drum stud, the plate of an output tube was at +120 volts for "0" and at +50 volts for "1." During contact, however, these voltages had to be +30 volts and −40 volts, respectively. Atanasoff's boost circuit worked as follows. One side of every drum capacitor was connected to a stud (or a commutator in the case of a carry-borrow capacitor); we call this the *outer* side because it went out to the surface of the drum. The other, *inner,* sides of all the capacitors were connected together and then to a slip ring at the end of the drum that was in constant (moving) contact with a brush from the boost voltage source.

The boost voltage was actually a timing or clock signal that changed every pulse time (every 6 degrees of slow drum rotation, i.e., every sixtieth of a second). It was usually at ground (0 volts) but was raised to +90 volts at the time the writing brushes were in electrical contact with the outer sides of the drum capacitors. Hence either +120 volts ("0") or +50 volts ("1") from an output triode was recorded on a capacitor, but with respect to the +90 volts on the inner side of the capacitor. When the boost line lowered the voltage of the inner side back down to ground, the voltage of the outer side went down to +30 volts ("0") or −40 volts ("1"). These were the approximate voltages read from the capacitors into the triodes at the next drum cycle.

The boost signal itself was generated in the following way. The power supply had terminals for +120 volts, +90 volts, 0 volts, and −120 volts. A wire went from the +90-volt terminal to a brush in constant (moving) contact with a slip ring on the fast drum. This slip ring was connected to alternate segments of a fast-drum commutator, the other segments being grounded. A brush contacting this commutator fed the boost line, which was connected with brushes to slip rings on all three read-write drums—including the fast drum, where the thirty carry-borrow capacitors had also to be boosted.

Atanasoff's system for lowering voltage levels from the output triodes of his add-subtract mechanism to the levels required by his drum capacitors was an ingenious method of recording bits on capacitors without loss of signal. Moreover, it was the first timing system to produce a clock pulse for each bit position of a binary number. It did, of course, rely on mechanical motion, commutation, and slip rings for its operation.

We look next at his system for lowering voltage levels from one triode to the next. He lowered those from the driving tube to the driven tube by means of a resistor network, a method that did diminish the signal. That is, the voltage difference between a "0" and a "1" supplied to a grid was less than the voltage difference between a "0" and a "1" produced at the plate. But this loss was immediately compensated for by the signal amplification of the triodes.

Let us use figure A.11 as our example here. It is an elaboration of figure 10, drawn to show how the basic circuit constituents of a one-input switch operate separately and how they interact. As we noted earlier, the circuit was always operated so that a vacuum tube was either fully conducting or essentially nonconducting; this end was achieved by always operating the grids (α and γ) at 0 volts or higher or at −10 volts or lower.

Look first at the driving triode by itself, that is, with points β and β' unconnected. As indicated in the figure, the plate resistor is 50K ohms. When the driving triode is off, no current flows through its plate resistor, and so the plate voltage β is +120 volts. When the driving triode is on, a current of about 1.7 milliamperes flows from the plate supply through the plate resistor and on through the tube to ground, so that β is at about +37 volts.

Consider next what happens when β is connected to β'. Then current flows from the plate supply through the plate resistor, on through the dropping resistor

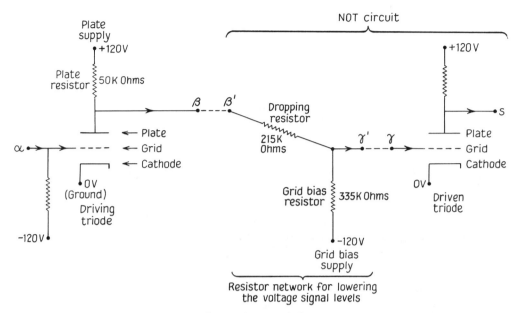

Fig. A.11. **Basic circuit constituents of a one-input switch**

and the grid bias resistor to the grid bias supply. This current is in addition to whatever current flows through the vacuum tube when it is conducting. The voltages at β are now different from the earlier ones, +100 volts when the driving triode is off and +33 volts when it is on. Correspondingly, the voltages at γ' are +14 volts and −27 volts. When γ is connected to γ', these become the voltages on the grid of the driven triode and are more than adequate to turn it on and off, respectively.

Actually, the situation is still more complicated when the grid of the driven tube is positive with respect to its cathode. For then the grid acts as a plate and draws current; this current causes the voltage drop across the plate and the dropping resistors to increase so that the positive voltage is near ground, rather than +14 volts. Such a decrease does not affect the switching action, however, because a positive voltage on the grid of the driven tube causes it to conduct. Having mentioned this point for the sake of completeness, we ignore it in explaining Atanasoff's two- and three-input switches (e.g., in figs. A.13 and A.14). We do so partly for simplicity, and partly because the voltage calculated by ignoring grid current shows how much safety factor the circuit has, that is, how much the voltage at the grid can vary because of variations in the vacuum tubes and resistors before the switch fails to work correctly.

We now explain how these voltages and those for figures A.12, A.13, and A.14 were calculated, in order to indicate what was involved in circuit design with vacuum tubes. There are two principles here. The first is Ohm's law, $E = IR$, which states that the electromotive force (in volts) across a resistor equals the current intensity (in amperes) times the resistance (in ohms). The second is Kirkhoff's law, which states that the sum of the currents flowing into a junction or point equals the sum of the currents leaving that point.

A resistor is a passive element, with a fixed resistance at any moment (though its resistance can change with temperature and aging). A vacuum tube is an active element, using the heat in its cathode to amplify electrical signals applied to its grid, and its behavior is much more complicated. As we mentioned earlier in this section, the current through a vacuum tube is a nonlinear function of the voltages at its grid and plate (relative to the voltage level of the cathode). To design a vacuum-tube circuit, one could obtain an estimate of this nonlinear function from a standard tube manual or measure it at the laboratory bench. For the 6C8G triodes used in Atanasoff's add-subtract mechanism, however, it is reasonable to treat each triode as having a fixed resistance of 22.5K ohms when it was conducting and as having an infinite resistance (being an open circuit) when it was not conducting.

(This is only an approximation, and it was a serious problem for computer design that the function describing the behavior of a vacuum tube not only was nonlinear but also varied from tube to tube and over the life of the tube. We return to this point in our discussion of the reliability of Atanasoff's circuits in the last section.)

The NOT circuit, which we have used to introduce the triode and ways of reconciling its output and input voltage levels, was, of course, the simplest of Atanasoff's atomic switches. His design for the switches that had two or three inputs (NOR and NAN) was more complicated. For those, he invented a circuit that did two things: (1) it added the binary voltage signals from the driving tubes to produce a multivalued voltage signal, and (2) it established a threshold level for the cathode relative to this multivalued voltage signal so that the tube switched on or off according to the desired switching function.

Figure A.12 shows how resistor networks can be used to add voltage signals, and why they lower the voltage levels of the signal and decrease its amplitude. The resistor values in this figure are chosen so as to make the arithmetic simple; but resistor networks like those of figure A.12a–c are used in figures A.11, A.13, and A.14, respectively. We start with Atanasoff's method of designing switches for the two-input case (figs. A.12b and A.13), in terms of conditions (1) and (2) in the preceding paragraph.

1. In Figure A.12b, a network with one lower and two upper resistors adds two independent binary signal inputs to produce a three-valued voltage sum at the output. At the inputs, "0" is represented by +240 volts and "1" by +120 volts. Logically speaking, there are four possible input states, "00," "01," "10," and "11." But because of the symmetry of the NOR and NAN functions, these four cases reduce to three: no "1," one "1," and two "1's"; or, if we add the "1's," to the ternary values "0," "1," and "2." Correspondingly, there are three electrical states at the output, for by Ohm's law the output voltage is one-half the sum of the input voltages, and hence is +120 volts, +90 volts, or +60 volts. Notice that this output voltage range can be shifted up or down by changing the ratio of the lower to the upper resistances, although the magnitude of the range will be altered in the process. For example, if the 100K resistor were replaced by a 200K resistor, the output voltage levels would be +160 volts, +120 volts, and +80 volts.

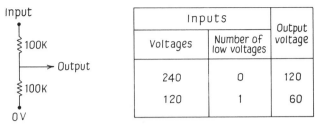

Inputs		Output voltage
Voltages	Number of low voltages	
240	0	120
120	1	60

(a) Lowering signal levels

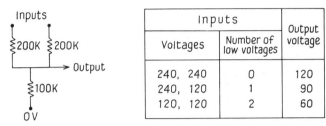

Inputs		Output voltage
Voltages	Number of low voltages	
240, 240	0	120
240, 120	1	90
120, 120	2	60

(b) Adding two signals

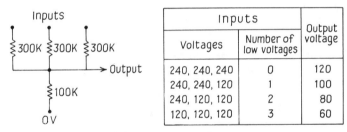

Inputs		Output voltage
Voltages	Number of low voltages	
240, 240, 240	0	120
240, 240, 120	1	100
240, 120, 120	2	80
120, 120, 120	3	60

(c) Adding three signals

Fig. A.12. Adding signals with a resistor network

2. Figure A.13 shows how the Atanasoff network with three resistors could be used to make a NOR switch or a NAN switch. The two driving tubes produce a signal at the grid g of the driven tube that has three voltage levels about 20 to 25 volts apart. By choosing the resistance values R_1 and R_2 appropriately, the designer controls where the three voltage levels at the grid fall with respect to the cathode voltage, and thus can choose the switching function NOR or NAN, as in figure A.13b. The design of Atanasoff's three-input switches was similar; see figure A.14.

We have now completed our explanation of the electronics of the six atomic switches Atanasoff defined in his 1940 manuscript: NOT, two- and three-input NORs, two- and three-input NANs, and the threshold switch. It is a significant aspect of his switching achievement that he designed these atomic circuits so that they could be compounded easily. The procedure for compounding is illustrated in figure A.15. In the next section we study the compound switch that was one of Atanasoff's two most important legacies to computing technology, namely his add-subtract mechanism.

He did not actually use the three-input NOR in his add-subtract mechanism—

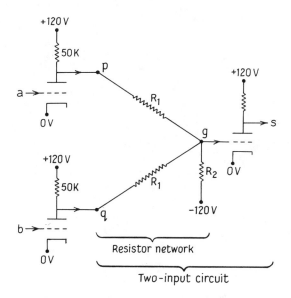

(a) NOR and NAN switches

Logical function	Resistor values		Input voltages at p and q How many are low?	Grid voltage g	Output voltage s
	R_1	R_2			
NOR At least one input voltage must be low to make the output voltage high	492K	304K	None	7	Low
			One	-14	High
			Two	-34	High
NAN All input voltages must be low to make the output voltage high	364K	368K	None	34	Low
			One	8	Low
			Two	-16	High

(b) Table of voltage levels

Fig. A.13. Atanasoff's two-input switching circuits

none is indicated on his circuit diagram. But, as we noted in chapter 1, neither did he use the threshold switch (3,2), two instances of which he did indicate. Instead, he connected the plates of these two threshold switches together, so as to use one switch during addition to generate the carry bit, the other during subtraction to generate the borrow bit. This circuit is represented logically by figure A.16, which is a subnet of figure A.17 discussed in the next section.

As we turn to the add-subtract circuit Atanasoff designed from his list of atomic switches, let us note that because this list included both NAN and NOR, it was complete (see fig. A.7). Consequently, for any logical version of a compound switch composed of these atoms, there was an isomorphic vacuum-tube circuit that

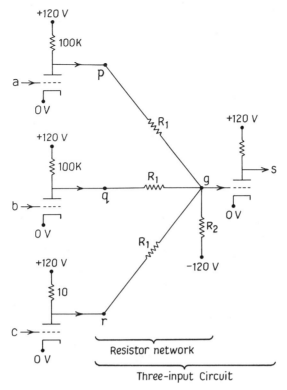

(a) NOR, "Threshold," and NAN switches

Logical function	Resistor values		Input voltages at p, q, and r. How many are low?	Grid voltage g	Output voltage s
	R_1	R_2			
NOR At least one input voltage must be low to make the output voltage high	1092K	436K	None	6	Low
			One	-11	High
			Two	-27	High
			Three	-43	High
"Threshold" At least two input voltages must be low to make the output voltage high	873K	509K	None	26	Low
			One	7	Low
			Two	-12	High
			Three	-30	High
NAN All input voltages must be low to make the output voltage high	576K	608K	None	55	Low
			One	30	Low
			Two	9	Low
			Three	-13	High

(b) Table of voltage levels

Fig. A.14. Atanasoff's three-input switching circuits

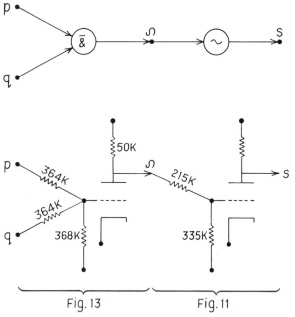

Fig. A.15. Isomorphism of logical and electronic switches

would realize it. But while this was true in principle, there were limits in practice. We discuss both the theoretical principle and its practicality in connection with adding and subtracting circuits, especially Atanasoff's.

Structure of Atanasoff's Add-Subtract Mechanism

Atanasoff's circuit diagram of his add-subtract mechanism, complete with (p,q) designations, is reproduced in chapter 1 as figure 12. We show its logical structure in figure A.17, where $\sigma_i, \alpha_i, \gamma_{i-1}, \beta_i, \gamma_i, f, g$ (in that order) correspond to his input and output points A through G, and where the variables p_1 through p_8, and p_{10} through p_{13} and σ_i correspond to the switching points driven by his tubes 1 through 8, and 10 through 14, respectively (his tube 9 was the nonfunctional half of one of his seven envelopes). His NOT (1,1), two-input NOR (2,1), and two-input NAN (2,2) switches, discussed earlier, are shown in figures 10, 13, and 14, respectively, and also in the corresponding appendix figures A.11 (NOT) and A.13 (NOR, NAN); his three-input NAN (3,3), three-input NOR (3,1), and threshold (3,2) switches, also discussed earlier, are shown in figure A.14.

The unit delay depicted in figure A.17 is not a switch, but is a logical representation of the delay inherent in the switching circuit plus the further delay accomplished by the storage of the carry or borrow bit on the fast drum from one bit cycle to the next (see brushes B_2'' and B_1'' in fig. 7). Our formal analysis of addition and subtraction made explicit the need for clearing the carry-borrow drum between

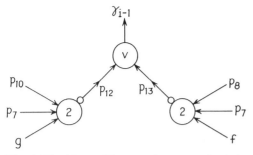

Fig. A.16. Atanasoff's carry-borrow output circuit

addition times. The rightmost carry input c_{49} should be "0," but if the carry-borrow drum were not cleared after the previous addition it might set c_{49} to "1." On Atanasoff's machine, signals from the timing drum cleared all the carry-borrow capacitors during the 60-degree blank period. It was not necessary to clear the capacitors of the keyboard and counter drums because the signals from the vacuum tubes writing on these capacitors overrode the charges stored on them.

We first analyze the switching net of figure A.17 and show that it performed the logical functions required of it. This is best done by treating addition and subtraction separately, just as Atanasoff did in his table of high and low voltages. The control lines f and g were used to select between the two operations, the settings $f \equiv 0$ and $g \equiv 1$ causing addition and the settings of $f \equiv 1$ and $g \equiv 0$ causing subtraction. Recall that these lines came from electromagnetic relays, which were fast enough for this application.

For addition, the truth table of figure A.18a traces the truth values through figure A.17 for each of the eight input states (the possible values of α_i, β_i, and γ_i) for

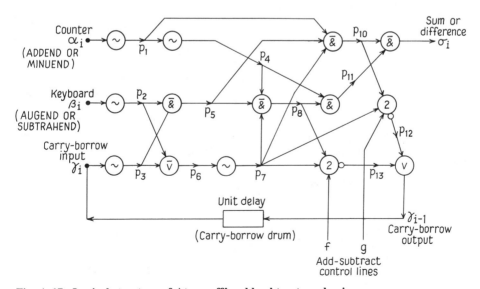

Fig. A.17. Logical structure of Atanasoff's add-subtract mechanism

Variable Inputs			Control Settings		Intermediate Switching Points													Outputs	
Addend α_i	Augend β_i	Carry In γ_i	f	g	P1	P2	P3	P4	P5	P6	P7	P8	P10	P11	P12	P13	Sum σ_i	Carry Out γ_{i-1}
0	0	0	0	1	1	1	1	0	0	0	1	1	1	1	0	0	0	0
0	0	1	0	1	1	1	0	0	1	0	1	1	0	1	0	0	1	0
0	1	0	0	1	1	0	1	0	1	0	1	1	0	1	0	0	1	0
0	1	1	0	1	1	0	0	0	1	1	0	1	1	1	0	1	0	1
1	0	0	0	1	0	1	1	1	0	0	1	1	1	0	0	0	1	0
1	0	1	0	1	0	1	0	1	1	0	1	0	1	1	0	1	0	1
1	1	0	0	1	0	0	1	1	1	0	1	0	1	1	0	1	0	1
1	1	1	0	1	0	0	0	1	1	1	0	1	1	0	0	1	1	1
Height: 0					1			2			3	4	5			6		

(a) Addition

Variable Inputs			Control Settings		Intermediate Switching Points													Outputs	
Minuend α_i	Subtrahend β_i	Borrow In γ_i	f	g	P1	P2	P3	P4	P5	P6	P7	P8	P10	P11	P12	P13	Difference σ_i	Borrow Out γ_{i-1}
0	0	0	1	0	1	1	1	0	0	0	1	1	1	1	0	0	0	0
0	0	1	1	0	1	1	0	0	1	0	1	1	0	1	1	0	1	1
0	1	0	1	0	1	0	1	0	1	0	1	1	0	1	1	0	1	1
0	1	1	1	0	1	0	0	0	1	1	0	1	1	1	1	0	0	1
1	0	0	1	0	0	1	1	1	0	0	1	1	1	0	0	0	1	0
1	0	1	1	0	0	1	0	1	1	0	1	0	1	1	0	0	0	0
1	1	0	1	0	0	0	1	1	1	0	1	0	1	1	0	0	0	0
1	1	1	1	0	0	0	0	1	1	1	0	1	1	0	1	0	1	1
Height: 0					1			2			3	4	5			6		

(b) Subtraction

Fig. A.18. Truth table for the add-subtract mechanism

the control settings $f \equiv 0$ and $g \equiv 1$. The switching points are grouped according to their heights, so that the truth values in each column depend only on the truth values in earlier columns. (Note that Atanasoff's subscripts satisfy this condition.) It is apparent from this table that Atanasoff's add-subtract mechanism did add correctly, for the sum and carry outputs σ_i and γ_{i-1} satisfy the evident rules (see chap. 1): an odd number of inputs of "1" yielded a sum of "1" $[\sigma_i \equiv (\alpha_i \not\equiv \beta_i \not\equiv \gamma_i)]$; and at least two inputs of "1" yielded a carry of "1" $[\gamma_{i-1} \equiv ② (\alpha_i, \beta_i, \gamma_i)]$.

For subtraction, the truth table of figure A.18b traces the truth values through figure A.17 for these same input states for the control settings $f \equiv 1$ and $g \equiv$ 0. Again, the outputs of σ_i and γ_{i-1} satisfy the rules: an odd number of inputs of "1" yielded a difference of "1" [$\sigma_i \equiv (\alpha_i \neq \beta_i \neq \gamma_i)$]; and at least two inputs of "1" yielded a borrow of "1" *if* every counter bit was replaced by its complement [$\gamma_{i-1} \equiv$ ② $(\bar{\alpha}_i, \beta_i, \gamma_i)$].

Having completed our demonstration that Atanasoff's add-subtract mechanism did indeed work correctly, we turn to the question of whether its design was reliable. There are two main issues: first, the reliability of his method of switching by adding voltages in resistor networks; second, the matter of time delay through the circuit. (Note here that these same considerations apply to the reliability of the much simpler associated restore-shift mechanisms.)

Both issues involve the relation of logical structure to its electronic realization. Let us begin, then, by defining three parameters of logical nets (e.g., fig. A.17) that are of special relevance: *fan-in* to switches, *fan-out* from switches, and *height* of switching points. The number of sources driving an atomic switch of a circuit is called the fan-in of that switch, and the largest number of sources driving any atomic switch of a circuit is called the fan-in of that circuit. Likewise, the number of sinks (circuit triodes and drum capacitors in the case of the ABC) driven by an atomic switch is called the fan-out of that switch, and the largest number of sinks driven by any atomic switch of a circuit is called the fan-out of that circuit. The height of a switching point in a circuit is the largest number of switches between that point and an input to the circuit, and the height of the circuit is the greatest height of all its switching points.

Now we have made our calculations of voltages in Atanasoff's atomic switching circuits assuming exact values of resistances of resistors and conducting vacuum tubes (figs. A.11, A.13, A.14). The actual situation is more complicated, both initially and over time. The resistances of resistors vary initially and as the temperature changes in an operating computer, and the effective resistances of conducting vacuum tubes vary from tube to tube and with the actual operating voltages of the grids. Moreover, in both cases the resistance values change with use, sometimes significantly.

The inexactness of resistances was a problem that early computer designers had to face; indeed, it is a general problem that still exists. Initial variations among both resistors and tubes could be handled by measuring each item, a painstaking task that Berry did for the add-subtract mechanisms. Variations among resistors over the life of a computer were more difficult to manage, however, because these—unlike tubes—were not easily replaced.

These variations do clearly constitute a limitation on Atanasoff's method of adding voltages with resistor networks. Correct logical switching by this method depends on establishing certain voltage levels and positioning the on-off operation of the driven vacuum tube in a specified region (cf. figs. A.13, A.14). Variations in resistors and tube characteristics may shift the critical voltage levels so far out of the specified range that the output of the switch will be incorrect. It is important here that the maximum fan-in of Atanasoff's circuit was three, requiring four voltage levels (fig. A.14); a fan-in of four with five voltage levels would have been unsafe.

We give here two extreme examples of the possible effects of parameter variations. Consider first the two-input NOR of figure A.13. When inputs p and q are both high ("0") the grid voltage at g should be +7 volts, causing the output at s to be low ("1"), which is correct. But suppose that the resistance R_1 is 10 percent *higher* than the stated value, while R_2 is 10 percent *lower* than the stated value. Then the voltage at g will be −5 volts, the driven tube will be off or nearly so, and the output at s will be high ("0") and incorrect.

Second, consider the three-input NAN of figure A.14, and assume that after aging the internal resistance of each of the three driving tubes has doubled to 45K ohms. Then when all three driving tubes are conducting, switching points p, q, and r will be low ("1"), but not as low as they should be, and the voltage at g will be −2 volts, rather than the desired −13 volts. A voltage of −2 volts is not low enough to hold the driven tube off, and the output at s will not be high ("0"), as required by the logic.

As we have already mentioned, these are extreme cases; actual variations were normally much less, although of course variations in resistor values and tube characteristics would occur together. We think Atanasoff's electronic arithmetic circuits were reliable in design, for two reasons. First, they were modular, each of the thirty add-subtract mechanisms occupying its own chassis, which could be removed independently of the others, replaced by a module known to work, and then tested and repaired at the bench. Second, the electronic computing circuits of the computer taken as a whole (including the restore-shift mechanisms) were not very complex; they involved only about 280 envelopes (double triodes). The variations of parameters in an electronic system, such as resistance values, are statistical, and other things being equal, the probability of failure of the system increases with size. The arithmetic electronics of the ABC was simple enough to be reliable.

It is an interesting sidelight to compare this reliability of Atanasoff's computing circuits with that of his sparking method of recording and reading binary digits. There, for one basic arithmetic operation such as eliminating a variable between two equations, the number of bit operations carried out by the arithmetic circuits far exceeded the number of bit operations carried out by the binary-card mechanisms. And yet he found the former reliable, the latter unreliable! Let us explore the reason for this seeming anomaly.

Hardware errors are associated with components: vacuum tubes, soldered connections, resistors, recording components, reading components, and so on. A component can fail *permanently,* that is, stop working. A component can also fail *intermittently,* making an error once or a few times, then working correctly for a while, and then malfunctioning again. Intermittent errors are much more difficult to locate than permanent errors, since by the very nature of statistical inquiry much more data are needed to certify rare events than regular repetitions.

Thus the difference between the reliability of Atanasoff's electronic arithmetic and his binary input-output comes down to this. Vacuum tube failures are usually permanent, and when resistances drift far enough to cause logical error such an error is likely to be repeated as long as the same temperature conditions prevail. Errors in the ABC's binary input-output were transient and rare, and therefore much harder to deal with. Indeed, the only satisfactory solution to such hardware errors came with the discovery and employment, much later, of error-correcting codes.

This concludes our discussion of the reliability of Atanasoff's method of switching by adding voltages in resistor networks. We look next at the matter of time delay through his add-subtract mechanism.

An electronic circuit takes time to respond after input signals are applied, and the output will not be correct until the response is complete. The delay through the circuit is due largely to the time required to charge or discharge the capacitance of the circuit, that is, the capacitances of the wires, the vacuum tubes, and any capacitors there are. (The capacitance of a wire depends mainly on its length.) The logical parameters of fan-in, fan-out, and height are rough indicators of the delay. In current practice, none of these parameters ordinarily exceeds ten. The fan-in of Atanasoff's add-subtract mechanism was only three, the fan-out four, and the height of the circuit just six (path $\gamma_i,p_3,p_6,p_7,p_8,p_{11},\sigma_i$).

Actually, as we noted in chapter 1, Atanasoff was operating his electronic circuits at a rate of only sixty pulses per second, far below their speed capability. The fundamental constraint on the speed of his computer derived from his use of standard sixty-cycle alternating current to punch binary digits on cards and read them from cards. It is nevertheless of interest that Atanasoff's add-subtract mechanism illustrates a principle generally followed in the interplay of switching with memory: information flows from memory into a switching circuit with a height of about one order of magnitude, and then flows back into memory. Of course, Atanasoff had only one level of internal memory (his drums), whereas a modern computer has several levels: main memory, buffer memory, registers, flip-flops (or latches), etc.

We close our analysis of Atanasoff's adding and subtracting circuits by observing how his electronic switching theory was applied in subsequent computers; indeed, how it has come down to us today via its use in the EDVAC and the IAS classes of first-generation computers.

Atanasoff stored binary numbers serially and so used his add-subtract switches serially. This was also done in the EDVAC design and the succeeding machines of this general type (see fig. 28). The machines of the IAS class stored binary numbers in parallel, a common practice today. Figure A.19 shows the arrangement for parallel use of adder-subtracter switches. Although fifty stages are indicated (to provide the parallel equivalent of an add-subtract mechanism), there can be any number of them.

Different adder-subtracter switches can be used, and different complement systems. One can do subtraction by borrowing, and use the two's complement system, which is what Atanasoff did (cf. figs. A.10 and A.16). In this application of figure A.19, the carry-in wire γ_{49} is not needed, and the add-subtract control is set at "0" for addition, "1" for subtraction. (This is input r_i in fig. A.10; in fig. A.17 it is input f, with the proviso that $g \equiv \sim f$.)

One can also do subtraction by complementing, and use either the two's or

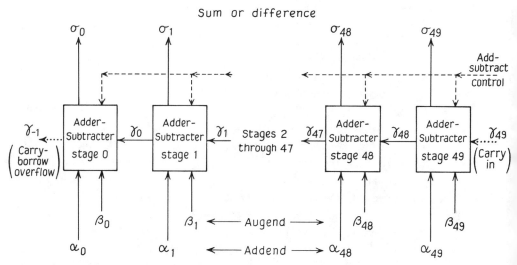

Fig. A.19. Fifty-bit parallel adder-subtracter

the one's complement system. For the two's complement system, figure A.9 can be applied at each stage of figure A.19, with p_i being the add-subtract control and q_{49} being γ_{49}, the external carry input needed for subtraction. The same circuits can be used for the one's complement system, but in this case the carry-borrow overflow from γ_{-1} is connected back to the carry-in γ_{49}. The add-subtract control is held at "0" for addition and at "1" for subtraction, as before.

Atanasoff's Place in the History of Computer Switching

In this section we review the history of logical switching, covering hardware as well as theory, and the interrelation of the two. Our purpose is to demarcate Atanasoff's contribution to computer switching; we will show that it was both important and influential.

What is a switch? Generalizing from the materials of the preceding sections, we can say that a *switch* is a finite device that transforms an input state into an output state in one basic time-step. It is the physical realization or automation of a finite function table, such as the multiplication table or a logarithm table. Each input state represents an argument value, and the output state produced by the table represents the corresponding function value.

The mechanical multiplication table invented by Ramon Verea (1878) and Leon Bollée (1889) to expedite multiplication is a good example. It represented the multiplication table by bars of lengths 0 through 9. There were ten inputs, the digits 0 through 9. A single digit (e.g., 7) was entered and the table mechanism then emitted the product of that digit by all other digits (i.e., 00, 07, 14, 21, 28, 35, 42, 49, 56, 63). Using such a table, a calculating machine could multiply one digit of the multiplier by the whole multiplicand in one addition time. Similar tables were used in electromechanical machines and in the electronic ENIAC.

A switch may be analog or digital, and historically the two types are interrelated, as when Atanasoff suggested that his computer could be transformed to do the work of the differential analyzer. We are, of course, interested mainly in the digital case. A switch may operate in any number base, or even in a mixture of bases. The mechanical multiplication table just described was decimal. Most electronic switches are binary, however, and we ultimately concentrate on the binary case.

From the logical point of view, switching circuits constitute one "half" of a computer in the sense that any computer is in principle reducible to a switch interacting with a memory. Moreover, switches play essential roles in memories. Storage is usually accomplished by means of switches plus delay. A flip-flop, for example, consists of two switches interacting through capacitor delays for stable storage, with other switches provided for access. Large memories also need switches for access, such as an address switch (see fig. A.5).

Lull and Leibniz

We begin our story by mentioning a very interesting precursor of switching. Ramon Lull (ca. 1232–1316), invented a combinatorial procedure (the "Great Art") and mechanical devices for computing it. One of Lull's simpler devices consisted of a fixed disk with sixteen of God's attributes (goodness, greatness, wisdom, . . .) inscribed around its circumference, and a smaller duplicate disk rotating around the same center. By rotating the smaller disk with respect to the larger, the user could form all the conjunctive pairs of God's properties.

Lull exhibited many such combinatorial generating devices, some made of parchment, some of metal. Most used circles, but some had squares, equilateral triangles, and other figures rotating on them. Some were brightly colored. One device formed combinations of different possible characteristics of the soul, another of various virtues and sins. Lull used his "Great Art" to argue for and promote the unity of religion, politics, and science.

An atomic theory of a subject matter or discipline holds that the objects to be studied by the discipline are compounded of atoms interconnected by certain rules or laws. The atomic theory of matter and the kinetic theory of gases are examples from physics. The switching theory presented earlier in this section is an abstract example of an atomic theory. The general metaphysical theory of atomism is ancient, going back to Plato, Democritus, and Euclid. Various forms of this theory hold that objects, systems, languages, disciplines, the whole universe—even all of knowledge—are compounds or wholes composed of atomic parts, concrete or abstract. Lull's contribution to atomism was to invent the first device for mechanically calculating combinations of a few basic terms. We will see shortly that, naive as it may seem, this idea was influential in an important way.

It is generally a useful discovery heuristic to consider different combinations of a set of basic ideas. But Lull's choice of primitives was unoriginal, and his method was too simple to generate anything useful. He did, however, form different combinations by moving pieces of information relative to one another, and this is a precursor to digital switching.

Lull's most important historical influence was on Gottfried Leibniz (1646–1716). Starting from Lull's method, Leibniz conceived of a universal language in which ideas and statements would be expressed algebraically. Since prime numbers are not decomposable, Leibniz proposed to represent every basic concept with a prime. Thus "man is a rational animal" was expressed by the equation $6 = 3 \times 2$. This algebraic-arithmetic language was universal in the sense that any concept or statement could be expressed as a formula in it. Moreover, these formulas could be transformed computationally, so that logical reasoning would be reduced to arithmetic computation (Leibniz 167?/1977, p. 337).

> All our reasoning is nothing but the joining and substituting of characters, whether these characters be words or symbols or pictures.

> . . . if we could find characters or signs appropriate for expressing all our thoughts as definitely and as exactly as arithmetic expresses numbers or geometric analysis expresses lines, we could in all subjects *in so far as they are amenable to reasoning* accomplish what is done in Arithmetic and Geometry.

> For all inquiries which depend on reasoning would be performed by the transposition of characters and by a kind of calculus, which would immediately facilitate the discovery of beautiful results. . . .

> Moreover, we should be able to convince the world what we should have found or concluded, since it would be easy to verify the calculation either by doing it over or by trying tests similar to that of casting out nines in arithmetic. And if someone would doubt my results, I should say to him: "Let us calculate, Sir," and thus by taking to pen and ink, we should soon settle the question.

Leibniz's plan for a universal language was an anticipation of modern computing, long before the general-purpose computer was even conceived. Leibniz expected the calculations to be done by humans working with pen and paper, perhaps aided by a mechanical calculator like one he invented as an extension of Pascal's adding machine.

It is easy to see now that Leibniz greatly underestimated both the difficulty of developing a universal computational language and the complexity of the computations required to settle interesting questions (Leibniz 167?/1977, pp. 395–96).

> Nothing is needed, I say, but that philosophic and mathematical procedures, as they call them, be based upon a new method, which I can prescribe and

which contains nothing more difficult than other procedures, or more re-mote from our usual practices, or more foreign to our habits of writing. It will not require much more work than we see already being spent on a good many procedures and on a good many encyclopedias, as they call them. I believe that a number of chosen men can complete the task within five years; within two years they will exhibit the common doctrines of life, that is, metaphysics and morals, in an irrefutable calculus.

Once the characteristic numbers of many ideas have been established, the human race will have a new organon, which will increase the power of the mind much more than the optic glass has aided the eyes, and will be as much superior to microscopes and telescopes as reason is superior to vision.

Practically speaking, these matters came only with the development of the electronic stored-program computer. This in turn required logical switches of substantial complexity, so that Leibniz's vision was not achieved until these were developed.

Leibniz's improved form of Pascal's adding machine had an accumulator, a keyboard, a crank for adding the contents of the keyboard into the accumulator, and a crank for mechanically shifting or displacing the decimal position of the keyboard relative to the accumulator. This machine incorporated mechanical switching, for the carry mechanism at each position was a binary switch and the shifting mechanism was a decimal switch.

Mechanical Scanning Switches

We go back now about two hundred years, to the pegged drum (or barrel), which also switched by mechanical displacement, but in a somewhat different way. The original pegged-drum automaton consisted of a rotating drum with pegs mounted on its surface. As the drum rotated, the pegs switched sound-producing mechanisms on and off (e.g., they twanged the reeds of a music box). These drum devices arose from the mechanical clock invented a few centuries earlier. The first mechanical clocks controlled the ringing of church bells. Pegged drums are more digital than mechanical clocks (although the escapement of a clock is binary [tick-tock] in its operation), and produce more complicated output patterns.

Drum automata were the first switching mechanisms of significance, and they left a profound practical legacy. Rotating pegged drums were used not only for music boxes, organs, and automated glockenspiels, but also for writing and acting automata. The epitome of the development of pegged-drum automata was the programmable writing automaton built in the last half of the eighteenth century. One such device was made by Knaus in 1760 and is now in the Vienna Technical Museum. There was a writing arm holding a pen, and there were analog mechanisms that could dip the pen into an inkpot and direct it to write any of thirty different characters. The automaton wrote a sequence of sixty-eight letters under the direction of the (read-only) program pegged into holes in its drum.

The writing of each letter resulted from the simultaneous rotation of three cams, each controlling a lever arrangement. The first cam and lever arrangement moved the arm to and from the paper, the second moved it up and down, and the third moved it right and left. Note that each cam was the analog version of a ring of pegs around a drum.

The sequence of letters to be written was programmed by placing a sequence of pegs around the drum. For each possible letter there was a ring of sixty-eight holes around the drum, positioned so that a peg in one of these holes would trigger the levers and cams that controlled the writing of the designated letter. Thus a complete message was programmed by placing a sequence of up to sixty-eight pegs around the drum, one peg at most per row. For each letter, the drum would advance a step, a peg would trigger the operation of the levers and cams corresponding to the peg position, the letter would be written, and the drum would advance another step. A message of up to sixty-eight characters would be written each time the drum rotated a full revolution.

The programmable writing automaton was a truly marvelous mechanism, the most complicated computing device to appear for a long time and a tribute to both the mechanical technology and the skilled craftsmanship of its day. Yet it was a toy, albeit a toy for royalty. Music boxes and glockenspiels had a wider audience and were more useful, but the first really practical application of the principle underlying these devices was the Jacquard loom.

What we call the Jacquard loom was actually conceived and used long before Jacquard created his version; the invention came to be named after him because he made it practical. Some earlier versions had used pegged drums, but a sequence of punched cards chained together in a loop had proved superior. These cards were scanned in sequence, and the loop of cards cycled over and over. Each card controlled the binary relations of a woof (crosswise) thread to all of the warp (lengthwise) threads: a hole in the card caused the corresponding warp thread to be raised so that the woof thread went under it, while the absence of a hole caused the warp thread to remain below the woof thread.

The underlying principle of the punched-card loom is the same as that of the pegged drum and is important enough in modern applications to merit an abstract formulation. There is a cyclic recorded medium with many parallel channels that moves under a row of reading heads, one for each information channel. The reading heads are levers (for the pegged drum) or hooks connected to levers (in the original Jacquard looms). The reading heads scan the information on the medium as it moves past them and transmit it to the device that uses it (music device, automaton, or loom).

Pegged-drum devices and punched-card looms embody switches of a kind, because they switch from one word (e.g., row of pegs versus no peg) to another. Atanasoff's drum with capacitors, studs, and brushes was an electrical/electronic variant, which he adopted because it provided a simple, inexpensive way of switching access to different pieces of information. Yet while these devices clearly do switch, they are quite different from the switches discussed in the previous sections of this appendix, and so do not fall directly under the definition given at the beginning of the present section.

That definition characterized a switch as a finite physical device that transforms an arbitrarily given input state into a specified output state in one basic time-step. Mathematically speaking, a switch is the physical realization of a finite function or map from argument values $0, 1, 2, \ldots, N - 1$ to function values $F(0), F(1), F(2), \ldots, F(N - 1)$. When a physical representation of argument n is entered into it, the switch produces a physical representation of the function value $F(n)$.

Keeping this definition in mind, consider again a pegged drum rotating in a music box and a chain of punched cards controlling the pattern woven by a loom. In each case information is recorded on a solid medium and moved past a reading mechanism. Formally speaking, the medium stores a finite sequence of function values $F(0), F(1), \ldots, F(N - 1)$ and these values are enumerated over a succession of basic time-steps, each function value resulting in an action of some kind. Such a switch could be called an enumerative switch, but we prefer the term *scanning switch*. Where we wish to stress the difference between these switches and switches proper as originally defined, we call the latter *function switches*. In brief, when given an arbitrary n as input, a function switch produces the value $F(n)$ as output in one time-step; whereas a scanning switch successively scans the values $F(0), F(1), \ldots, F(N - 1)$ and emits these as outputs.

Now, the scanning switches we have mentioned thus far employ memories that are digital, mechanical, and read-only. But the mechanical scanning principle has been generalized along each of these three dimensions: (1) analog-digital representation, (2) mechanical-electromechanical-electronic storage and access, and (3) write-read-erase storage. Let us look at these in turn.

1. The Knaus writing automaton illustrates the use of analog representation as well as digital, for the cams and templates that controlled the shape of the letters were analog and were scanned. Early mechanical phonographs and dictating machines were analog. In the analog case, both the variable x and the function value $F(x)$ range over a finite and approximate continuum rather than over sets of discrete values.

2. Although some punched-card machines were mechanical, the most successful ones were electromechanical. Electric motor commutators and IBM timing drums were electromechanical scanning switches. The recording and reading of magnetic tapes is done by means of electronics, for both the analog and the digital forms.

3. The Knaus automaton wrote characters after reading coded representations of them. Teletype machines punched and read paper tape electromagnetically, and card calculators punched and read cards. Magnetic tapes and disks for computers are good examples of write-read-erase storage. In a read-only memory the words scanned are unchangeable and thus are constants, logically speaking. But in the erasable form, word positions are scanned and these function as variables, logically speaking. Controls in the drive units select the regions of the memory to be used.

With this conceptual background, we are now in a position to see an underlying unity in the evolution of scanning switches. Mechanical scanning switches produced a tremendous legacy, which can be divided roughly into two major streams. The entertainment-cultural stream includes the glockenspiel, various automatic music machines (including the hand organ and the player piano), toy automata, phonographic records and tapes, and videotapes. The practical-control-computing stream includes the Jacquard loom; punched-card accounting machines; dictating machines; punched paper tape for teletype, computer data, and computer programs; and computer drums, tapes, and disks.

The idea of using mechanical scanning motion to accomplish switching has an important history of its own, a history we trace briefly so as to demarcate Atanasoff's contribution to scanning switches. Mechanical technology gave way to electromechanical, which in turn was gradually replaced by electronic and magnetic technologies (now being supplemented by video technology). Important new forms of scanning switches have appeared. There were noncyclic forms of storage: paper tapes that could be moved back and forth, packs of cards that could be rearranged. Punched paper tapes holding programs for relay computers were later developments; they employed controls for branching and read-only indexing. Magnetic drums and tapes were essential to early electronic computers, and current computers make essential use of magnetic disks.

This last is an anomaly in current computers, which are based on the latest technology of high-density, solid, electronic circuits. But magnetic disks employ the oldest switching technology, that of mechanical scanning. The reason is, of course, that mechanical motion of the storage medium relative to the input-output is still the cheapest way to switch large amounts of information. Moreover, just as the density of electronic circuits has increased, so has the density of magnetic information storage.

Thus scanning switches are ubiquitous in the history of computing. Atanasoff's work stands out in this evolution. He conceived and built the first scanning switches for electronic computing. These were of all kinds: read-only, write-only, and write-read-erase. His decimal-binary and timing drums worked well. His timing drum led to the ENIAC cycling unit and to successor electronic clocks for timing digital signals. His spark method of recording on cards and his electrode method of reading them, although not ultimately successful, constituted the first attempt to write on and read from external media by electronic means.

Most important, Atanasoff's erasable counter and keyboard drums were the first memories for electronic computers of *any* type. Moreover, these led to Eckert's mercury-delay-line memory for the EDVAC. By substituting acoustic scanning for mechanical scanning and piezoelectric crystals for electrical brushes, Eckert's memory attained electronic speed. The greater storage capacity that resulted made possible the modern stored-program computer. The regeneration principle invented by Atanasoff to make his capacitor memory drums feasible was used by Eckert to make the mercury-delay-line memory workable, and is used now in DRAMs (dynamic random access memories).

Mechanical Function Switches

Atanasoff also stands out in the history of switching for having the first electronic function switch, his add-subtract mechanism (figs. 12 and A.17). Note that this switch is "holistic" in the sense that each output bit depends on all three input bits (fig. A.18). We will call a function switch *holistic* if most output digits depend on most input digits. An address switch (e.g., fig. A.5) is holistic. A holistic function switch is intrinsically complicated because it has many branching and merging connections (remember the switching concepts of fan-in and fan-out). The term "holistic" is appropriate because "the whole is greater than the sum of its parts" in that an output digit is a nonlinear function of the input digits.

Because holistic function switches are essential to modern computers, let us trace their history. We believe that Charles Babbage was the first to conceive such a switch, although he never built one. For his analytical engine, he planned to separate the memory (the "store") from the arithmetic unit (the "mill"). Instructions were to be punched on cards and the cards chained together to make a read-only program. Babbage called the decimal registers in which numbers were to be stored "axes," because they were axles with one gear wheel for each decimal position. (Actually, axes came in pairs, so that a number could be saved during read-out, which was destructive, and so that a number could be shifted.)

Babbage considered stores of various sizes and worked on the design of one having 100 pairs of axes. This would have required an address of two decimal digits and one bit, and an address decoding switch in which an address would activate any one of 200 axes in the store.

We turn next to the formal contribution of mathematician/logician George Boole (Boole 1847, 1854). For the first holistic switch to be built came not from the computing tradition, but from a theoretical advance in symbolic logic, namely, Boolean algebra, developed by Boole around 1850. During the next few decades, his work led to the design of mechanical logic machines by William Stanley Jevons and Allan Marquand, then to the suggestion by Charles S. Peirce that Boolean algebra could be applied to switches, and finally to the design by Marquand of the first electrical function switch.

George Boole created a logic of classes. At the base there were primitive classes A, B, C, D, etc., the null or empty class (0), and the universe class (1). Compound classes could be made from these by the operations of complementing a class with respect to the universe class (e.g., $1 - A$), intersecting or multiplying two classes (e.g., AB), and adding exclusively two classes (e.g., $A + B$). There were equations between class terms (e.g., $AB = A$ and $BC = 0$).

At the time of Boole, logic consisted mostly of the theory of syllogisms, formulated by Aristotle and refined over the intervening centuries. We use the following two syllogistic arguments as examples:

(1) All A are B, All B are C, therefore All A are C [Valid]
(2) All A are B, All B are C, therefore No C are A [Invalid].

In Boole's algebra these could be represented as sequences of equations, whereby questions of validity were reduced to questions of mathematical proof.

In writing these equations, it is convenient to represent the complement of a primitive class by the corresponding lowercase letter. Thus "$1 - A$" is written as "a," "$1 - B$" as "b," etc. Our two arguments may then be expressed in equational form as

(1') $Ab = 0$, $Bc = 0$, $\therefore Ac = 0$ [Valid]

(2') $Ab = 0$, $Bc = 0$, $\therefore CA = 0$ [Invalid].

The conclusion of (1') is derivable from the premises by the laws and rules of Boolean algebra, but this is not so for (2').

(Some syllogistic statements involved inequalities; e.g., "Some A are B" becomes $AB \neq 0$. We will ignore these, since they did not play a significant role in the history of switching.)

The first step in the mechanization of Boolean algebra was taken by William Stanley Jevons, who designed a wooden logic machine for evaluating syllogisms (Jevons 1870, 1874). He called his machine, constructed in 1869, a *logical piano,* because it had a row of keys below and a logical display above (fig. A.20). Note that Jevons's logic switch preceded Verea's decimal multiplication table by a decade.

In testing a syllogism on this machine, one first pushed a special key to initialize the display, so that it showed all sixteen multiplicative combinations of the primitive class variables and their complements:

Display α
$$
\begin{array}{cccccccccccccccc}
A & A & A & A & A & A & A & A & a & a & a & a & a & a & a & a \\
B & B & B & B & b & b & b & b & B & B & B & B & b & b & b & b \\
C & C & c & c & C & C & c & c & C & C & c & c & C & C & c & c \\
D & d & D & d & D & d & D & d & D & d & D & d & D & d & D & d
\end{array}
$$

Jevons called these sixteen *basic class combinations* the logical alphabet, because any class describable by means of the variables A, B, C, D and the Boolean operations (other than the null class) could be expressed as a sum of these basic class combinations. For example, the class $1 - (A + b)$ is the same as the class $aBCD + aBCd + aBcD + aBcd$. It will be convenient to refer to several basic class combinations at once by replacing some of the variables by blanks. Thus AB_d will refer to $ABCd$ and to $ABcd$.

One then entered premises into Jevons's machine by pressing the appropriate keys. The effect of entering a premise or equation onto the keyboard was to eliminate from the display those basic class combinations that were false according to the equation. Consider, for example, the statement "All A is B," which Jevons wrote as $A = AB$. To initialize the machine, one pressed FINIS, causing all sixteen basic conjunctions to be displayed. Then, to enter "$A = AB$," one pressed the following keys from left to right: A, COPULA, A, B, FULL STOP.

Now $A = AB$ is equivalent to $Ab = 0$, that is, to "No Ab's exist." And so, after $A = AB$ was impressed on the keyboard, all basic class combinations of the form Ab__ __ disappeared from the display, leaving:

	A	A	A	A		a	a	a	a	a	a	a	a
Display β	B	B	B	B		B	B	B	B	b	b	b	b
	C	C	c	c		C	C	c	c	C	C	c	c
	D	d	D	d		D	d	D	d	D	d	D	d

Similarly, entering the premise "All B are C" $(Bc = 0)$ caused all combinations of the form __Bc__ to disappear, leaving:

	A	A		a	a		a	a	a	a
Display γ	B	B		B	B		b	b	b	b
	C	C		C	C		C	C	c	c
	D	d		D	d		D	d	D	d

After the premises were entered into the keyboard, the operator inspected the display to see if the conclusion of the argument was true. Because no combination of the form A__c__ was left, "All A are C" proved true and argument (1) valid. But $ABCD$ and $ABCd$ were left, so that "No C are A" proved false and argument (2) invalid.

As figure A.20 shows, Jevons's logical piano had twenty-one inputs (keys) and sixteen outputs (basic class combinations). (The keys next to each end of the keyboard signified inclusive OR.) Thus it was a holistic function switch of substantial complexity. Apparently the first complex holistic function switch ever built, it led to some interesting developments before its influence died out. We noted earlier that a holistic function switch, with its many branching and merging connections, is intrinsically complicated. But mechanical technology is ill suited to such complex interconnections, so that the design of Jevons's logical piano was truly ingenious (see Jevons 1870).

We turn next to Allan Marquand's contributions to logic machines. Marquand was a Ph.D. student in logic under Peirce at Johns Hopkins University. At that time logicians often tested syllogisms by means of diagrams, following John Venn and, originally, Leonhard Euler. These were for just the three primitive classes involved in a syllogism (e.g., A, B, C), but Marquand extended them to cover any number of primitive classes. They are shown for the four primitive classes A, B, C, D of Jevons's piano in figure A.21, marked to represent the equations of arguments (1) and (2) (cf. the Jevons displays α, β, γ).

By using the square format of his diagrams, Marquand designed a simpler version of Jevons's logical piano. This was built in 1881; see figure A.22 (Marquand 1886; Peirce 1887). There was a two-position pointer for each basic class combination, pointing to the left when that combination was present, down when it was absent. Figure A.23 shows the three successive displays on Marquand's machine

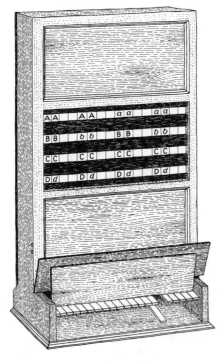

(a) Machine

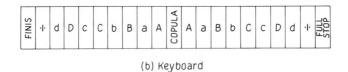

(b) Keyboard

Fig. A.20. Jevons's logical piano

that corresponded to the displays α, β, γ of Jevons's machine and of Marquand's class diagrams (fig. A.21).

Whereas Jevons's piano had twenty-one keys, Marquand simplified the design to ten keys. As can be seen in figure A.22, the keys were actually protruding bars. There were long bars for *A, B,* 0, 1, *C, D*. Interspersed among these were short bars for *a, b, c, d*.

To initialize Marquand's display, the operator pressed the 0 and 1 keys simultaneously, then released them in that order. This left all sixteen pointers horizontal, signifying that all basic class combinations were present (fig. A.23α). To enter premise $Ab = 0$, the operator pressed keys *A* and *b*, then the 0 key. The *A* key held up all pointers for combinations containing *a*, while the *b* key held up all combinations containing *B;* when the 0 key was pressed, all other pointers fell to 0, giving figure A.23β. The premise $Bc = 0$ was entered similarly, giving figure A.23γ.

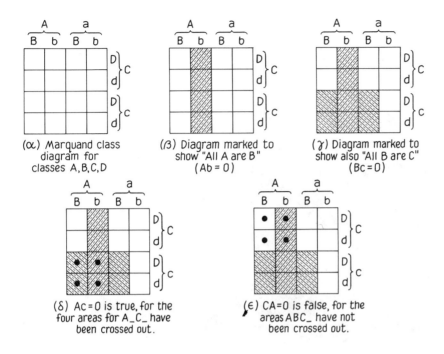

(α) Marquand class
diagram for
classes A,B,C,D

(β) Diagram marked to
show "All A are B"
(Ab = 0)

(γ) Diagram marked to
show also "All B are C"
(Bc = 0)

(δ) Ac = 0 is true, for the
four areas for A_C_ have
been crossed out.

(ε) CA = 0 is false, for the
areas ABC_ have not
been crossed out.

Fig. A.21. Marquand's class diagrams for testing argument validity

To determine whether the argument being tested was valid, the operator looked at the display to see if the conclusion was true, as for Jevons's machine (figs. A.23δ and A.23ε).

We have presented these logic machines as syllogistic devices, involving the logic of classes, because this is how Jevons and Marquand thought of them. But it is simpler to view them as truth-table machines, calculating the values of truth functions. From this perspective, the display showed at the outset all basic conjunctions of the two-valued variables A, B, C, D and their negations. As each premise was entered certain basic conjunctions were removed as being false, so that after all the premises had been entered the display showed those basic conjunctions that were still true.

Though it is simpler to view these machines as truth-functional calculators than as syllogistic machines, the connection between the two viewpoints is not obvious. Interpreted as a logic of classes and their interrelations, Boolean algebra seems quite remote from the propositional calculus and switching theory. This disparity can be shown by comparing the traditional interpretations of the two systems. Each class expression of Boolean algebra (A, a, AB, etc.) represents a class that can have *any number* of members. In contrast, a formula of the proposi-

Fig. A.22. Marquand's mechanical logic machine

tional calculus or truth-function theory (p, q, p & q, $p \lor q$, etc.) can have only the two values "0" (false) and "1" (true). Similarly, in a computer switch each point of the switching network is in one of two states (on or off, high or low, closed or open, etc.).

Even though a class can have any number of members, and in that sense is multivalued, two values for each basic class combination were sufficient for testing the validity of those syllogistic arguments that could be tested on these machines. For the validity status of these arguments depended only on whether or not each basic combination $ABCD$, $ABCd$, . . . , $abcD$, $abcd$ was asserted to be empty. Thus we saw that the result of entering "All A are B" into Jevons's or Marquand's machines was to strike out all basic class combinations of the form $Ab\_\_\_$.

Basically, syllogistic logic did not adequately distinguish quantifiers from truth functions. When both are used explicitly, it becomes clear why Jevons's and Marquand's logic machines were essentially truth-table machines. Symbolized in this way, arguments (1) and (2) are:

(1″) $(x)(Ax \supset Bx)$, $(x)(Bx \supset Cx)$, $\therefore$ $(x)(Ax \supset Cx)$
(2″) $(x)(Ax \supset Bx)$, $(x)(Bx \supset Cx)$, $\therefore$ $(x)(Cx \supset Ax)$.

These reduce to the truth-functional arguments

(1''') $Ax \supset Bx$, $Bx \supset Cx$, $\therefore Ax \supset Cx$ [Valid]

(2''') $Ax \supset Bx$, $Bx \supset Cx$, $\therefore Cx \supset Ax$ [Invalid].

The validity of each of these arguments depends only on the truth values of Ax, Bx, and Cx for representative x, and not on the sizes of classes A, B, C.

Thus it is conceptually much simpler to think of Jevons's and Marquand's logic machines as truth-table machines rather than class-algebra machines. The basic class combinations ($ABCD$, etc.) of the displays are better viewed as *basic conjunctions* of primitive statements that are either true or false. Similarly, sums of basic class combinations correspond to disjunctive normal forms of truth-function theory. Since an individual switch can be in only one of two states at any moment, this two-valued truth interpretation of the variables of Boolean algebra is much closer to switching theory than the class interpretation is. Compare Atanasoff's table specifying the add-subtract mechanism (see table 1 and figs. A.2 and A.18).

Electromechanical Function Switches

Charles Peirce was the first to see clearly that Jevons's and Marquand's machines were essentially truth-table machines. This semantic and syntactical insight led to a novel suggestion in the history of switching, which we will present after a brief review of Peirce's logical background.

Peirce's most important contribution to logic was the development of the logic of relations, starting from Augustus De Morgan's work. (The logic of relations was developed independently by Gottlob Frege in Germany at about the same time. Compare the independent development of relay and vacuum-tube computers in the United States and Germany in the late 1930s, and in England a little later.)

In the early 1880s, Peirce discovered that the NAN (NOT-AND) and NOR (NOT-OR) functions were sufficient to express all truth functions (cf. fig. A.7). He employed partial truth tables, and he knew that every truth function could be expressed either as a disjunction of basic conjunctions, e.g.,

$$(A \neq B) \equiv (Ab \lor aB),$$

or as a conjunction of basic disjunctions, e.g.,

$$(A \neq B) \equiv (a \lor b)(A \lor B).$$

(See fig. A.6 for a switching application of disjunctive normal form, and note that in principle every possible switching function [other than the constant function 0] can be realized in this way.)

Peirce knew of Babbage's plan to build a difference engine and then an analytical engine, and of the Scheutzes' difference engines based on Babbage's (see

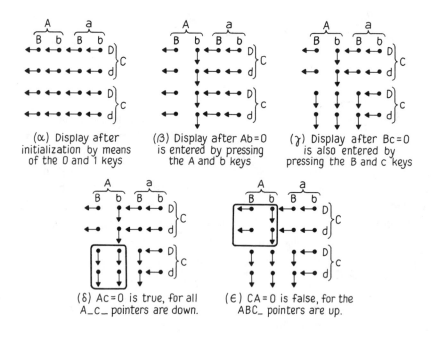

(α) Display after initialization by means of the 0 and 1 keys

(β) Display after Ab=0 is entered by pressing the A and b keys

(γ) Display after Bc=0 is also entered by pressing the B and c keys

(δ) Ac=0 is true, for all A_c_ pointers are down.

(ε) CA=0 is false, for the ABC_ pointers are up.

Fig. A.23. Testing syllogisms on Marquand's logic machine

chap. 5). He followed his student Marquand's work on logic machines, and in a letter to him dated December 30, 1886, wrote:

> You spoke, when I saw you, as if disappointed with the reception your machine had met with. I wish I could see it. [Peirce made some minor suggestions about it, and then went on] though not absolutely required, it would be well to have it capable of adding. I think you ought to return to the problem, especially as it is by no means hopeless to expect *to make a machine for really very difficult mathematical problems*. But you would have to proceed step by step. I think *electricity* would be the best thing to rely on. [Then he drew fig. A.24 and went on] Let A, B, C be three keys or other points where the circuit may be open or closed. As in [fig. A.24a] there is a circuit only if *all* are closed; in [fig. A.24b] there is a circuit if *any one* is closed. This is like multiplication & addition in logic.

Thus Peirce saw that mathematical logic could be applied to the design of relay switching circuits and that one could thereby build a general mathematical machine.

His letter stimulated Marquand to design a relay version of his logical machine (fig. A.26). To explain it, we need first to describe how an electromagnetic relay switch works (fig. A.25). The relay has an electromagnet consisting of an iron

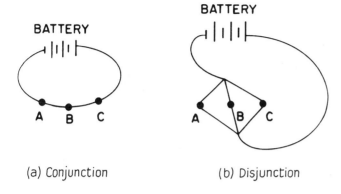

(a) Conjunction   (b) Disjunction

Fig. A.24. Peirce's relay switching circuits

core (shaded) with a many-turned coil around it. It also has an iron armature held away from it by a spring. The armature controls the electrical contacts of the relay by a lever action, though it is insulated from them electrically. Only two pairs of contacts are shown, one forward pair and one backward pair, but a single relay can control many pairs and hence many circuits.

The relay is normally open, in which state the forward contacts are open and the back contacts closed (fig. A.25a). When the relay is energized by passing current through its coil, the iron becomes magnetized and closes the armature against the force of the spring. Then the forward contacts are closed and the back contacts are opened (fig. A.25b). When the current is removed, the magnetic field falls off and the spring opens the armature. The normally open or forward contacts are also called "make-circuit contacts," while the normally closed or back contacts are also called "break-circuit contacts." Notice that the latter contacts give negation.

The figure shows the relay being controlled by a hand-operated key, but in general relays are controlled by the contacts of other relays. As Peirce pointed out, relay contacts in series realize AND, and contacts in parallel realize OR. Note that p and q in series paralleled by $\sim p$ and $\sim q$ in parallel yields EQV. Since AND, OR, and NOT can all be realized by relay contacts and can be compounded, any switching function can in principle be realized by relays.

Computing requires memory as well as switching. Memory is usually accomplished with relays by means of contacts that hold the current in a relay coil on. These holding contacts can be on the relay itself, provided that they are in series with independently controlled contacts, since otherwise a relay once closed could never open again.

Let us now see how a machine built according to Marquand's circuit diagram of figure A.26 would have worked. (Actually there are slight wiring errors in the windings of the $aBCD$ and $abCD$ relays in this figure, but they are easily corrected.) We note first that the wooden logic machines of Jevons and Marquand both in-

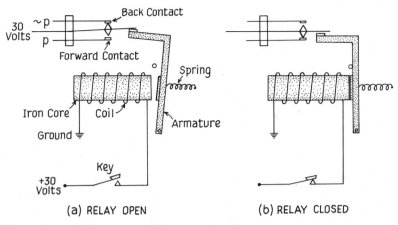

Fig. A.25. Switching with electromagnetic relays

volved memory as well as switching. Had Marquand done his switching with relay contacts in the way Peirce suggested and also used relays for memory, he would have needed a very large number of relays indeed. Instead, Marquand ingeniously planned to use relays with many coils working disjunctively. One coil held a relay closed as well as initially closing it; other coils controlled the relay according to the four basic variables. Because *any* one of these coils could operate a relay, they were disjunctive. And so, by cleverly combining memory and switching in a single relay, Marquand had designed a logic machine with only one relay per basic conjunction.

The keyboard of the electrical machine had the keys *A, B, C, D, a, b, c, d*, as before. The control keys 0, 1 of the wooden machine were replaced by a rotary switch with positions 1, 2, and 3. The display on the front would have looked somewhat like the display of Marquand's wooden machine, with each of the sixteen relays controlling a pointer of the display. When a relay closed, it raised its attached pointer.

These truth-table machines contrast with hand-calculated truth tables in an important respect. For the latter, one begins with a tabular form that has blanks where the calculated truth values are to be written, and then writes in "0's" and "1's" as the table is filled out; thus these forms are essentially three-valued (blank, 0, 1). But in the logic machines, each basic conjunction was displayed for *truth* (1) and was changed to a blank for *falsity* (0) if a premise so required.

This difference explains why figure A.26 has two electrically independent circuits, one for initializing the machine and the other for premise entry. The *initializing and holding* circuit is connected to a coil on each relay and is powered by battery B_1. When the rotary switch is in position 1, sufficient current flows through each coil to close the relays and thereby raise all pointers to true (1). When the rotary switch is in position 2, enough current flows through each coil to hold the raised pointers up, but not enough to raise any lowered pointer. The rotary switch is moved to position 3 to transfer control to the premise entry circuit.

The *premise entry circuit* is powered by battery B_2 and contains the eight

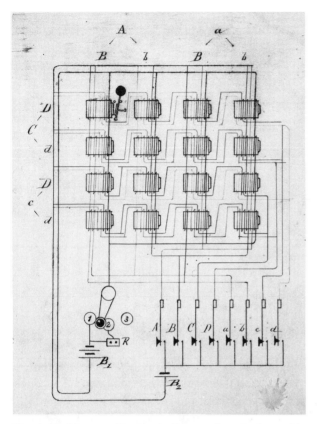

Fig. A.26. Marquand's circuit diagram for a logic machine

variable keys, each of which is connected to coils on eight relays. When a variable key is held closed while the rotary switch is moved to position 3, enough current passes through it and on to each of its coils to hold its closed relays closed, though not enough to close its open relays. To enter a premise the operator performed this sequence of steps: the appropriate premise entry keys were closed, the rotary switch was moved to position 3 and held there long enough for all closed relays not held closed to fall open, the rotary switch was returned to position 2 to hold the state of the machine, and, finally, the premise entry keys were released.

Marquand's premise entry circuitry was neatly color coded. All wires from the *A* and *a* keys and around the *A* and *a* coils on the relays were drawn in red. The *B* and *b* circuits were in blue, the *C* and *c* circuits in green, and the *D* and *d* circuits in yellow. The rest of the wiring was in black.

Figure A.27 shows the circuit for a representative relay, *ABCD,* in more detail. It has five coils: a coil for each of the logical variables *a, b, c, d;* and the initiating and holding coil. If the relay was already closed, current through *any one* of these five coils would keep it closed, and so the coils functioned disjunctively. More magnetic field is required to close an open relay than to hold a closed relay

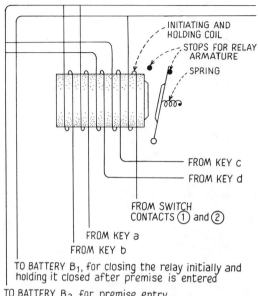

Fig. A.27. Relay for basic conjunction *ABCD* in Marquand's electrical logic machine

closed, and Marquand was using threshold effects in his relay coils to differentiate among closing open relays, opening closed relays, and leaving open relays open. Thus the initiating and holding coil received enough current from switch contact 1 to close the relay, but received only enough current from switch contact 2 to hold the relay closed, not enough to close it.

The four coils connected to keys *a*, *b*, *c*, and *d* are the premise entry coils. This is the *ABCD* relay, but its coils are connected to keys *a*, *b*, *c*, *d*, rather than to keys *A*, *B*, *C*, *D*. For although the entry of a premise causes the appropriate relays to open, this result is accomplished indirectly by holding closed all those relays that should remain closed while allowing all other closed relays to fall open. Therefore, when relay *ABCD* is closed, a current from any of the keys *a*, *b*, *c*, *d* can hold it closed. On the other hand, if relay ABCD is already open, the currents in its coils caused by closing switches *a*, *b*, *c*, or *d* will not be sufficient to close it.

Marquand's circuit was initialized by moving the rotary switch to position 1. Current from B_1 then flowed through the initializing and holding coil of every relay, closing them all and setting all pointers to "1." The switch was then moved to position 2 for holding. The current from B_1 had now to flow through resistance R. This reduced the strength of the current through each coil so that it was sufficient to hold a closed relay closed but not enough to close an open relay. As we have mentioned, position 2 was also used after each premise was entered to hold the state of the truth table.

The premise entry circuit was quite complicated, but was systematically organized. To see how it worked, suppose the premise $Ab = 0$ was entered by closing keys A and b and holding them closed while the rotary switch was moved to position 3, held there while the relays operated, and then returned to position 2. While key A was closed, current from battery B_2 passed through coils on all a___ relays; and while key b was closed, current from B_2 passed through coils on all _B__ relays. The joint effect was that all a___ relays *and* all _B__ relays already closed remained closed, but all relays other than the a___ and _B__ relays not already open did open. Since $\sim(a \lor B) = Ab$, at the end of premise entry all Ab__ relays were open, as the premise $Ab = 0$ requires. The rotary switch was then moved back to position 2 to hold the state of the machine.

After the premises of an argument were entered, the truth value of the conclusion (and hence the validity status of the argument) was decided by inspection of the display, as in the case of the wooden machines.

We saw earlier that Atanasoff employed several voltage levels to do electronic switching. It is clear from the preceding description that Marquand was using several levels of magnetic field intensity to do relay switching. These were: essentially no field (with the relay open), sufficient field to hold a closed relay closed but not enough to close an open relay, and sufficient field to close an open relay. Moreover, when a premise was entered, two of the premise coils on an open relay could receive currents from variable keys, and for the machine to work correctly these currents could not produce a total magnetic field strong enough to close the relay. Thus both Marquand's and Atanasoff's machines employed threshold effects for switching.

Figure A.28 shows Marquand's proposed electrical machine drawn in the symbolism of switching theory, with flip-flops for memory. This illustrates the logical complexity and clever circuitry of Marquand's design, and hence indirectly of its predecessor wooden machines. In this figure, the output of an unset flip-flop displays "0." The set key is numbered "1" because it performs the same function as contact 1 in Marquand's circuit. Pushing the set key causes pulses to go into the set inputs (S) of all flip-flops, so that they are turned on and display "1's." Thus these flip-flops perform the memory function accomplished in Marquand's circuit by the relays and switch position 2.

Each OR of figure A.28 receives inputs from four premise entry keys, and corresponds to Marquand's disjunctive use of the four coils around a relay that are connected to premise entry keys. The reset key of figure A.28 is labeled 3 by analogy to the switch contact 3 of figure A.26, which is used for premise entry. The premise $Ab = 0$ would be entered in figure A.28 by holding keys A and b closed while pushing the reset key. As a consequence, all Ab__ flip-flops would receive reset signals and thus end up displaying "0's," while all other flip-flops would be unaffected.

With this technical background, we are ready to interpret Peirce's very suggestive statement "it is by no means hopeless to expect *to make a machine for really very difficult mathematical problems*" in his letter to Marquand. Although Peirce nowhere systematically addressed this question, he made many scattered comments in his writings that bear on it, and there is a good deal of relevant contextual information (Peirce 1931–58, 1982–).

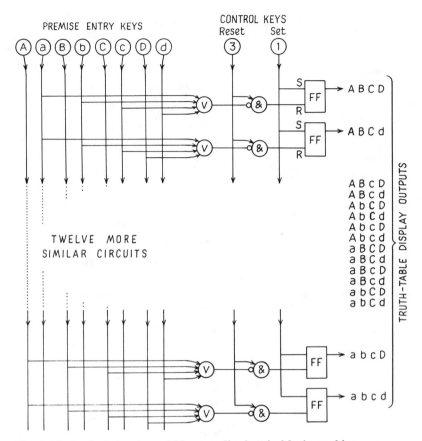

Fig. A.28. Logical structure of Marquand's electrical logic machine

At various places in his writings and book reviews, Peirce referred to Lull's combinatorics and to Leibniz's idea of reducing all thought to mathematics and determining by hand calculation whether a mathematical statement is true or false. Members of Peirce's family were influential in bringing a Scheutz version of Babbage's difference engine to the United States. In a memorial notice on Babbage, and elsewhere, Peirce showed a deep understanding of Babbage's proposal of an analytical engine, the first formulation of the concept of a general-purpose programmable computer. He saw that Babbage's analytical engine could be used for the computation Leibniz envisaged in connection with his universal language. Moreover, Peirce was interested in logic machines and was aware of the close connection between mathematical logic and mathematics proper. He realized that his own relational logic was a bridge between truth-functional logic and mathematics, and that one could grind out or enumerate theorems from an axiomatic system by enumerating proofs.

The most plausible interpretation of Peirce's suggestion to use electromagnetic relays for a machine to solve very difficult mathematical problems is that he saw that one could use electromechanical technology to replace the mechanical technology of Babbage's planned analytical engine. One would then have an *electrical analytical* engine—in modern terminology, an electrical general-purpose programmable computer!

This was an important original concept. Although only that, a concept that had yet to be developed, it did contain the basis for such a computer, because Marquand's circuit diagram included both switching and memory. Moreover, electricity is a better technology for computing than mechanics, for two reasons. First, electrical signals are easily amplified, since the contacts of a relay can be used to control a separate power source. Indeed, the electromagnetic relay was invented for the purpose of amplifying telegraphic signals after they had degenerated over distance. In contrast, mechanical power amplifiers are large and clumsy—the clutch and power steering of an automobile being examples. Second, electrical signals are carried on wires, which are easily bent, whereas changing the direction of mechanical signals is relatively difficult.

And yet Peirce's and Marquand's ideas on the logic of electrical computing had no influence; they did not even become known until after World War II. (The fate of the first calculator, created by Wilhelm Schickard some 250 years earlier, was similar but worse; it too was not "rediscovered" until after World War II!) Jevons and Marquand built their logic machines because of intellectual curiosity, and their machines were not of practical value, for one could test the same arguments by means of Venn's and Marquand's paper diagrams about as fast. Peirce, following Babbage, had a vision of practical applications, but there was no pressing need at that time for a general-purpose electromechanical computer, since the slide rule and desk calculator seemed adequate for scientific calculations.

There *was* a clear need for electrical calculation in a very different area, that of census data processing. In 1880, John Shaw Billings, a physician associated with the United States Census Bureau, suggested to engineer Herman Hollerith that the census could best be processed by using punched cards (Truesdell 1965). Hollerith then developed an electromagnetic punched-card reader, tabulator, and sorter for this purpose. His equipment was used in the 1890 U.S. census and later came into widespread governmental and commercial use.

Hollerith used relays to detect simple truth-functional properties. For example, there were positions on a card representing *white, native,* and *male.* Each card hole passed an electrical signal that operated a corresponding relay. The contacts of these three relays were connected in series to operate a counter. When a card had holes in all three positions, all three relays closed, causing the counter to advance one step. Hollerith used electromagnetic stepping switches that counted to 100, with carrying circuits between successive counters. Stepping switches were the electromechanical equivalent of mechanical toothed-wheel or gear counters, and of the later electronic counters. In contrast, as we have seen, Marquand used a single relay for binary storage, comparable to the electronic flip-flop used later.

Hollerith's work ultimately led to what Peirce had envisaged, a programmable electrical computer, but only after a gap of fifty years. Hollerith had founded a punched-card machine company that eventually became IBM, and Howard

Aiken's proposal to IBM in the mid-1930s led to the IBM-Harvard Mark I. By this time, however, Konrad Zuse was building his relay machines and Bell Laboratories were building theirs. Also by then Atanasoff had begun his work, which led directly to the ENIAC and on to the stored-program computers, which in turn made relay computers obsolete.

The story of delay in applying Boolean algebra to switch design is similar to that of delay in building complicated relay computers. The connection of Boolean algebra to relays was mentioned in a published review of a logic book in the early 1900s. But a systematic account of how to realize this connection was not developed until the mid-1930s, and then simultaneously by engineer Claude Shannon in the United States (Shannon 1938) and others in Japan and Russia. The resultant theory was then used for designing relay switches, having been explicitly created for this purpose.

Electronic Switching

Because the focus of this history of switching is on Atanasoff's place in it, we turn now to electronic switching. We first discuss vacuum-tube switching, then transistor and solid-state switching.

Vacuum-Tube Switching

Vacuum tubes were originally invented for amplifying and transferring continuous (analog) signals. In the early 1930s, physicists used them to make electronic switches for detecting the paths of subatomic particles. Two Geiger counters were placed in a row, each supplying a pulse input to an AND. A particle that passed through both counters would produce an output from the AND, but a particle that passed through just one counter would not.

The physicists had only atomic switches. Atanasoff was the first to design a compound vacuum-tube switch, his holistic add-subtract mechanism. Helmut Schreyer worked on electronifying Zuse's relay calculator a little later, and the World War II Colossi had holistic electronic switches, but neither Schreyer's work nor the Colossi had much influence. On the other hand, as we have said, Atanasoff's work led directly to the ENIAC and on to the first stored-program computers, all of which had complicated holistic electronic switches.

Although Atanasoff had heard of Boolean algebra, he did not know that it was relevant to his design of the add-subtract mechanism. Indeed, Boolean algebra was not used in the design of the ENIAC or of the first stored-program computers; it became useful only later, when circuits were standardized and many circuits were being designed.

It was a serious limitation of Atanasoff's electronic switches that they were not completely binary. Even though each vacuum tube operated in only two states (off and on), its single grid received signals from as many as three other tubes. These signals were added together in a resistor network, fed by driving tubes that usually drove other such networks as well (see figs. A.14 and A.17, and also fig. 12 in chap. 1). Because the characteristics of tubes and resistors vary with age, this design would not have been satisfactory for the fast and complicated switching circuits needed for general-purpose programmable computers.

A completely binary vacuum-tube switch uses one grid per input. The ENIAC switching system, designed mainly by J. Presper Eckert, improved on the ABC in this respect. Its most common types of switches were NORs and NANs. Expressed in terms of positive logic, n triodes connected together with a common plate resistor constituted an n-input NOR (for $n = 1$, a NOT), as in figure A.29. A multigrid tube with two grids used as logical inputs constituted a NAN, as in figure A.30. A single tube envelope contained either one multigrid tube or *two* triodes, hence two control grids or switch inputs.

The ENIAC had 18,000 vacuum tubes (envelopes), some used as switches and some as storage elements in counters. To ensure that this huge electronic system worked reliably at the chosen speed of 100,000 pulses per second, Eckert had the circuits designed with very large safety factors. Each vacuum-tube circuit was designed to produce at least three times as much voltage swing as was needed to switch the tubes it drove, and to effect this switching action in half the time allowed by the electronic clock that synchronized all digital actions.

Another safety feature was the following. Notice that in figures A.29 and A.30 the plate resistor of each driving tube is connected to a power supply voltage that is higher than the supply voltage of the cathode of the driven tube. Thus in figure A.29 the 10K plate resistor of the driving triodes is connected to -360 volts and the cathode to -375 volts. This "overdrive" feature provided an extra safety factor by counteracting the cumulative effect of small currents flowing through the driving triodes when they were all supposed to be off. In the worst case (the program control circuits of the high-speed multiplier), there was a twenty-four-input NOR (a fan-in of twenty-four)!

It was an important feature of the ENIAC that when stopped the whole system held its state. This was useful for debugging both the circuits and the program. The state of each counter and each flip-flop could be read from neon bulbs, and the on-off state of each electronic switch could be checked with a voltmeter. This advantage was achieved by having the ENIAC's switching circuits directly coupled. That is, the plate of each driving tube was wired directly to the grid of the driven tube, rather than being coupled through a capacitor. This procedure eliminated the possible unreliability of a voltage-lowering network, such as Atanasoff used. The cost was an extraordinarily complex direct-current power supply system: there were seventy-eight different voltages running around the machine, produced by twenty-eight large power supplies and a large resistor network!

Although the ENIAC's method of direct coupling made it a complex and clumsy switching system, the aggregate effect of this and other reliability features introduced by Eckert was to produce a huge and unprecedentedly fast system that was reliable enough to establish the preeminence of electronic computing.

The ENIAC's direct-coupling system was eliminated in later computers, by

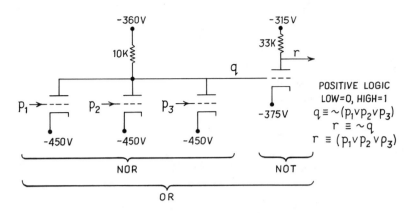

Fig. A.29. ENIAC NOR, NOT, and OR switches

various techniques, so that only a few different direct-current power supplies were required thereafter. One technique was to take the switch output from the cathode rather than from the plate. (Indeed, the ENIAC made some use of this principle, in the OR circuits driving the communication lines of the trunk system.) Another was to use resistors to lower the voltage level from driving plate to driven grid (cf. fig. A.11) while *not* adding voltages in resistor networks, and thereby retaining the valuable one-grid-per-switch-input principle. In pulse-based systems, such as the EDVAC and its successors, successive switching tubes were interconnected by capacitors that passed the switched pulses but blocked the direct currents.

We have now completed our account of switching through the vacuum-tube era. Much further engineering work was done to design still more complicated switching circuits that would be both faster and more reliable, but these developments go beyond the scope of our survey.

Transistor and Solid-State Switching

Let us now summarize the advances in switching technology that—along with corresponding advances in memory technology (which used similar principles)—drove the computer revolution.

Vacuum tubes had been the *active* components of electronic computers: they amplified electronic signals and transformed them nonlinearly in order to make digital switching and memory possible. To do so, they were supplied with energy from outside—the filament current that heated the cathode. Wires, resistors, capacitors, and sometimes inductances were the *passive* components: they communicated signals and transformed them linearly, degrading them in the process.

Correlated with these fundamental differences between the active and passive components of the first electronic computers were economically important physical and engineering differences. Wires, resistors, and capacitors were solid, small, simple, reliable—and so relatively cheap to install and maintain. In contrast,

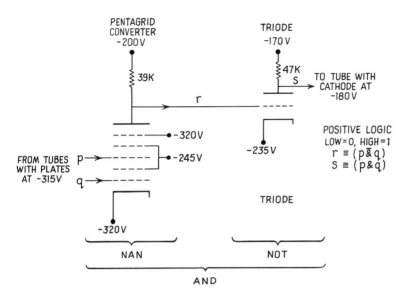

Fig. A.30. ENIAC NAN, NOT, and AND switches

vacuum tubes were "empty," large, complicated, much less reliable—and so much more costly to install and maintain. Recall Atanasoff's economic constraints on how many vacuum tubes he could afford, which led to his development of the capacitor regenerative memory to obtain a lower per-bit cost.

As a consequence, the next really big advance in computer switching (as well as in computer memory) came when the *vacuum* tube was replaced by a *solid* component, the transistor. A transistor is essentially a solid-state triode, with parts analogous to the three parts of the triode: (*a*) cathode, (*b*) grid, and (*c*) plate. In a field effect transistor, (*a*) the *source* emits electrons, (*b*) with a small voltage difference the *gate* controls their flow, and (*c*) the *drain* collects the electrons. In a bipolar transistor, (*a*) the *emitter* emits electrons, (*b*) the *base* controls their flow, and (*c*) the *collector* collects them. Because it is much smaller than a vacuum tube, a transistor is much faster. Because it does not have a separate heating element and is simpler in internal structure than a vacuum tube, it is much cheaper. Thus the invention of the transistor has led to the rapid evolution of ever smaller, ever faster, ever cheaper modules.

The transistor was invented by William Shockley, John Bardeen, and Walter Brattain. This invention required both a basic research study of the movement of electrons (and their opposites, "holes") in semiconducting materials and the practical application of this knowledge. As soon as it became practical, the transistor replaced the vacuum tube in all its uses, analog as well as digital, in communications as well as computing. As time has gone on, these different uses have become more and more closely linked, so that communications is now closely interrelated with computing.

It will be recalled that Shockley's invention of the acoustic delay line, along with Atanasoff's memory regeneration principle, led to Eckert's invention of the mercury-delay-line memory, which in turn led to the modern stored-program computer (see chap. 5). Thus Shockley made important contributions to the development of the computer, a fact not often noted.

With the availability of reliable transistors, the "solidification" of computing components was complete. This solidification had two important consequences for the revolution created by electronic computers. First, it connected the computer revolution to another important technological revolution, the materials revolution. (The nuclear energy revolution is the third technological revolution with which we now live.) Although the first transistors were made from germanium, most are now made from silicon; the search for other suitable semiconductors is now on, however, gallium arsenide being the most promising. Moreover, there is a widened horizon as to the possible material bases of digital computing: optical computers are being studied and artificial organic computers are being considered.

Second, the replacement of the vacuum tube as the active element of a computer by the transistor made possible integrated circuits. This development may be viewed as the extension of the solidification process from computer components to computer modules. An integrated circuit is a multilayered chip composed of conducting, semiconducting, and insulating materials. It is made by a complicated process involving photolithography, etching, and diffusion. Present chips have as many as 100,000 transistors, and chips with 1 million transistors are being produced. Ultimately, there will be three-dimensional integrated circuits.

We have seen that advances in electronic switching technology (and in memory technology) have followed the pattern of producing computers that are smaller and cheaper than their predecessors, and yet faster. The integrated circuit has continued this pattern at an ever-accelerating pace. We can now see that the decrease in component size is the key factor. As components become smaller, more components can be fabricated on the same chip, at a smaller cost per component. And smaller components, being closer together, can both compute and communicate faster.

Conclusion

We have now reviewed the history of logical switching, with appropriate references to the contributions of Atanasoff, the subject of this book. We end with a brief summary and some reflections.

A function switch is a finite device that transforms an input state into an output state in one basic time-step. More technically, when an input state n is supplied to a *function switch* it produces an output state $F(n)$. The first six sections of this appendix gave the logic of electronic function switches, but the present section covers their predecessors in mechanical and electromechanical technology as well. The present section also covers *scanning switches,* which use mechanical motion to scan a sequence of words or memory positions.

The first switches of importance were mechanical scanning switches: rotating pegged drums for music boxes and automata, and punched-card looms. Punched-card and paper-tape machines came later. Atanasoff introduced the first scanning switches for electronic computing: his read-only timing drum and base-conversion drum, and his read-write counter and keyboard drums.

Babbage conceived of the first function switch, an address switch for his planned store, although he never built a switch. Stimulated by Boole's logic of classes, Jevons built a wooden logic machine for testing syllogisms; this had twenty-one inputs and sixteen outputs. Marquand built a simpler wooden machine that performed the same function.

In the 1880s, Peirce saw that these were essentially truth-table machines and suggested building them out of electromagnetic relays. Marquand then designed a relay version of his logical machine. Peirce also suggested that one could build an electromechanical version of Babbage's analytical engine, but nothing came of his idea. At about the same time, Hollerith used relays for simple truth-functional operations, and relays continued to be used in punched-card machines thereafter.

Although physicists used vacuum-tube switches, Atanasoff's add-subtract mechanism was the first substantial electronic function switch. The ENIAC improved on his electronic switching circuits, and later computers improved on the switching circuits of the ENIAC.

We conclude this appendix by drawing some generalizations about the evolution of computer technologies and noting Atanasoff's role in that evolution. We observed earlier that electromagnetic computer components were superior to mechanical components in two crucial respects: they both *amplify* and *communicate* more easily. Electronic computing components also have both advantages. These are the underlying technological reasons why the first complex digital computing machines were not mechanical, but were either electromagnetic or electronic.

The first electronic computers (the ABC, the Colossi, and the ENIAC) were developed at about the same time as the first relay computers (Zuse's Z3, the IBM-Harvard Mark I, Bell relay computers). But electronic computing components were not much more expensive than electromechanical, and they were orders of magnitude faster. Hence electronic computers were destined to win out over relay computers from the start. Moreover, the vacuum tube was solidified into the transistor; and ultimately transistors, capacitors, and wires were fabricated as integrated circuits, thus spurring the computer revolution on to its present state.

The development of integrated circuits has interesting implications for our analysis of Atanasoff's contributions to the history. His principle of regenerative capacitor memory has been revived in the form of DRAMs (dynamic random access memories). The capacitors are now parts of integrated circuits, along with a large random-access address switch for access (writing, regenerating, reading). Thus they are not mechanically dynamic, for Atanasoff's mechanical scanning switch for accessing memory has been replaced by an electronic address switch, which is a holistic function switch.

Our last implication is an irony. For Atanasoff, vacuum tubes (the active elements) were expensive, but capacitors (passive elements) and mechanically rotating drums were cheap. These circumstances provided the technological motivation for his invention of the capacitor drum memory, involving the important principle of regeneration and the use of a scanning switch. With the development of integrated circuits, however, the technological economics changed dramatically. In an integrated circuit, it is the area that counts, and so transistors and capacitors are of comparable cost. Yet Atanasoff's regenerative principle lives on.

Response to Kathleen Mauchly

Kathleen Mauchly's Advocacy

At the beginning of chapter 2, we observed that a main reason why the Mauchly-Atanasoff controversy has remained unsettled was that the rich store of ENIAC trial data had not been brought to bear on it. We invited public debate, on the basis of that wealth of validated information, as we set out to argue the Atanasoff side through relevant exhibits and testimony. At the same time, we acknowledged the contribution of Kathleen R. Mauchly in her article, "John Mauchly's Early Years" (Mauchly 1984), as a "best case" for her late husband's side of the dispute and said we would address her arguments in this appendix.

Kathleen Mauchly's arguments are important not just as those of one interested party. As we said in chapter 2, we believe the case she has made against the ENIAC trial findings on derivation from Atanasoff is essentially the one Mauchly himself was developing in the last years of his life. She had the benefit of hearing his views directly from him, she was present for much if not all of his testimony in the trial, she now has access to the papers he left behind, and she possesses artifacts that he had preserved from his Ursinus College days.

Perhaps even more importantly, Mrs. Mauchly in her article criticizes the Sperry Rand side for what she considers its poor conduct of the defense; that is, she maintains that the trial itself was so deeply flawed as to render Judge Larson's finding incorrect. Moreover, she bolsters her case by quoting from several old Mauchly letters discovered by her after his death.

Thus, with her 1984 article, Kathleen Mauchly has come to be viewed not only as an authority on the issues of the ENIAC trial, but as an authority in a position to reexamine those issues and refute the outcome. For this reason, and also because her article was actually prompted by our own article on the ENIAC (Burks and Burks 1981), we feel constrained to supplement our presentation of the suit in the main text with this direct response to her retrying of it.

Main Claims

Two passages in the preface of Mrs. Mauchly's article encompass her overall thesis. These are (p. 117, col. 1):

The development of the ENIAC (Electronic Numerical Integrator and Computer) flows in a simple, logical, natural way from the work of John William Mauchly while he was at Ursinus College in Collegeville, Pennsylvania, and from the genius, inventiveness, and experience of his partner, J. Presper Eckert, Jr. Mauchly's background and work prepared him for his part in the conception, design, and construction of the first electronic digital computer.

and

. . . we have the physical components of the electronic computer that Mauchly was building during the time he was teaching at Ursinus College. These components alone are evidence that Mauchly's concept of an electronic "computer-calculator" predated any association with John V. Atanasoff and led directly to the design of the ENIAC.

Now there is a difficulty at the very outset with this formulation of events. Mauchly taught at Ursinus College until mid-June, 1941, but he had met Atanasoff over five months earlier. This need not concern us greatly, since he did not really accomplish anything in electronic computing in those months anyway. As we shall see, however, the author does claim much for him in that last spring term, partly on the basis of her newfound letters, and does claim that the "components" she attributes to him then were free of Atanasoff influence.

There is a second, more serious, difficulty with the two passages, a difficulty that lies in Mrs. Mauchly's misleading use of language. One would take *the physical components* of an electronic computer to be its basic parts. But we find, as we read her presentation of them, that these "components" were only experimental contrivances that were not even models of parts of a machine except in the most rudimentary sense of the word *model*. In every instance the contrivance remains far short of the component it is said to be. (We explain these deficiencies as we examine each of them in turn.)

One would also take the statement that Mauchly *was building* an electronic computer at Ursinus College to mean that he had a machine under construction there. Yet we discover later that the "components" she presents were never interconnected in any way and lacked any indication of how they were to be interconnected to perform the functions she attributes to them.

Lastly, the references to Mauchly's *concept* of an electronic "computer-calculator" prior to any association with Atanasoff, a concept that "led directly to the design of the ENIAC," would suggest that he had executed some preliminary design in connection with some calculating machine. But no written description, no diagram—no sketch—by Mauchly before the end of 1940 has been produced, either at trial or by the author, for any aspect of any electronic calculator, much less the complex one she claims for him. Even for the balance of his Ursinus period, after he had met Atanasoff and while he was corresponding with him, there were no descriptions; there were just the barest sketches depicting far less sophisticated ideas than those she attributes to him as of the end of 1940. Indeed, written descriptions and diagrams—cited and quoted at length by her!—date only from August, 1941, two months after Mauchly's visit to Atanasoff in Iowa to see and learn about his computer.

The reader will have to decide, of course, between the author's case for Mauchly's early computing activities and our countercase. We want it to be clear from the beginning, however, that the two passages from her preface that we take to encompass her overall thesis are only contentions. She is only *contending* that objects she has from Mauchly's Ursinus years were the components of an electronic computer, and that he was building such a computer at Ursinus. She is only *contending* that he had the concept of an electronic computer-calculator in the pre-Atanasoff portion of his Ursinus period, and that that concept led directly to the design of the ENIAC. Finally, she is only *contending* that Mauchly and Eckert conceived and developed the ENIAC without the example and ideas of Atanasoff, and that the ENIAC, not the ABC, was the first electronic digital computer. At this stage of her article, she has yet to prove every one of these assertions.

Main Claims Reformulated

It is helpful to reformulate Mrs. Mauchly's statement of her thesis, as quoted above, in an ordered list of (seven) claims:

1. Mauchly was building an electronic computer during his Ursinus period.
2. Mauchly built the physical components of that computer in that period.
3. Mauchly had the concept of an electronic computer or calculator before his association with Atanasoff began.
4. Those components, still in existence, are evidence for Mauchly's having had that concept prior to that association.
5. That concept led directly to the design of the ENIAC.
6. Atanasoff had no influence on the conception of the ENIAC by Mauchly and Eckert.
7. The ENIAC was the first electronic digital computer.

Now claims 1 through 5, inasmuch as they are taken to exclude any Atanasoff influence, rightfully fall not just in Mauchly's Ursinus period but in his *pre*-Atanasoff period as we have distinguished the two. Claims 6 and 7, on the other hand, are two years *post*-Atanasoff. Mrs. Mauchly has, then, set herself a two-pronged task: first, the positive one of establishing significant progress in electronic digital computing by Mauchly prior to any input from Atanasoff; second, the negative one of establishing a lack of Atanasoff influence once that input began. The first prong is the more crucial for her, of course, in that the more she can show Mauchly achieved before he met Atanasoff, the less she has to discount Atanasoff's influence in the later period.

Sources

The author provides a list of sources for her article in the preface (p. 117, col. 1):

> The material I present here is based partly on my own recollections of discussions with my husband, supported and extended by conversations with Eckert and by a study of depositions and testimony from various trials, of Mauchly's own written articles, of tapes of his talks and videotapes of interviews (particularly an 18-hour videotape made by Esther (Mrs. John) Carr in 1977), of documents, letters, and scientific magazines of the 1930s, and of other artifacts that Mauchly saved during his lifetime.

It will be seen at once that one of her chief sources is, as we have already suggested, John Mauchly, and, in fact, John Mauchly in recent years—or, more precisely, in the years after the ENIAC patent trial decision of October 19, 1973. Now of course Mauchly's own account of the events addressed in that decision, even though written or spoken later, is a legitimate source for the present article. We find it very disturbing, however, that the author makes two serious omissions, from a research point of view, in her use of that account.

One, she never tells the reader when she is relying on it; many strong assertions remain completely unsubstantiated and can only be assumed to be from Mauchly's recent account. Two, she does not acknowledge in any way, much less attempt to reconcile, the many serious discrepancies between that account and Mauchly's own ENIAC trial testimony of 1971.

Certainly it is helpful to have Mrs. Mauchly's collection into a single article of all the salient features of her late husband's case, as construed by her and apparently by him. One can also sympathize with her effort to make this presentation of it. Nevertheless, one cannot accept without further documentation this new version of critical events, especially where it differs from an earlier version given by him in court, under oath and subject to cross-examination. It must be kept in mind that John Mauchly was a principal to the disputed patent suit, and was also the present author's husband; and that this article is offered as serious research intended to demonstrate that a federal court decision of tremendous import was incorrect.

Rationale for Disputing the Trial Outcome

Before we proceed to our critique of Mrs. Mauchly's main claims, let us look into her grounds for retrying the ENIAC case: first, that previously overlooked letters have come to light; and second, that the Sperry Rand defense was inadequate. The latter charge is of particular significance, because it serves to shift the blame for loss of the suit away from the merits of the case and onto the attorneys who conducted it.

The author has found a "letter file from the Ursinus period . . . [with]

carbons of letters showing that Mauchly had been actively working on a computer at Ursinus" (p. 117, col. 1). We examine these in due course, but we would note here that Mauchly had been subpoenaed to produce all relevant documents, not only for the ENIAC case but for other court cases dating back to 1967 or earlier. It was his responsibility, his legal obligation, to locate them, and he had ample time to do so.

Moreover, now that Mrs. Mauchly has uncovered this file, she again errs on the side of omission. In her entire article, she presents just five Mauchly letters that were not introduced at the trial, and only selected portions of these. We submit that at the very least she should have reproduced them in full, as would have been required in court, and assured her readers that nothing else in this old file pertained to Mauchly's computing interests and endeavors.

It also turns out that only two of the five letters fall in the pre-Atanasoff period; indeed, they fall at the very end of that period (November 15, 1940, and December 4, 1940). Two other letters fall in the post-Atanasoff Ursinus period (April 26, 1941, and June 7, 1941). Oddly enough, the fifth newfound Mauchly letter postdates the entire Ursinus period and falls fully two months after Mauchly's highly instructive visit to Atanasoff's laboratory (August 18, 1941).

The author next represents the Sperry defense against the Honeywell case for derivation of the ENIAC from Atanasoff as inadequate, or even remiss, on three scores. First, she alleges that the devices Mauchly had built at Ursinus, which in her view constituted strong evidence for his independent progress toward the ENIAC, were not adequately exploited in the trial (p. 118, col. 1). Second, she alleges that there were witnesses from the Ursinus period who could have related those devices to the ENIAC, but were not called (p. 118, col. 1). Third, she alleges that the Sperry defense team failed to respond to the Honeywell presentation of its case for derivation of the ENIAC from Atanasoff—in fact, that they chose not to respond because they believed no evidence had been introduced by Honeywell to substantiate such derivation (p. 117, col. 2). Let us look at each of these allegations.

Mrs. Mauchly acknowledges that "*pictures* of nearly all the devices Mauchly had built at Ursinus were introduced, and Mauchly described them for the court" [her italics] (p. 118, col. 1). We think the reader will agree, on the basis of Mauchly's own depiction of his early computing efforts in court, as presented in chapter 2 of this book, that the Sperry attorneys promoted these objects as fully as prudence allowed. (Recall that the Sperry side did have Mauchly describe in detail the one item that represented his greatest pre-Atanasoff achievement in digital electronic computing, his two-neon device, only to have the Honeywell side reduce it to a toy crossing light that could not in any realistic sense be viewed as a binary counter.) We will, nevertheless, consider her new interpretation of these Ursinus devices as possible steps by Mauchly toward an electronic digital computer.

As to her second objection, that witnesses were not called who could have related the Ursinus devices to the ENIAC, she herself cites only one such person by name in her entire article. This is John DeWire, a former Ursinus student of Mauchly's. She quotes from DeWire's 1982 letter to her in which he recalled

Mauchly's work; his account, however, does not go beyond Mauchly's interest in flip-flops, with a *motivation* to develop a binary electronic arithmetic device (p. 120).

On the other hand, the Sperry attorneys did call Joseph Chapline, another former Ursinus student who had spent much of his free time in Mauchly's laboratory and was familiar with his projects and devices in the 1939–41 period, and who had also worked at the Moore School (in the differential analyzer room) throughout the ENIAC period. Mauchly himself testified (ENIAC trial transcript pp. 11,878–79), that it was Chapline who had called his ideas on electronic computing to the attention of Goldstine, who had then proposed the ENIAC project to the military. Chapline identified the photographs of Mauchly's harmonic analyzer and his cipher machine for the court; he also identified the photograph of Mauchly's experimental circuit boards explicitly as a "five-stage counter" and testified that he had seen Mauchly demonstrate it (transcript pp. 12,591–93; see our chap. 2). He said nothing, though, to link it to the ENIAC any more than Mauchly had done at trial.

As to Mrs. Mauchly's third charge, that the Sperry team did not respond to the Honeywell case for derivation of the ENIAC from Atanasoff, we have already seen in chapter 3 that they did respond—and vigorously. Attorney H. Francis DeLone, in his direct examination of Mauchly as a defense witness, educed Mauchly's own detailed version of all his contacts with Atanasoff, both in person and through correspondence, and of all his own independent accomplishments in computing.

The author's stance here is poor, indeed, and does grave disservice to the Sperry side. She simply overlooks the volumes of trial testimony from which we have quoted and relies instead on her interpretation of the opening statements of both sides to establish, first, that Honeywell had no intention of pursuing the Atanasoff derivation, and, second, that Sperry accordingly would not counter it. Worse yet, she quotes so selectively as to convey impressions totally inconsistent with their true import.

Mrs. Mauchly observes that the Sperry attorneys "may have been thrown off guard" by the opening statement of Henry Halladay (p., 117, col. 2). She quotes Halladay (trial transcript pp. 111–12):

> The plaintiff's case, your honor, does not deny that the ENIAC machine was a pioneering effort. Indeed some of the witnesses whom the plaintiff intends to call to testify will so say, that it was the premier machine, that it was a break-through. Indeed, we and Honeywell do not deny that the ENIAC machine, the ENIAC system, may well be considered a pioneering invention.

But she has cut Halladay off at midsentence! He actually said:

> . . . pioneering invention, but that does not constitute a confession that the ENIAC machine, as it existed in Philadelphia . . . , at any time constituted a

patentable invention which shall be subjected to monopoly power by [Eckert or Mauchly] or any of the others who worked on the research and development of that machine . . . .

In short, the Honeywell side, while acknowledging much that was novel in the ENIAC, nevertheless meant to challenge its originality in essential respects. And it proceeded to do just that, in terms of derivation from Atanasoff and others, over the next five months of the trial.

Mrs. Mauchly then quotes attorney DeLone of the Sperry side from *his* opening statement (transcript pp. 11,724–25), to the effect that in the defense view plaintiff Honeywell had not presented any evidence beyond a great many patents for its derivation claims, and that he, therefore, would not be offering any evidence on that issue. Instead, he added—and Mrs. Mauchly omits this factor—he would proceed with the Sperry charge of patent infringement against Honeywell.

DeLone had, in fact, just prior to this statement, made a more general assertion as to the emptiness of the Honeywell presentation not only on derivation, but on coinventors, public use, on sale, and prior publication; that is, on Honeywell's entire Count 2 requesting a judgment of invalidity of the ENIAC patent. His purpose was to have Count 2 dismissed, leaving Honeywell with only Count 1, its antitrust charge. (See chap. 4.) Indeed, he had also tried at the outset of the trial, after Halladay completed his opening statement, to have this invalidity count dismissed, arguing that Honeywell did not *intend* to pursue it seriously. He had failed then, as he did now, because Halladay each time reaffirmed Honeywell's seriousness about the charge.

(It should be noted that it was in this interim period between presentations of their cases by the two sides that DeLone did succeed in having Honeywell's third count, violation of the Clayton Antitrust Act—as distinguished from the first count, violation of the Sherman Antitrust Act—dismissed by Larson [again, see our chap. 4]. We mention this as one of many actions by Sperry Rand attorneys that demonstrate their keen pursuance of the defense in the ENIAC trial.)

What this Sperry posturing amounted to was, first, an attempt to disparage the Honeywell case for invalidation, including its argument for derivation from Atanasoff, and, second, a shifting of emphasis from the Honeywell count of invalidity to its own countercharge of infringement. There was really no question of not responding. The issues of invalidity and infringement were intertwined, since infringement of a patent depends on that patent's being valid in the first place. The Sperry team would simply make its courtroom presentation against the Honeywell case for derivation *in the context of* its own infringement case, thereby having the matter both ways: maintaining that the Honeywell case for derivation did not even merit a response, yet in fact responding through the infringement issue.

Thus, as the Sperry attorneys conducted their infringement case, they examined Mauchly—their chief witness—at length about his computing activities at Ursinus College and about his relationship with Atanasoff, including what he had learned from him in their two meetings and what was conveyed in letters to, from, or about him. Former Ursinus student Joseph Chapline followed him on the stand; somewhat later, Eckert, too, underwent questioning on the Atanasoff connection. (The Sperry side had, of course, already cross-examined Atanasoff and other Honeywell witnesses as to his "prior art.")

Let us sum up our review of Mrs. Mauchly's preface to her *Annals* article with the observation that she manages to mislead her readers rather badly from the very start. First, she blurs Mauchly's pre-Atanasoff period into his Ursinus period. Second, she falsely charges the Sperry Rand attorneys with neglecting to capitalize on Mauchly's Ursinus devices as computer components that led to the ENIAC, overlooking key witnesses who could have established that relationship, and deliberately electing not to respond to the Honeywell case for derivation of the ENIAC from Atanasoff.

Third, and most serious of all, she makes a number of strong contentions that she fails to prove in the body of her article. These are based partly on her own interpretation of the so-called computer components and partly on her newfound Mauchly letters (the omission of which from the trial can be blamed only on Mauchly himself). We turn now to her presentation of these devices and these letters.

The Pre-Atanasoff Years

Although Kathleen Mauchly has stated that she possesses electronic computer-calculator components not properly presented at trial, or not presented at all, she herself presents only six "devices," of which she actually identifies only three as the neglected "components."

These three are: a two-neon device said to constitute not only the "control switches" of the calculator Mauchly was supposedly building, but its "binary counters" and "flip-flops" as well; a "decade [ring] counter" with gas triodes replacing the standard vacuum triodes; and a "pulse former" for use with the decade counter. The reader will recognize the first as the two-neon relaxation oscillator Mauchly called a binary counter in court (but not a flip-flop or a switch) and the second as one of the gas-triode scaling circuits with which he said he had experimented on circuit boards (see chap. 2). The third, as we shall see, was actually a standard laboratory source of pulses.

The three "devices" Mrs. Mauchly describes but does not cite as computer components are: a "cipher machine" or "cryptographic device"; a "harmonic analyzer"; and a "voltage regulator" to be used with the harmonic analyzer. We have seen the first two examined thoroughly in the ENIAC trial (chap. 2). The third did not appear there, presumably because Mauchly did not consider it worth bringing in; like the pulse former, it was a standard laboratory instrument. (Mrs. Mauchly seems to elevate both the pulse former and the voltage regulator to special status because of their hand construction, even though scientists in those days often built their own equipment.)

We need not concern ourselves further with the author's account of devices unrelated to Mauchly's design and construction of an electronic digital computer in his pre-Atanasoff years. We must, however, examine the three items she construes as digital computer components. We must also examine the two Mauchly letters of late 1940 recently uncovered by her, letters she contends prove a long-standing plan to build an electronic desk calculator. The reader has seen the pertinent passages of these, too, as quoted from other sources (see chap. 2), but Mrs. Mauchly finds new meaning in them. We look first at the components and then at the plan.

The Three Components

The author leads up to her computer components in three steps: by characterizing an electronic "desk calculator with storage facilities" she says Mauchly aspired to build in 1936; by stressing his familiarity with the physicists' scaling circuits described in the professional journals; and by concluding that he had still to develop "the control circuits" to use with the "various types of ring counters" being built at the time (pp. 119–21). Notice the strong implication here that those ring counters were *computing* (accumulating) counters, rather than the simple rough *counting* counters they really were. We will return to this difficulty as we discuss the author's presentation of Mauchly's decade counters, her second computer component, below.

Before we turn to her first component, let us quote one short passage from this preliminary account to illustrate the degree to which she undertakes to reproduce Mauchly's very thoughts, all those years ago, without source and without evidence (p. 120, col. 2):

> . . . Mauchly began to think of ways he could utilize such a [binary scaling] circuit. He reasoned that if one could generate pulses in his own way, then one could, with proper controls, use vacuum tubes to do arithmetic. All of the foregoing contributed to Mauchly's thoughts about electronic computing. The year was 1936.

Control Switches

Mrs. Mauchly introduces the two-neon device (PX 21,374 and PX 21,375) as follows (p. 121):

> The computer he envisioned had decade counters controlled by electronic switches for the transfer of data from one counter or storage element to another. He first attacked the problem of electronic switching, the control of one set of signals by another. Because he couldn't afford the cost of vacuum tubes for these experiments, he substituted neon bulbs. Although neon bulbs are not functionally interchangeable with vacuum tubes, the logic was the same. The neon bulbs worked electronically, and they gave a visible

indication. They were slow compared with vacuum tubes, which the cosmic-ray scientists were using, operating at only a few hundred pulses per second, but they were cheap—a good tradeoff.

We would note, first, that there is again neither source nor evidence for this depiction and, second, that the author continues her implication that the physicists' scaling circuits were suited to use in a computer without modification. Her more immediate offense, however, is that she is promoting Mauchly's little relaxation oscillator to a *switch,* as well as to the binary counter Mauchly had testified it was, *and* to a flip-flop in the bargain. She states flatly (p. 121, col. 2):

> . . . With these neons he built his control devices. He wired up two neon bulbs with three resistors and one capacitor (condenser) and created, by interrupting the current through the circuit, the off-on action of a switch—one of the elements needed for a computer. This device with one input that would throw it into an alternate state is called a binary counter; with two inputs the device is a flip-flop. . . .

Yet Mauchly, demonstrating this device from the witness stand, had made clear that it was the current interruptions themselves that were counted—and that these interruptions were actually administered by hand, though they could have been administered from a hand-operated switch had he cared to *add* a switch! Moreover, he had avoided saying just how he might have interconnected two of these devices, for a count above "1," because the device as it stood had no input at all, let alone one for a counter, two for a flip-flop! (See our chap. 2.)

(Mrs. Mauchly says that he actually connected ten of these devices in a ring and "built for himself a trial decade counter." She provides a photograph captioned "Remains of an early attempt at a flip-flop counter using neon tubes, built around 1936" that is beyond recognition. Significantly, Mauchly did not bring this item into court. Rather, he testified only that several of the two-neon devices *could have been* connected to form a *binary* counter to count, say, to "8" [chap. 2].)

Quite incredibly, then, Mrs. Mauchly has taken this crude device, which was truly an embarassment at the ENIAC trial, and stretched it from Mauchly's "binary counter" pretense into a flip-flop, and—even more fantastically—into the control circuits of an electronic computer. As we pointed out in the text, this device was incapable of functioning in any computing context. It would have required additional circuitry to function as a counter or a flip-flop, and it was not even a starting point for a switch.

Switches, of course, are not memory devices at all, but the basic *logical* devices of a computer. They interact with counters and flip-flops and other memory elements to impose NOTs, ANDs, ORs, NANs, NORs—and combinations thereof—both on signals entering and on signals emanating from those elements. Moreover, they are not nearly as easily characterized as these other two devices; they range from simple one-triode circuits to extremely complex circuits that transfer information in very intricate ways. (The ENIAC, Mrs. Mauchly's goal in this part of her presentation, had some highly complicated switching circuits.)

Nevertheless, with this single stroke, the author moves on to the counters this little neon device was to control.

Decade Counters

This second "component," a biquinary form of decade counter (PX 21,382), was not a computing component either. It was exactly what Mauchly testified it was: an experimental ring counter copied from the literature, with gas tubes replacing vacuum tubes. Mrs. Mauchly, however, now shifts from her earlier implication that scaling counters were the same as computing counters to an outright assumption that such was the case. (See chap. 2 for the distinction between, on the one hand, counters that merely cycle as a stream of pulses enters at the ones place, with appropriate carries passed along, and, on the other hand, counters that are capable of receiving streams of pulses in parallel at all places and performing various arithmetic operations on them, that is, counters that make up an accumulator.)

The author completes this shift by relating Mauchly's circuit board experiments first to the ENIAC and then to the desk calculator he was supposedly building, or planning to build—she dates these counters from 1937 to 1941 (pp. 122–23):

> . . . Mauchly bought many of these 0A4G tubes and began experimenting with circuits that would advance in response to pulses the same way the ENIAC ultimately advanced to pulses using vacuum tubes. . . .

and

> These cold cathode gas counters were satisfactory in that they showed Mauchly at that time the possibility for building a small, cheap electronic computer for his own use at Ursinus, even though the speed would be inferior to what could be expected from vacuum-tube counters or scaling circuits described in the literature or in use in the nuclear laboratories that he had visited.

But Mauchly did not even get so far as to interconnect two of his ring counters in series, and he certainly did not learn how to compute with counters by virtue of substituting poorer tubes in the physicists' cosmic-ray counters! The great challenge for designing a counter that could compute reliably lay ahead, to be met for the ENIAC, as we have said, by Eckert, not by Mauchly.

Thus we have a second completely unjustified inflation of a simple device, one that the author herself does not claim was original with Mauchly, into a sophisticated computing component that was in fact only invented some time later, and then by others.

Pulse Former

The trouble with Mrs. Mauchly's third computer component is that it was just a standard laboratory instrument (PX 21,384) that Mauchly testified he had built to test his experimental scaling circuits. She, however, claims much more for it, likening it to what was really a pulse *standardizer* in the ENIAC. We quote her one-paragraph presentation (p. 123):

> A counter is of no use unless it is reliable, so to ensure reliability of the ring counters, some reshaping of the pulses going into the counter was needed. Mauchly built what he called a "pulse former," similar to what was later used in the ENIAC, to precede the counter. He needed to know what kind of pulses were going into the circuits and what was happening to them in the circuits. The pulse former . . . reshaped the pulses that entered the counters. It . . . contained a choke, an 884 gas tube, and a 6J7G (G stands for glass) power tube with screen grid. Simple manual switches regulated the height, width, and duration of the pulses. The pulses entering the pulse former were generated by a low-frequency oscillator. A small 3-inch oscilloscope from Culpman and Romander monitored the pulses for frequency and reliability. Mauchly was able to make the gas-tube counters, with the reshaped pulses, operate reliably at 500 pulses per second.

In direct examination, Mauchly said only that this instrument was a "pulse generating device which I used in making my experiments with these various counters" (transcript p. 11,803). Under cross examination, he said it was a "pulse former," which he defined as "essentially a test device which creates pulses of various sizes and length and spacing," but added that his was "not very complicated" and "the change in shape or the variety of shape possible was not great." For use in connection with it, he mentioned only a power source and an oscilloscope on which he monitored the slightly reshaped "sawtooth pulse" it generated. Finally, he said that it was not original with him, but that he had referred to the work of others before him in building it. (Transcript pp. 12,173–79)

Now Mauchly's characterization of PX 21,384 and the photograph supplied by Mrs. Mauchly both indicate that this device was what was better known as a pulse generator, that is, an instrument incorporating a pulse former *and* an oscillator (which she does mention, as a separate entity). It was the oscillator that provided the basic sawtooth wave and the pulse former that reshaped it into a succession of "somewhat distinct" pulses of "rather rapid rise and rapid descent," as Mauchly testified to Honeywell attorney Halladay. (His early 1941 drawings of ring counters actually specified pulse generators, rather than pulse formers.)

In any case, it is the grossest exaggeration for Mrs. Mauchly to claim that this instrument was "similar to what was later used in the ENIAC" to reshape pulses entering its counters. As Mauchly had said in court, he built it to test his gastriode scaling circuits, that is, to observe the shapes of their pulses at various points and to measure their counting speed. The ENIAC, by contrast, had a pulse reshaping problem of immense magnitude, for three reasons. First, its (accumulator) decade counters were to operate 200 times faster than the 500-pulse-per-second rate of Mauchly's counters. Second, the ENIAC's pulses were to be transmitted among 18,000 vacuum tubes, over a distance of up to 80 feet, with the result that both their shapes and their magnitudes could deteriorate before they reached their destinations. Third, these pulses were not equidistant, but varied in their spacing with the information being transmitted.

And, again, it was Eckert who both recognized and solved this difficult reliability problem for the design of the ENIAC, in 1943. His pulse standardizer took the distorted pulses at input and reformed them into the sharp, strong standard pulses needed to operate his counter reliably. Plainly, it was many orders of magnitude beyond the simple testing instrument Mauchly described in court, or anything yet devised in that day for use with counters.

And so the author has again inflated a relatively simple device, built by Mauchly but not original with him, into the conscious antecedent of a complex and highly sophisticated computer component—a component whose need arose only years later with the ENIAC and whose creation was the work, not of Mauchly, but of Eckert. Once again she has provided no documentation for her version of it, and has made no reference to Mauchly's court testimony belying that version. As with her other two so-called computer components, she has also moved this device back in time; she specifies 1937–38 in her figure legend, whereas Mauchly had testified to "about 1939" for this piece of test equipment.

The Plan

Having developed her three basic "computing components" ("control switches," "decade counters," and "pulse former"), none of which *was* a computing component, the author declares Mauchly "ready to build" as of November, 1940, and discloses his plan (pp. 124–25). In her preface she claimed that he had conceived a "computer-calculator" before he began his association with Atanasoff in December, 1940, and that these alleged components were evidence for that claim. In the body of her article, as we just saw, she dated the concept itself back to 1936, with the components following in 1936 (controls), 1937–41 (counters), and 1937–38 (pulse former). (Notice that the date for the counters somehow goes beyond Mauchly's first encounter with Atanasoff, in which they did discuss digital computing with vacuum tubes [see chap. 2].)

If the author's account of Mauchly's early plan and his progress toward it has been highly creative up to this point, it now becomes utterly fanciful. For we are told that this plan encompassed the concept and the components of a calculator that would use "only electronic elements" and have "no moving parts." Indeed, its components were "the seeds of an ENIAC"!

Mrs. Mauchly bases this tremendous escalation of events on the first two of the five letters she found in a sealed file in 1980. As we noted earlier, portions of both were reproduced, from publication elsewhere, in our chapter 2. We now present them as offered by her.

The First Two Letters

The first letter, dated November 15, 1940, was from Mauchly to meteorologist H. Helm Clayton, with whom he had worked the previous summer in Canton, Massachusetts. Mrs. Mauchly acknowledges its prior publication, in full, in the *Annals of the History of Computing* (Mauchly 1982) She reproduces the same paragraph we did in chapter 2, to the effect that he was about to start to build his second (twenty-seven-ordinate) harmonic analyzer, and that he was also considering the construction of a calculator. The germane part for this latter aspiration is:

> In addition, we are now considering the construction of an electrical computing machine to obtain sums of squares and cross-products as rapidly as the numbers can be punched into the machine. The machine would perform its operations in about 1/200 second, using vacuum tube relays, and yielding mathematically exact, not approximate, results. That is, its accuracy would not be limited to the accuracy with which one can read a meter scale, but could be carried to any number of places if one cared to construct the machine with that many parts. With conventional tubes, it would be rather bulky, but special tubes could be designed to make it very compact.

The second letter, dated December 4, 1940, was to former student John DeWire. Mrs. Mauchly does *not* say that part of it, too, had been published earlier, in a letter James McNulty wrote to the editor of *Datamation* (McNulty 1980). She quotes just two sentences of one paragraph, as follows (p. 125, col. 1):

> For your own private information, I expect to have, in a year or so, when I can get the stuff and put it together, an electronic computing machine, which will have the answer as fast as the buttons can be depressed. The secret lies in "scaling circuits," of course.

These sentences read the same as they did in McNulty's version, except that she has inserted the word *can* in the first one. But she has also omitted entirely a third sentence included in his version of this same paragraph:

> Keep this dark, since I haven't the equipment this year to carry it out and I would like to be the first.

A number of points come to mind concerning these two letters, vis-à-vis Mrs. Mauchly's ongoing account of Mauchly's pre-Atanasoff activities. First, they do not support her claim that he had conceived of an electronic desk calculator in 1936, but only at the time of writing, late 1940. Nor do they support her claim that he had already built and tested its components. The letter to Clayton said he was *considering* building the calculator—after he had reported progress toward a second harmonic analyzer. The one to DeWire said he *expected* to get the materials and put it together in a year or so. And surely if he had had this goal back in 1936 or 1937, when DeWire was his student, he would not have waited until late 1940 to reveal it to him, as he clearly was doing in this letter.

Second, the "concept" expressed in these letters is one of function, not sub-stance, and the only projected "components" mentioned are vacuum-tube relays.

Third, the comment quoted by McNulty but omitted by Mrs. Mauchly, "I haven't the equipment this year to carry it out," flatly contradicts her statement that he was now "ready to build."

Fourth and last, Mrs. Mauchly's very withholding of a sentence contrary to her thesis is disturbing. As we pointed out earlier, had these letters been brought into court, they would have been disclosed in full and Mauchly subjected to examination on them. Readers of this *Annals* article, however, have only selected passages from them, with no assurance that unfavorable passages were not deleted—as, indeed, one was from the letter to DeWire.

It is all the more remarkable, then, that these two letters, which constitute the earliest documentation of any plan by Mauchly to build any electronic computer, come down to so little. For they establish only that he intended, as of the end of 1940 and just prior to meeting Atanasoff, to build a keyboard-operated electronic calculator to perform sums of squares and cross-products, and to do so by the end of 1941. Nevertheless, having quoted from them, Mrs. Mauchly forges ahead with her fantasy that the components she has been conjecturing were actually the starting point of the ENIAC, a mammoth, highly complex, revolutionary general-purpose computer that was actually proposed more than two years later.

Seeds of an ENIAC

Mrs. Mauchly's four "seeds" turn out to be just the three earlier ones that have been growing all along in her fertile imagination. The first is the biquinary ring counter of the physicists (still with gas triodes), now fully realized as accumulating counters or "registers" that "perform the arithmetic and store the numbers." The second is Mauchly's simple two-neon relaxation oscillator now further extended to "additional vacuum tubes for control and circuit switching" in the transfer of results "from one register to another." (This extension is undoubtedly taken from Mauchly's testimony that the vacuum tubes interspersed among the neon counters in his 1941 drawings were "a switch to do that [current] interruption" that was to be counted [transcript p. 12,125].)

The third and fourth "seeds" are just Mauchly's laboratory pulse generator, still presented as a pulse former but with its oscillator now given equal status, and still performing for the author in the fashion of the ENIAC's pulse standardizer.

Of course, as we saw in chapters 2 and 3, John Mauchly had not, by late 1940—or by mid-1941, for that matter—built or even designed any electronic accu-

mulator or register, any electronic control or switch, any electronic pulse standardizer. And Mrs. Mauchly, for her part, has introduced no new evidence.
Rather, she has created these basic elements of ENIAC design entirely from the
selfsame items that had been examined in court—items that Mauchly himself never
so much as hinted were presagers of the ENIAC. Yet she continues undaunted,
summarizing her listing of the "seeds of an ENIAC" as follows:

> In his careful, methodical way, Mauchly had thought out, built, and
> tested the components of an automatic electronic general-purpose calcula-
> tor. . . .

While it is hard to find a single word of truth in this statement, the author's
injection of "general-purpose" into Mauchly's thinking at this juncture is perhaps
the greatest aggrandizement of all. She herself has been maintaining that the projected computer was explicitly intended to replace mechanical calculators in his
analyses of weather data. In fact, Mauchly's characterization of it to Clayton as a
machine to obtain sums of squares and cross-products not only related to those
same analyses, but depicted a special-purpose machine that was probably even less
general than the mechanical version it was to replace. Moreover, with cost an
overriding factor, the notion of building a general-purpose computer would have
been out of the question.

Let us close our critique of the author's presentation of Mauchly's pre-
Atanasoff period by checking it against the claims of her preface, the contentions
her retrying of the ENIAC patent case was to substantiate. Of the five claims in this
crucial positive prong of her task, we find only one established—and that rather
weakly. She has in fact failed to show that he was building an electronic computer
before he met Atanasoff (claim 1); she has failed to show that he had the physical
components of that computer at that time (claim 2); she has failed to show that her
still-existing "components" constitute evidence for that concept at that time (claim
4); and she has failed to link that concept directly to the design of the ENIAC
(claim 5). She has shown only that he had the concept of an electronic desk calculator before he met Atanasoff (claim 3)—something on which Mauchly's word was
never challenged in court—but only about one month before, and only in terms of
function, not substance; this one "established" claim is diminished, too, by the high
probability of a Travis influence (see chap. 3).

Mrs. Mauchly devotes the balance of her article to the last two of her seven
claims, what we have called the negative prong of her argument: that Atanasoff had
no influence on the conception of the ENIAC once their association began (claim
6); and, moreover, that Atanasoff's machine was not an electronic digital computer, so that the ENIAC was actually the first such machine (claim 7). There, for
the first time, she addresses the trial testimony, as well as Mauchly's notes and his
considerable correspondence, including the remaining three letters not produced
for the trial. She does not really have anything new to offer for this post-Atanasoff

period either, however, but continues to concentrate on her own biased selection and interpretation of material already examined in court.

The Post-Atanasoff Years

We have seen thus far that Mrs. Mauchly's physical evidence from her late husband's pre-Atanasoff period has not produced the promised results. Likewise, her presentation of the first two of the five letters she found after his death in 1980 has done her cause more harm than good, showing only a hope, in late 1940, of building an electronic calculator "in a year or so,"—*and* revealing her willingness to edit a document to suit her purposes. We turn now to the last three of these letters, and then to a sampling of her treatment of courtroom testimony, Mauchly notes, and other correspondence that we have already covered—unselectively, we believe—in chapter 3. Our response to Mrs. Mauchly's *Annals* article concludes with a quotation from her own testimony, given in pretrial deposition.

The Last Three Letters

The third and fourth letters fall in the period between the first Atanasoff-Mauchly meeting on December 28, 1940, and Mauchly's visit to Iowa from June 13 to June 18, 1941. The third was dated April 26, 1941, and, like the one of November 15, 1940, was to meteorologist H. Helm Clayton. Mrs. Mauchly quotes from it as follows (p. 127, col. 2); ellipsis and italics are *hers:*

> As our work progresses, there will be more need for reducing long series of daily values to some standard form by multiplication or division. Hence we are still thinking about and working on electrical computing circuits, as I wrote in a previous letter. Some of the *simple components of an electronic computer have been constructed, and we find they work,* so we are going on to the real job of putting the components together . . .
>
> In June I am thinking of doing two things. One is to go to Ames, Iowa, to see an *electric* computing machine being developed there in the Physics Dept. by a friend of mine, and the other is to *further the construction of my own devices.*

As in her presentation of the first two letters, Mrs. Mauchly provides only selected passages, with no assurance that omitted passages are inconsequential. What had Mauchly said, for example, before her opening sentence, and did he go on at the end to cite any specifics of the "simple components" he (and his students) had built, or of what would be entailed in the "real job of putting the components together"?

Mrs. Mauchly asserts that these components were "a number of decade counters," the "biquinary counters using gas tubes." And possibly they were, but it is more probable that they were the neon counters depicted in Mauchly's four April, 1941, notes (see chap. 3). It would seem significant, too, that Mauchly had written to Atanasoff less than a month earlier that he had not "been able to do more than just paper-work" on his proposed calculator and hadn't "yet found out how practical my ideas are" (March 31 letter, PX 744, omitted by Mrs. Mauchly; see chap. 3).

We should perhaps remind the reader once more, too, that anything Mauchly did after he had met Atanasoff cannot be assumed to be entirely his doing. While it is certain that Atanasoff did not advocate the use of neons for computing, the fact remains that Mauchly raised that issue with him and that Atanasoff told him of his own success with vacuum tubes.

The fourth letter also conveys only an expectation to build an electronic computing "circuit." Written on June 7, 1941, to George Bailey, chairman of the Radio Section, Office of Scientific Personnel, National Research Council, it was a follow-up to an employment application. We know a little more about it, because, as with Mauchly's December 4, 1940, letter to DeWire, a fuller excerpt was published by McNulty in his letter to *Datamation* (McNulty 1980). Mrs. Mauchly quotes it as follows (p. 128); again, the italics are *hers:*

> I have already filled out the Specialized Personnel check lists in Physics, Statistics, and Meteorology; in each of these I indicated my interest in electrical computing devices, and electronic devices particularly. I don't think I qualify as an "expert" in radio engineering, but my deep *interest in electronic computing* and *control devices* is accompanied by practical laboratory work, leading now to the *construction of a high-speed electronic computing circuit.* Incidentally, during the next week I am going to Ames, Iowa, to visit with Dr. Atanasoff of the Iowa State Physics Dept. and to discuss with him the pros and cons of his electronic computer versus mine.

Now a comparison with the McNulty excerpt reveals, again, some selective editing. A sentence is omitted after her first one, to the effect that Mauchly now wished to place more emphasis on his electronic activities than he had in his formal application. And two changes appear in the wording: a "with" has been inserted in her last quoted sentence, so that Mauchly was going to Iowa *to visit with* Dr. Atanasoff instead of *to visit* Dr. Atanasoff, and the second occurrence of "Dr. Atanasoff" has been changed to "him."

Although these changes are not of great moment, they do shift the tenor of the letter in Mauchly's favor. What is more disturbing about them is that they constitute a second clear instance of Mrs. Mauchly's editing of her new evidence without using the conventional devices to alert the reader. We are left wondering what else she has deleted or altered.

Her fifth and last letter is one Mauchly wrote to his first wife, Mary, on August 18, 1941, while he was living in Philadelphia (taking the summer defense course at the Moore School) and she and the children were still in Collegeville (pp. 132–33). Strangely enough, not only was this letter written a full two months after Mauchly's visit to Iowa, but it had nothing to do with computing! It concerned his employment prospects, including a possible instructorship at the Moore School for the coming academic year. It discussed matters of teaching load, the chance for promotion if he should be retained after a trial year, and some of the school's policies. It did contain one sentence, however, that would have interested Honeywell attorneys investigating the Mauchly-Eckert quarrel with the University of Pennsylvania over ENIAC patent rights:

> The school doesn't restrict your publication in any way, and although it has rights to patents, it always releases engineering patents to the individuals anyway, so that's that.

Thus it turns out that Mrs. Mauchly has not five, but four letters on which to base her retrial of the ENIAC case vis-à-vis Atanasoff's influence, and that not one of them furthers her cause in the least.

ENIAC Trial Testimony

Mrs. Mauchly turns to the trial testimony for the first time in her coverage of the Iowa visit, as she seeks to promote her running argument that the ABC was of limited potential and that Mauchly was greatly disappointed in it. She quotes seven passages from Atanasoff and one from Mauchly. But now, even though presenting material that is in the public record, she boldly deletes certain phrases, sentences, paragraphs, even pages that detract from her case. And again she fails to indicate her deletions—indeed, effectively disguises them by citing a single transcript page for passages that occupy three or more pages, by citing an incorrect page, or by citing only a volume number.

We take as our example Mrs. Mauchly's treatment of a portion of Atanasoff's testimony on the state of his computer during Mauchly's visit. Her single-paragraph reduction of a *five-page* description is as follows (p. 129, col. 1):

> I had asked Mr. Berry to get the machine in as complete as possible form while Dr. Mauchly was visiting. I remember the main shaft was off the machine when I spoke to Clifford Berry about it, so he assembled the parts. The base 10 "card in" device was on the machine while he [Mauchly] was there. The "base 2 in" and "base 2 out" units had temporary connections while he was there. They could be mounted on the machine to indicate the mechanical arrangements here, but they were not connected underneath. We had no ability to read from a base 2 punch card at that time. [Vol. 17, p. 2,434]

374 The First Electronic Computer

Atanasoff gave the description from which Mrs. Mauchly has constructed this passage during cross-examination by Sperry attorney Ferrill (transcript pp. 2,429–34), in the presence of drawings to which both witness and counsel referred repeatedly (see chap. 3). Her first three sentences (about half of her paragraph) occur on transcript page 2,429, in response to a question by Ferrill, "Were these base 2 in and base 2 out card units that you show on the picture there in place on the machine when he was there?" Atanasoff replied (transcript pp. 2,429–30):

> Yes, they were. They were, and this was part—you remember I testified that I had asked Mr. Berry to get the machine in as complete as possible form while Dr. Mauchly was visiting. He had things kind of disassembled. I remember the main shaft was off the machine when I spoke to Clifford Berry about it for some reason so he assembled the parts. The base 10 card in device was on the machine while he was there. The transformer bank was not complete. We had a single transformer which is used for punching a single line, a single series of lines. . . .

A comparison of this passage with Mrs. Mauchly's first three sentences shows that she has effectively altered the impression of a temporarily disassembled machine to that of a machine being assembled for the first time. (It will be recalled that Berry was working feverishly to complete his master's thesis on the binary-card system by an early July deadline, and that in any case the machine was designed to be taken apart for ease of access to all systems.)

Mrs. Mauchly's remaining three sentences suffer from similar distortion. She omits entirely a clarifying question-answer exchange, and then she quotes—or, rather, paraphrases—her fourth sentence from an exchange on transcript page 2,430. She next skips forward to page 2,434, to form her fifth sentence from a longer one in the middle of that page and her sixth from one at the top. The omitted four pages had been devoted to Atanasoff's explanation of how the base-two units could be removed from the machine and replaced as desired, and also to where his one-cycle switch was located and how it operated in conjunction with an oscilloscope during demonstrations.

Mauchly Notes

The author's presentation of the Mauchly notes of August and September, 1941 (pp. 131–34), is both highly selective and difficult to sort out. She paraphrases and reorganizes in outline form; she combines quotations from more than one file, citing only one as a reference; she supplies dates where none were given; she pushes on with her conjectures on Mauchly's anticipation of the ENIAC; and she completely ignores his trial testimony in interpreting her versions of these very tentative notes.

We take as our example here her summary paragraph (pp. 133–34):

Other items referred to in these early notes indicate clearly that the basic units of computation that were ultimately used in the ENIAC had already taken shape in Mauchly's mind: a multiplier, subtraction by complement, carry-over, a master pulser, and a pulse generator. We can see these reflected in the final ENIAC. Subtraction by complement was the method used for subtraction in the decade counters of the accumulators. The "master pulser" at this time was a pulse scheme between units for controlling the sequence of events. In later work with Eckert, the "master pulser" developed into the "master programmer" for initiating programs, for subroutines, and for branching. The "cycling unit" of the ENIAC was a linear descendant of Mauchly's "pulse generator." . . .

As we saw in chapter 3, Mauchly himself had testified that the two undated pages of block diagrams with "Pulse gen," "Accum.," "Multiplier," etc., written in their squares could have dated from his Ursinus period *or* from his Moore School period (up until 1943!). He had added that he could be certain of only one thing and that was that they were "absolutely unrelated to anything that might be proposed to Aberdeen, or anything that was suggested and proposed in that August '42 memorandum" of his. They were, he said, "the kind of thing I was hoping to do with small statistical machines." (Transcript pp. 12,229–32.) And, of course, these *were* only crude block diagrams, not electronic designs.

The page with the phrase "master pulser" was such a tentative effort that it left Mauchly blank at trial (transcript pp. 12,227–28). His single page on "subtraction by complement" featured a mechanical switch. And his page on "carrying" was only a question, "Can carrying be done with little delay (or at least without complex equipment)," which remained unanswered by the sketches of counters that occupied the rest of the space. (See chap. 3.)

Other Correspondence

Mrs. Mauchly's presentation of correspondence examined at trial—again, documents in the public record—follows a similar pattern. In quoting Mauchly's letters to Atanasoff and Clayton, she omits parts that express frustration with his digital efforts or enthusiasm for his analog harmonic analyzer. And, in her treatment of a critically important letter from Atanasoff to Mauchly, we find perhaps our most extreme example of her unorthodox editing of quoted material.

This is the letter of May 31, 1941, in which Atanasoff disclosed his idea for converting the ABC into an electronic digital integraph (see chap. 3). Her reconstruction of this key piece of trial evidence features: (1) an untagged omission of a vital clause, (2) a very damaging "typographical error" *not* in the original but tagged with *sic,* and (3) an interpretive footnote that contradicts the essential burden of the letter. We first quote and comment on the relevant passage from the trial exhibit, with our own *sic* at "excell":

. . . The figures on the electronic differential integraph seem absolutely startling. During Dr. Caldwell's last visit here, I suddenly obtained an idea as to how the computing machine which we are building can be converted into an integraph. Its action would be analogous to numerical integration and not like that of the Bush Integraph which is, of course, an analogue machine, but it would be very rapid, and the steps in the numerical integration could be made arbitrarily small. It should therefore equal the Bush machine in speed and excell [sic] it in accuracy.

Now, two points about the computer Atanasoff was here envisaging are apparent from this passage, without resort to any courtroom testimony concerning it. First, the proposed computer was to be *digital:* the digital ABC was to be "converted into" it; it was to solve differential equations by "numerical integration" using "arbitrarily small" steps; and "its action" was *not* to be like that of the Bush integraph, which was "of course, an analogue machine." Second, the computer was to be *electronic:* again, the electronic ABC was to be "converted into" it; and its "action" was to be "very rapid," so much more rapid than the action of the mechanical integraph that it would equal the integraph in speed, that is, equal it in solution time despite the large number of "arbitrarily small" steps required.

In her article, Mrs. Mauchly creates a different impression of Atanasoff's meaning by "editing" the last two sentences. Her version of them reads as follows (p. 128, col. 1):

. . . Its action would be analogous to numerical integration, and not like that of the Bush Integraph which is, of course, an analogus [sic] machine, and the steps in the numerical integration could be made arbitrarily small. It should therefore *equal* the Bush machine *in speed** and excel it in accuracy.

*He is talking about a mechanical device, not an electronic one.

Her two serious text alterations compromise, in turn, Atanasoff's two cardinal points, the one that his computer was to be *digital,* the other that it was to be *electronic!* First, the replacement of "analogue" with "analogus [sic]" effectively denies him his direct statement that his computer was *not* to be analog. Second, the omission of the clause, "but it would be very rapid," (after "machine,"), effectively cancels his stated advantage of the machine's electronic speed in the accurate performance of each digital step; the placement of "equal" and "in speed" in italics enhances this impression, as does the footnote, which from its placement seems to refer to Atanasoff's proposed "device," labeling it "mechanical."

Now the first change has the further complication that "analogus" would very probably be taken as a mistyping of *analogous,* not of *analogue,* both because it is more similar to *analogous* and because *analogue* is today an outmoded spelling. And to readers who indeed saw the word as "analogous," Atanasoff might actually seem to be acknowledging that the computer was to be *analog!* (Though now quite garbled, the passage would seem to state that, whereas his machine's *action* would not be like that of the Bush integraph, the machine itself was still *analogous to* that integraph, that is, was still analog.) Indeed, as if to ensure this effect, Mrs. Mauchly adds a footnote to the earlier (January 23, 1941) Atanasoff letter about this same idea; that footnote says: "He is talking here about an analog machine."

A Question of Innocent Error

The question naturally arises as to how that word "analogue" came to appear in Mrs. Mauchly's article as "analogus." Did she (or her typist) simply misread the original "e" for an "s," after which either she or the editor conscientiously inserted the "*sic*"? We ourselves have seen far too many changes, all favoring her case, to believe they could be inadvertent. And, in this particular instance, we *know* that it was not. For we happen to have read an earlier draft of this article in which the word appeared baldly as "analogous," with the "o" (and without the "*sic*")! At some point before publication, then, the "o" was removed, necessitating the editorial "*sic*." (In mid-September, 1982, we were sent the first draft of Mrs. Mauchly's article by an *Annals* editor, who wrote that he thought it might be at least part of the anticipated response to our own 1981 article; Eckert had been expected to respond also.)

And so it is that, just as her presentation of Mauchly's pre-Atanasoff devices and the five newfound letters could not have helped the Sperry Rand case, neither could her handling of the post-Atanasoff letters, notes, and testimony. Indeed, the very suggestion that her reconstruction of those devices, those letters, or this trial material might have seen the light of day in a court of law is ludicrous.

In terms of the author's original claims, she has now also failed to prove the last two, those relating to Mauchly's post-Atanasoff period. That is, she has failed to refute Judge Larson's findings that Atanasoff influenced the conception of the ENIAC and that the ABC was a prior electronic digital computer.

Kathleen Mauchly's Own Testimony

As we remarked earlier, Mrs. Mauchly was present in court during her husband's testimony, and she has also represented herself in her article as informed on the trial proceedings. She must have known, then, as she wrote of Mauchly's activities prior to "any association" with Atanasoff, that he had himself, at trial, disavowed even an *attempt* to invent an electronic digital computer before his exposure to Atanasoff's computer in Iowa (see chap. 3). But what is more shocking, she too had testified, not only that he had not *conceived* or *invented* a computer before that exposure, but that he had not wanted even to *build* one himself.

We quote her answer to Halladay's question as to whether she had ever heard of Atanasoff (p. 291 of Kathleen Mauchly's ENIAC trial deposition):

. . . His original desire was to find a computer, not to have to build one. Any time he heard of anyone who might have a computer he was going to build, John tried to get by to see how and find out when the person would have something that he could use or if they would.

And (p. 292):

. . . He indicated that he went to see Atanasoff, but Atanasoff didn't have what he wanted.

References

Atanasoff, John V. 1940/1973. Computing Machine for the Solution of Large Systems of Linear Algebraic Equations. Originally unpublished. In *The Origins of Digital Computers,* pp. 305–25. *See* Randell 1973.

———. 1984. Advent of Electronic Digital Computing. *Annals of the History of Computing* 6:229–82.

Boole, George. 1847. *The Mathematical Analysis of Logic, Being an Essay toward a Calculus of Deductive Reasoning.* Cambridge: Macmillan, and London: George Bell. Reprint. Oxford: Basic Blackwell, and New York: Philosophical Library, 1948.

———. 1854. *An Investigation of the Laws of Thought, on Which Are Founded the Mathematical Theories of Logic and Probability.* London: Walton and Maberly. Reprint. New York: Dover Publications, 1951.

Burks, Arthur W. 1946. Super Electronic Computing Machine. *Electronic Industries* 5:62–67, 96.

———. 1947. Electronic Computing Circuits of the ENIAC. *Proceedings of the Institute of Radio Engineers* 35:756–67.

———. 1980. From ENIAC to the Stored Program Computer: Two Revolutions in Computers. In *A History of Computing in the Twentieth Century,* pp. 311–44. *See* Metropolis, Howlett, and Rota 1980.

Burks, Arthur W., and Burks, Alice R. 1981. The ENIAC: First General-Purpose Electronic Computer. *Annals of the History of Computing* 3:310–99. With comments by John V. Atanasoff, J. G. Brainerd, J. Presper Eckert and Kathleen R. Mauchly, Brian Randell, and Konrad Zuse, together with the authors' responses.

Burks, Arthur W.; Goldstein, Herman H.; and von Neumann, John. 1946. *Preliminary Discussion of the Logical Design of an Electronic Computing Instrument.* Princeton: Institute for Advanced Study. Reprinted in *Papers of John von Neumann on Computing and Computer Theory,* pp. 97–142. *See* von Neumann 1987.

Eckert, J. Presper. 1953. A Survey of Digital Computer Memory Systems. *Proceedings of the Institute of Radio Engineers* 41:1,393–1,406.

———. 1980. The ENIAC. In *A History of Computing in the Twentieth Century,* pp. 537–39. *See* Metropolis, Howlett, and Rota 1980.

ENIAC Trial Records. Pretrial depositions, affidavits, complaints, transcripts, exhibits, briefs, and decision. Honeywell, Inc. v. Sperry Rand Corp. et al. No. 4-67 Civ. 138. D. Minn. Filed May 26, 1967, decided October 19, 1973. General Services Administration, Federal Records Center, Chicago. Decision published in *U.S. Patent Quarterly* 180:673–773.

Gardner, Martin. 1982. *Logic Machines and Diagrams.* 2d ed. Chicago: University of Chicago Press.

Goldstine, Herman H. 1972. *The Computer from Pascal to von Neumann.* Princeton: Princeton University Press.

Jevons, William Stanley. 1870. On the Mechanical Performance of Logical Inference. *Philosophical Transactions of the Royal Society of London* 160:497 (with plates).

———. 1874. *Principles of Science.* London: Macmillan.

Ketner, Kenneth L. 1984. The Early History of Computer Design: Charles Sanders Peirce and Marquand's Logical Machines. *Princeton University Library Chronicle* 45, no. 3: 186–211.

Leibniz, Gottfried. 167?/1977. Three quotations translated from the Latin in *Chance, Cause, Reason: An Inquiry into the Nature of Scientific Evidence,* by Arthur W. Burks, pp. 337 and 395–96. Chicago: University of Chicago Press. An analysis of what Leibniz intended is given on pp. 366–67 and 393–96.

McNulty, James. 1980. Letter to the editor. *Datamation* 26, no. 11: 23–26.

Marquand, Allan. 1886. A New Logical Machine. *Proceedings of the American Academy of Arts and Sciences* 21:303–7. Reprinted in *Princeton University Library Chronicle* 45, no. 3 (1984): 213–17.

Mauchly, John W. 1942/1973. The Use of High Speed Vacuum Tube Devices for Calculating. Originally unpublished. In *The Origins of Digital Computers,* pp. 329–32. *See* Randell 1973.

———. 1980. The ENIAC. In *A History of Computing in the Twentieth Century,* pp. 541–50. *See* Metropolis, Howlett, and Rota 1980.

———. 1982. Mauchly: Unpublished Remarks. *Annals of the History of Computing* 4:245–56. With foreword by Henry Tropp and afterword by Arthur W. Burks and Alice R. Burks.

Mauchly, Kathleen R. 1984. John Mauchly's Early Years. *Annals of the History of Computing* 6:116–38.

Mays, Wolfe. 1953. The First Circuit for an Electrical Logic Machine. *Science* 118:281–82.

Merrifield, C. W. 1879. Report of the Committee. British Government publication. In *The Origins of Digital Computers,* pp. 82–85. *See* Randell 1973.

Metropolis, Nicholas; Howlett, Jack; and Rota, Gian-Carlo, eds. 1980. *A History of Computing in the Twentieth Century.* New York: Academic Press.

Patterson, George W. 1960. The First Electric Computer: A Magnetological Analysis. *Journal of the Franklin Institute* 270:130–37.

Peirce, Charles Sanders. 1887. Logical Machines. *American Journal of Psychology* 1:165–70. Reprinted in *Princeton University Library Chronicle* 45, no. 3 (1984): 219–24.

———. 1931–58. *Collected Papers of Charles Sanders Peirce.* Vols. 1–6, ed. Charles Hartshorne and Paul Weiss, 1931–36. Vols. 7–8, ed. Arthur W. Burks, 1958. Cambridge, Mass.: Harvard University Press.

———. 1982–. *Writings of Charles S. Peirce, a Chronological Edition.* 3 vols. to date. Ed. Max H. Fisch, Christian J. W. Kloesel, and Edward C. Moore. Bloomington: Indiana University Press.

Randell, Brian, ed. 1973. *The Origins of Digital Computers: Selected Papers.* Berlin: Springer-Verlag. 2d ed., 1982.

Regenerative Memory Trial Records. Pretrial depositions, transcript, etc. Sperry Rand Corp. v. Control Data Corp. Civ. 15,823 and 15,824. D. Md. Filed April 1, 1964. An out-of-court settlement was reached in 1981.

Richards, R. K. 1966. *Electrical Digital Systems.* New York: Wiley.

Shannon, Claude E. 1938. A Symbolic Analysis of Relay and Switching Circuits. *Transactions of the American Institute of Electrical Engineers* 57:713–23.

Truesdell, Leon E. 1965. *The Development of Punch Card Tabulation in the Bureau of the Census, 1890–1940.* Washington, D.C.: U.S. Department of Commerce, Bureau of the Census.

von Neumann, John. 1945. First Draft Report on the EDVAC. Moore School of Electrical Engineering, University of Pennsylvania. In *Papers of John von Neumann on Computing and Computer Theory,* pp. 17–82. *See* von Neumann 1987.

———. 1987. *Papers of John von Neumann on Computing and Computer Theory.* Ed. William F. Aspray and Arthur W. Burks. Cambridge, Mass.: MIT Press.

Williams, Michael R. 1985. *A History of Computing Technology.* Englewood Cliffs, N.J.: Prentice-Hall.

Index